DISEASE CONTROL IN CROPS

Disease Control in Crops

Editor:

Dr. Nisha Dahiya, Ph.D.

Delve Publishing
www.delvepublishing.com

Disease Control in Crops
Editor: **Dr. Nisha Dahiya, Ph.D.**

708 3rd Avenue, 6th Floor
New York, NY 10017
United States of America

www.delvepublishing.com

ISBN: 978-1-68095-774-7

Library of Congress Control Number: 2016951530

About the Editor

Dr. Nisha Dahiya, Ph.D.

Dr. Nisha Dahiya is serving as a Research Scientist at the Indian Council of Medical Research (ICMR), New Delhi, India. She completed her PhD in 2014 from University of Camerino, Italy, under the Malaria and Human Development Program, which was supported by World Health Organization, Global Malaria Program. Her main interests are drug discovery for infectious diseases and surveillance and molecular epidemiology of rising drug resistant pathogens. Currently, she is assisting in the coordination of national network for surveillance of anti microbial resistance through out the country.

Contents

List of Contributors

Robert P. Larkin
USDA-ARS, Biocontrol of Plant Diseases Laboratory, Beltsville, MD 20705

Deborah R. Fravel
USDA-ARS, Biocontrol of Plant Diseases Laboratory, Beltsville, MD 20705

M. G. Cunha
Department of Plant Pathology, University of California, Davis 95616

D. M. Rizzo,
Department of Plant Pathology, University of California, Davis 95616

D. Sharma-Poudyal
Department of Plant Pathology, Washington State University, Pullman 99164-6430

X. M. Chen
Department of Plant Pathology, Washington State University, Pullman 99164-6430
U.S. Department of Agriculture-Agricultural Research Service, Wheat Genetics, Quality, Physiology, and Disease Research Unit, Pullman 99164-6430

Ayad Ismael Khaleel
Plant Protection Department, Faculty of Agriculture, Universiti Putra Malaysia, Serdang, Malaysia

Kamaruzaman Sijam
Plant Protection Department, Faculty of Agriculture, Universiti Putra Malaysia, Serdang, Malaysia

Tavga Sulaiman Rashid
Plant Protection Department, Faculty of Agriculture, Universiti Putra Malaysia, Serdang, Malaysia

Khairulmazmi Bin Ahmad
Plant Protection Department, Faculty of Agriculture, Universiti Putra Malaysia, Serdang, Malaysia

Anna A. Slavokhotova
Vavilov Institute of General Genetics, Russian Academy of Sciences, Moscow, Russia

Ekaterina A. Istomina
Vavilov Institute of General Genetics, Russian Academy of Sciences, Moscow, Russia

Emma N. Andreeva
Vavilov Institute of General Genetics, Russian Academy of Sciences, Moscow, Russia

Tatyana V. Korostyleva
Vavilov Institute of General Genetics, Russian Academy of Sciences, Moscow, Russia

Vitalij A. Pukhalskij
Vavilov Institute of General Genetics, Russian Academy of Sciences, Moscow, Russia

Alexander N. Shijan
Vavilov Institute of General Genetics, Russian Academy of Sciences, Moscow, Russia

Tatyana I. Odintsova
Vavilov Institute of General Genetics, Russian Academy of Sciences, Moscow, Russia

Omar Hammoudi
Agriculture Scientific Research Center, Lattakia, Syria;
Mazen Salman
Palestine Technical University-Kadoorie (PTUK), Tulkarm, Palestine;

Ruba Abuamsha
National Agriculture Research Center (NARC), Jenin, Palestine;

Ralf-Udo Ehlers
Department of Biotechnology & Biological Control, Institute for Phytopathology, Christian-Albrechts-University Kiel, Kiel, Germany.

Shabir H. Wani
Biotechnology Laboratory, Central Institute of Temperate Horticulture, Srinagar, Jammu and Kashmir, India;

Gulzar S. Sanghera
Rice Research & Regional Station (SKUAST-K) Khudwani, Anantnag Jammu and Kashmir, India;

N. B. Singh
Department of Plant Breeding and Genetics, COA, Central Agricultural University, Imphal, Manipur, India.

Akaamaa Clement Wada
National Cereals Research Institute, Badeggi, Nigeria;

Samuel Oladele Bakare
National Cereals Research Institute, Badeggi, Nigeria;

Mohammed Bashir
National Cereals Research Institute, Badeggi, Nigeria;

Mark Noye Ukwungwu
National Cereals Research Institute, Badeggi, Nigeria;

Jayakrishnan Saimandir
Department of Entomology, Ohio State University, Columbus, USA;

Madhuban Gopal
Madhuban Gopal, Division of Agricultural Chemicals, Indian Agricultural Research Institute, Pusa Campus, New Delhi, India.

Malkhan Singh Gurjar
Central Potato Research Station, Shillong, India

Shahid Ali
SNPG College, Azamgarh (VBS Purvanchal University, Jaunpur), India;

Masood Akhtar
SNPG College, Azamgarh (VBS Purvanchal University, Jaunpur), India;

Kangabam Suraj Singh
Central Potato Research Station, Shillong, India

Mark S. Sisterson
United States Department of Agriculture–Agricultural Research Service, San Joaquin Valley Agricultural Sciences Center, Parlier, CA 93648.

Drake C. Stenger
United States Department of Agriculture–Agricultural Research Service, San Joaquin Valley Agricultural Sciences Center, Parlier, CA 93648.

Judy A. Thies,
Research Plant Pathologist, U.S. Vegetable Laboratory, United States Department of Agriculture–Agricultural Research Service (USDA-ARS), Charleston, SC;

Richard F. Davis
Research Plant Pathologist, Crop Protection and Management Research Unit, USDA-ARS, Tifton, GA;

John D. Mueller
Professor, Clemson University, Edisto Research and Education Center, Blackville, SC;

Richard L. Fery
Research Geneticist, U.S. Vegetable Laboratory;

David B. Langston
Assistant Professor, Rural Development Center, University of Georgia, Tifton

Gilbert Miller
Senior County Extension Agent, Bamberg County, SC

T. C. Paulitz and K. L. Schroeder,
USDA-ARS, Root Disease and Biological Control Research Unit, Pullman, WA 99164-6430; and

W. F. Schillinger,
Department of Crop and Soil Sciences, Washington State University, Dryland Research Station, Lind, WA 99341

Steven A. Brooks,
United States Department of Agriculture–Agricultural Research Service (USDA-ARS), Stuttgart, AR 72160;

Merle M. Anders,
University of Arkansas, Rice Research and Extension Center, Stuttgart 72160;

Kathleen M. Yeater,
USDA-ARS, College Station, TX 77845

Preface

Plants are affected by fungi, bacteria, viruses and nematodes which cause a range of diseases in plants. Certain abiotic factors such as temperature, nutrition, moisture and toxicity also results in stunted growth, low productivity and death in crops. Globally, about 10-16% crop yield is lost to plant diseases which cost about US$220 billion. Demand for innovative methods of disease management is increasing due to increase in awareness for organic food, non-specificity of chemicals, environmental and health risks of the chemicals, along with the increasing fungicide resistance in pathogens. Management of plant diseases relies on a number of general strategies as fungicide spray, cultivating pathogen resistant varieties, adopting crop rotation and reduction of plant stress. Fungicides are majorly used for the management of plant diseases, which is leading to the increased incidence of fungicide resistance in pathogens. Novel biocontrol methods are being utilized for the control of plant pathogens. Non pathogenic strains of some bacterial species such as F. oxysporum and F. solani collected from a Fusarium wilt-suppressive soil have been demonstrated to be used for crop disesae management. Extracts from natural sources have reportedly been effective in vitro against Erwinia carotovorum subsp. Carotovorum, Ralstonia solanacearum, and Xanthomonas gardneri. Attenuated strain of green mottle mosaic virus (CGMMV) VIROG-43M strain has effectively been used as biocontrol agent for the protection of cucumber plant in greenhouses and laboratory conditions and in commercial greenhouses. Serratia plymuthica HRO-C48 and Gliocladium catenulatum J1446 have been reportedly used effectively as microbial antagonists against Blackleg disease caused by Phoma lingam. False smut (Ustilaginoidea virens) and kernel smut (Neovossia horrida) susceptible rice cultivars showed significantly reduced disease severity when rice was cultivated in 3-year rotation of soyabean and corn. Certain studies have observed that management practices such as management of straw treatments, tilling, and crop rotation of irrigated cropping systems such as wheat (Triticum aestivum) influences the severity of Rhizoctonia solani AG-8, Fusarium pseudograminearum and F. culmorum. In addition, RNAi is a potential molecular strategy for generating pathogen resistant plants by enabling them to combat fungal, bacterial, viral and nematod diseases. The strategy has been demonstrated to be successful in targeting various pathogenic fungi, bacteria and viruses with a possibility of potential for management of plant parasitic nematodes. The use of novel methods of plant disease management and control is expected to reduce the food safety and security concerns along with augumentation in improvement in crop yields.

Editor

Dr. Nisha Dahiya, Ph.D.

INTRODUCTION

Plant disease control is eminent process in crop yielding without complete or partial eradication of pathogen, or control measures of disease, one cannot get good crop yield results without proper control amendments of disease or pathogen. The major aspect of plant disease is to "eliminate or eradicate the pathogen from plant / crop, pre or post-harvest to control the disease and loss of crop yield by the pathogen." Chemical disease control employs the use of chemicals that are either generally toxic and used as disinfectants or fumigants or chemicals that target specific kinds of pathogens, as in the case of fungicides, bactericides (or antibiotics) and nematicides. No substances have yet been found that effectively control plant viruses. Ideally, a chemical control agent should be effective at concentrations that will not harm the plant, have low risk to humans and animals, and have minimal effect on the normal microflora on the plants and in the soil. Also, there should be little chance of the pathogen quickly developing resistance to it, and it should be suitable for long periods of storage in ambient conditions. These chemical agents can be sold as dusts, concentrated solutions, wettable powders, granules or emulsions.

Chemical control

A variety of chemicals are available that have been designed to control plant diseases by inhibiting the growth of or by killing the disease-causing pathogens. Chemicals used to control bacteria (bactericides), fungi (fungicides), and nematodes (nematicides) may be applied to seeds, foliage, flowers, fruit, or soil. They prevent or reduce infections by utilizing various principles of disease control. Eradicants are designed to kill a pathogen that may be present in the soil, on the seeds, or on vegetative propagative organs, such as bulbs, corms, and tubers. Protectants place a chemical barrier between the plant and the pathogen. Therapeutic chemicals are applied to combat an infection in progress.

Soil treatments are designed to kill soil-inhabiting nematodes, fungi, and bacteria. This eradication can be accomplished using steam or chemical fumigants. Soilborne nematodes can be killed by applying granular or liquid nematicides. Most soil is treated well before planting; however, certain fungicides can be mixed with the soil at planting time.

Seeds, bulbs, corms, and tubers are frequently treated with chemicals to eradicate pathogenic bacteria, fungi, and nematodes and to protect the seeds against organisms in the soil—mainly fungi—that cause decay and damping-off. Seeds are often treated with systemic fungicides, which are absorbed and provide protection for the growing seedling.

Protective sprays and dusts applied to the foliage and fruit of crops and ornamentals include a wide range of organic chemicals designed to prevent infection. Protectants are not absorbed by or translocated through the plant; thus they protect only those parts of the plant treated before invasion by the pathogen. A second application is often necessary because the chemical may be removed by wind, rain, or irrigation or may be broken down by sunlight. New, untreated growth also is susceptible to infection. New chemicals are constantly being developed.

Biological control

Biological control of plant diseases involves the use of organisms other than humans to reduce or prevent infection by a pathogen. These organisms are called antagonists; they may occur naturally within the host's environment, or they may be purposefully applied to those parts of the potential host plant where they can act directly or indirectly on the pathogen.

Although the effects of biological control have long been observed, the mechanisms by which antagonists achieve control is not completely understood. Several methods have been observed: some antagonists produce antibiotics that kill or reduce the number of closely related pathogens; some are parasites on pathogens; and others simply compete with pathogens for available food.

Cultural practices that favor a naturally occurring antagonist and exploit its beneficial action often are effective in reducing disease. One technique is to incorporate green manure, such as alfalfa, into the soil. Saprophytic microorganisms feed on the green manure, depriving potential pathogens of available nitrogen. Another practice is to make use of suppressive soils—those in which a pathogen is known to persist but causes little damage to the crop. A likely explanation for this phenomenon is that suppressive soils harbour antagonists that compete with the pathogen for food and thereby limit the growth of the pathogen population.

Other antagonists produce substances that inhibit or kill potential pathogens occurring in close proximity. An example of this process, called antibiosis, is provided by marigold (Tagetes species) roots, which release terthienyls, chemicals that are toxic to several species of nematodes and fungi.

Only a few antagonists have been developed specifically for use in plant-disease control. Citrus trees are inoculated with an attenuated strain oftristeza virus, which effectively controls the virulent strain that causes the disease. An avirulent strain of Agrobacterium radiobacter (K84) can be applied to plant wounds to prevent crown gall caused by infection with Agrobacterium tumefaciens. Many more specific antagonists are being investigated and hold much promise for future control of disease.

Therapy

Therapeutic measures have been used much less often in plant pathology than in human or animal medicine. The recent development of systemic fungicides such as oxathiins, benzimidazoles, and pyrimidines have enabled growers to treat many plants after an infection has begun. Systemic chemicals are absorbed by and translocated within the plant, restricting the spread and development of pathogens by direct or indirect toxic effects or by increasing the ability of the host to resist infection.

Antibiotics have been developed to control various plant diseases. Most of these drugs are absorbed by and translocated throughout the plant, providing systemic therapy. Streptomycin is used against a variety of bacterial pathogens; tetracycline is able to control the growth of certain mycoplasmas; and cycloheximides offer effective control for certain diseases caused by fungi.

Host resistance and selection

Disease-resistant varieties of plants offer an effective, safe, and relatively inexpensive method of control for many crop diseases. Most available commercial varieties of crop plants bear resistance to at least one, and often several, pathogens. Resistant or immune varieties are critically important for low-value crops in which other controls are unavailable, or their expense makes them impractical. Much has been accomplished in developing disease-resistant varieties of field crops, vegetables, fruits, turf grasses, and ornamentals. Although great flexibility and potential for genetic change exist in most economically important plants, pathogens are also flexible. Sometimes, a new plant variety is developed that is highly susceptible to a previously unimportant pathogen.

Variable resistance

Resistance to disease varies among plants; it may be either total (a plant is immune to a specific pathogen) or partial (a plant is tolerant to a pathogen, suffering minimal injury). The two broad categories of resistance to plant diseases are vertical (specific) and horizontal (nonspecific). A plant variety that exhibits a high degree of resistance to a single race, or strain, of a pathogen is said to be vertically resistant; this ability usually is controlled by one or a few plant genes. Horizontal resistance, on the other hand, protects plant varieties against several strains of a pathogen, although the protection is not as complete. Horizontal resistance is more common and involves many genes.

Fungicides

Fungicides are chemicals used in the control of fungal diseases. They are often classified as either protectant or systemic. Protectant fungicides are usually effective against a broad range of fungi and protect the plant against infection on the surfaces of the plant to which they are applied. Often, they require multiple applications during the growing season to maintain coverage as new growth emerges and weathering removes past coverage. Systemic fungicides can be absorbed by the plant without harming it, and transported to other tissues where they are toxic to fungi. These compounds can control and eradicate established infections, but they are also vulnerable to fungi developing resistance, as they generally only target one step in a biosynthetic pathway to kill the fungus. To minimise the development of resistance by chemical overuse, fungicides are classified into groups based on their chemical activity. By alternating between the different classes of fungicides, the fungal population has less opportunity to build up resistance to one chemical.

Antibiotics

Relatively few antibiotics are routinely used to control plant diseases. Antibiotics are chemical produced by micro-organisms, which destroy or injure living organisms, in particular, bacteria. Streptomycin is effective against a few fruit pathogens, such as blights and cankers, and cyclohexamine can be used to control some fungal pathogens of crops, particularly powdery mildews and rusts. Bacteria, as well as fungi, have the ability to develop resistance to antibiotics, which is a major disadvantage of using these compounds, and one of the reasons that they are not widely used.

Nematicides

The use of nematicides is confined largely to high-return horticultural crops, because they are expensive. Additionally, they are all highly toxic, and alternative measures for controlling nematodes are being investigated.

Issues Related To Chemical Disease Management

Pathogen resistance to chemical agents has been a problem since the 1970s. Throughout the 1980s and 1990s, public concern has grown about the use of agricultural chemicals in general. The main concerns are the risk of poisoning humans or animals, contamination of livestock products, harm to beneficial insects, and the contamination of food products, waterways and soil. When properly used, fungicides should not cause problems in any of these areas, but they do have the potential to alter the balance of diseases on certain crops.

The main risk to humans is during the preparation or application of these chemicals, when they can be inhaled, ingested or absorbed through the skin, and upon consumption of plants or their products. To overcome these risks, safety precautions must be followed when working with the chemicals, and various regulations exist with regard to consumable plant products. For example, some chemicals may not be applied within a certain time before harvesting (the with-holding period), thus ensuring that chemical residues have fallen to an acceptable level by the time they reach the market. Maximum residue limits have been established to allow monitoring of the chemical residue on plant products.

Environmental concerns focus mainly on protectant fungicides. For example, copper and sulphur sprays have the potential to affect a broad range of organisms if they are washed off the leaves and accumulate in the soil or are washed into the waterways.

In response to public concern about these issues, many countries have initiated programs to reduced pesticide use. They can involve education of farmers on best practices in pesticide application, the use of lower doses of chemicals, better understanding of the threshold level of infection that causes economic loss and developing disease forecasting systems to allow optimal use of chemicals.

What Are the Different Types of Crop Diseases?

Every year, agricultural producers lose vast quantities of crops to various crop diseases. To understand how and why certain diseases strike certain crops, plant pathologists have categorized all diseases into two general groups, infectious and non-infectious. Once segmented into one of the categories, diseases are then described by the characteristic of the offending pathogen or environmental cause, such as fungal, bacterial, extreme temperatures or soil deficiency. Finally, horticulturalists further classify a disease by the part of the crop that is affected, such as leaf diseases, root diseases and more. There are multitudes of different types of crop diseases, but all diseases basically fall under the category of infectious and non-infectious for further classification.

Non-infectious, or abiotic, diseases are those that are caused by unfavorable environmental circumstances and that do not result from living organisms. Generally, non-infectious diseases tend to weaken a plant, making the plant susceptible to the more serious infectious diseases. Examples of non-infectious diseases include extreme temperatures, excessive or lack of water, insects, poor soil nutrition content and toxic chemicals. These factors can weaken the entire plant or a certain part of the plant, making the plant susceptible to infectious stem diseases, seed diseases, pod diseases and more.

Infectious, or biotic, diseases make up the bulk of variety of crop diseases. These are caused by living organisms that infect the crop. Fungi, such as smuts, rusts, molds and blights, are the most common cause for crop diseases. Most fungi require very moist conditions in which to thrive and reproduce. Some fungi develop pustules on the crops, while others produce a colony of powdery, discolored soot.

Bacteria and viruses, which are other infectious hosts, are tiny organisms that infiltrate the plant organs and reproduce. Some bacteria do not harm crops, but others can produce toxins or attack the plants' cells for food. Viruses are usually spread by insects when an insect chews through the plant's protective outer layer and infiltrates the inner cells. Bacterial blights and wilts often plague wheat and barley crops as well as vegetables, such as cucumbers and beans. Crop diseases caused by viruses can cause considerable damage, such as wheat streak mosaic virus and potato virus diseases than can eradicate an entire crop very quickly.

Nematodes and protozoa are also classified as infectious diseases. These tiny worm-like organisms reside in the soil, waiting for an opportunity to attack a crop weakened by non-infectious causes. They may carry viruses, causing further damage to crops and complicating disease treatment. Examples of these biotic diseases include the potato cyst nematode, pine wilt and soybean cyst nematode. As ascertained from their names, these crop diseases invade the sensitive root or vascular systems of the crops and cause cysts that stunt plant growth and sharply decrease crop yields.

Curbing or curing crop diseases is big business. Scientists and agriculturalists have developed myriads of fungicides, insecticides, pesticides and antibacterial solutions for crop disease control. While advances have been made, pathogens have shown a tenacious propensity to adapt to repellants. As the agriculture industry learns more about the various types and causes of crop diseases it becomes better equipped to treat and prevent the many diseases that plague the world's crops.

Plant Diseases and their control

The word disease in connation with plants is considered to refer to any disturbance in the normal life processes which results in such things as:

- abnormal growth
- temporary or permanent check to the development or
- premature death of part or all of the plant.

Plant diseases can be divided into two sorts (i) Parasitic, where the trouble is due to attacks of parasites such as fungi, bacteria or viruses and (ii) Non-parasitic in which the trouble is a result of faults in the environment (soil, temperature, moisture etc). The biggest group of diseases is that caused by fungus parasites (including bacteria) which are spread about by spores produced in the fruiting bodies (equivalent to seeds of higher plants but, of course, microscopic in size). Viruses are incredibly small in size and in nature are carried from infected to healthy plants by insects, mainly aphids, and also by trimming knives and hands (in glasshouses), by knives etc, (in propagating houses) and in a very few instances by seed transmission. It will be noticed that the great difference between parasitic and non-parasitic disease is that the former is infectious while the latter is not.

Despite this, where a non-parasitic trouble begins, there is likely to be great loss unless the fault in environment is quickly corrected. The symptoms in either kind of trouble can be very similar and even almost identical so that to judge the cause it is often necessary to consult an expert to get microscopic examination and sound advice.

The control of plant disease depends on an accurate estimate of the symptoms shown by the affected plant so as to arrive at the exact cause of the trouble. Even with parasitic diseases the cause is usually microscopic, requiring a careful examination and often laboratory tests on the diseased tissues. Similarly with non-parasitic troubles the environment must be studied as well as the plants and the details of cultivation carefully considered. There may have to be a careful soil analysis as well as a study of the drainage and soil texture and there may also be an analytical test of some of the foliage or fruits which could reveal a shortage of some essential plant food.

Precautionary measures: Before considering measures which have to be taken to check diseases in plant crops we may look at some of the things which can be done to guard against disease ever appearing. These can best be termed precautionary measures. They aim at building up the vigor of the plants to help them to resist any possible attack by a parasitic disease and they also include various precautions which can be taken to eliminate the possibility of disease being in the neighborhood of the crop, more especially in the soil.

It is of the utmost importance to study the special requirements of any particular plant so that the soil can be prepared in order to ensure good drainage and also that it contains sufficient organic matter (humus) and the necessary plant foods. Where some plants are concerned, for instance camellias and rhododendrons it is important to ensure that there is no lime present in the soil. Everything should be done to get vigorous plants with robust stems and foliage. In glasshouses, proper light and ventilation must be arranged and in fruit trees skilful pruning helps to build up strong, healthy shoots and buds with good circulation of air among the branches.

Rotation of crops is intended to avoid growing the same kind of plant on the same spot year after year. If crops are grown on the same site year after year any disease of that crop is encouraged to build up and the soil can become heavily infected, addition to which the same plant foods are taken out of the soil. A different crop takes different amounts of the various chemical elements so that a balance can be easily kept. In glasshouses the same crop is very often grown each year but the disease build-up is checked by suitable methods of soil sterilization. Other precautionary measures are weed eradication (eg wild celery harbours celery leaf spot disease and the common plantain can carry the virus of spotted wilt). Careful spacing is also helpful so that diseases are not provided with the humid and moist conditions between plants which they need to germinate their spores and infect the leaves.

The protection of large wounds is another obvious precaution and one which is very important where large specimen trees or even expensive fruit trees and shrubs are concerned. It is not suggested that small pruning cuts need to be treated but where a large branch is broken down by wind or snow or cracked during severe frosts it is wise to try to protect the broken or cut places. After any branch is removed the cut should be painted over with a suitable protective paint such as Stockholm tar or Arbrex to prevent the entry of fungus parasites. Not only are plum trees likely to be infected by the silver leaf fungus but many fine ornamental trees and shrubs can be lost by neglecting this simple precaution. Even after cutting out a canker from apple trees the wound should be painted.

Resistant plants: An important method of avoiding plant disease is to use resistant plants. A plant immune to a particular disease is highly valued by the grower if its quality is as good as those which are susceptible to the disease. Growing immune varieties is the simplest way of avoiding disease and it is a pity that such varieties are so limited in number.

Soil sterilization: One of the most important precautions taken to avoid disease in horticultural crops is the practice of soil sterilization. In this the soil in a greenhouse intended for tomatoes, cucumbers, lettuces etc, is treated by passing hot steam through it (a commercial practice) or by watering with chemicals (eg formaldehyde or cresylic acid) before the crop is planted so that any dormant spores of disease are killed. However, at the temperature used, the spores of beneficial bacteria such as the nitrogen fixers are not killed, so that after the process is completed these can begin to enrich the soil without any immediate competition from other organisms. For use in small pots and seed boxes sterilized soil made after the John Innes formulae can be bought for use by gardeners. Similarly, sterile soilless composts are now available. Small glasshouses can be washed down inside with formalin or other disinfectants and the same sort of treatment can be given to garden frames, pots, boxes, seed trays and tools. Cresylic acid is one substance used for this purpose and in some instances also for sterilizing the surface of the oil, but there are several other good disinfectants available for the gardener.

Preventing the spread of disease When, despite the precautions referred to, a disease makes its appearance it is necessary to act quickly and to take direct action measures. The chief of these is to cover the plant with a protective film of a chemical which will kill the fungus or at least prevent the germination of its spores. Diseased parts can be removed before treatment, but it must be remembered that in most diseases an affected plant is often doomed (except where the disease is superficial or where affected parts can be cut away). The object of the treatment is to protect still healthy tissues, so that it is wise to spray or dust early.

Various chemicals, known as Fungicides, which will deal with all types of disease, are widely obtainable on the market. These are applied as a fine misty spray or as dust (most people hold that spraying is more efficient than dusting). Many types of spraying and dusting machines are in use. There are also smokes which are lit to fumigate glasshouses, which have previously been cleared of all plant life. Seeds, bulbs and corms are also treated by dusting or by immersion in a liquid fungicide to give them protection from soil-borne diseases, etc., after planting.

Fungus diseases: These are diseases which are caused through the attack of various parasitic fungi. Sime fungi obtain food by living on the decaying organic matter which we call humus in the soil, but some obtain their food by attacking living plants and injure or kill them in the process—these are called parasites. In general, the harmless saprophytes are large and easily seen but the parasites are very small and need a microscope for their proper identification, although their presence may be detected because of some whitish or greyish mould or furry growth (eg rose mildew). These parasitic fungi grow microscopically inside plants but some have also a smothering effect (mildews) and grow on the outside of leaves and stems etc. They reproduce and spread themselves by means of innumerable spores.

Fungicides: Fungicides are chemical substances which are used in the control of those diseases of plants which are caused by fungus parasites (see above). The ideal fungicide is a substance which will kill a fungus or prevent its spores from germinating, without doing any harm to the host plant. These substances are used in various forms and in various ways and many chemicals have been tried in the search for the most effective safe fungicide. Sulphur and copper are two of the oldest elements used for this purpose and in various forms are still used against some diseases. For instance copper sulphate is used to make the well-known Bordeaux mixture which has been in use for a century, while sulphur probably dates from Biblical times.

In recent years many of these fungicides have been replaced by more modern ones as the result of much research and there is now a much greater choice. In general the modern fungicide is more specific, that is to say it will prove very effective against a certain disease but is not so generally useful against many others, whereas the older kinds had an all-round fungicidal effect and exercised a check on many diseases likely to attack the plant.

- Fungicides are applied either as sprays or dusts on the foliage or in empty greenhouses as sprays or in the process of fumigation. They may also be applied as a 'smoke' from a special generator or tablet which is lit.
- But although fungus diseases are those we most frequently come in contact with in our gardens, the viruses are also important.

Virus diseases: A virus is a minute particle, visible only under the electron microscope, which causes disorders or diseases in living cells. The presence of viruses may result in leaves developing yellow or brown spots, streaks or ring patterns. Other symptoms affecting the leaves may include dark green areas along the veins (vein banding), a loss of color (vein clearing), complete yellowing, distortion and small outgrowths. Streaks and stripes may appear on stems. Other symptoms include witches' brooms (large numbers of side shoots), distortion, color breaks or white flecks on flowers. A condition where flowers become green and leaf-like is called `phyllody'. Fruits may be small, misshapen or bumpy. Less distinct symptoms of virus infection are a general stunting and reduction of cropping. Viruses can also change the internal structure and the metabolism of plants; for example, virus-infected tomatoes are said to have a better flavor.

- A plant may contain a 'latent' virus without any visible symptoms. There may also be interactions due to the hidden presence of viruses. Indicator plants, which give a quick and characteristic reaction when inoculated with the sap of a virus-infected plant, are widely used in the identification of viruses. Good indicator plants include Nicotiana species, beans and Chenopodium species.
- Non-persistent viruses adhere to the mouth parts of insects feeding on infected plants and are carried to healthy plants. The particle remains infective for up to about two hours.
- Persistent viruses are absorbed into the digestive system of the insect, and from there pass into the salivary glands where the virus multiplies. When feeding the insect injects saliva containing virus particles into plants and remains a disease carrier all its life.
- Virus diseases may be mechanically transferred from one plant to another by man. Plants propagated from infected material will contain virus particles. This is important in all vegetatively propagated crops such as potatoes, fruit and bulbs.

Control Heat therapy kills or inactivates the virus, and apical meristem culture is another method of approach. As a general rule avoid the spread of virus diseases by regular spraying against the insect vectors. Clean cultivation prevents weeds acting as carriers of infection. Remains of the previous year's crop are another source of infection. Plants susceptible to the same virus should not be grown close together. Regular inspection and rogueing of obviously diseased plants is often effective. Seed should not be sown near an old infected crop. Barrier crops are sometimes grown by planting immune plants around those liable to a particular virus infection. Early sowings may enable plants to become established before the activity of disease-carrying insects. Physiological disorders In the study of plant diseases it is usually found that most diseases are due to the attack of some small organism such as a fungus, a bacterium, or one of the microscopic viruses. The work of the plant pathologist when confronted with a disease is to try to identify the parasite which is causing the trouble and then to take steps to combat it. Long ago it was realized that plants sometimes become sick and show symptoms of ill-health although they are not infected by any parasite. This type of trouble in plants is referred to as non-parasitic disease or physiological disorder. It can arise from a multitude of factors such as unsuitable soil, lack of lime, excessive moisture, drought, lack of some essential food element, spray, fume or fumigation damage, and so forth. There is no parasite present so there is no infection to spread but it is obvious that unless something is done all the plants will become affected and weakened.

The identification of a physiological disorder is not easy and, in general, must depend on finding the symptoms as it does with the parasitic diseases which are microscopic. In both groups symptoms from different troubles are so similar that it requires an expert to recognize them.

Attempts have been made to form a rough classification of symptoms and the most study has so far been given to the effects of shortages of essential food elements which are often referred to as mineral deficiencies. These include shortages of the major elements such as nitrogen, potash, phosphate, magnesium and lime, which plants absorb in fairly large amounts and the lesser known minor or trace elements, such as boron, copper, zinc, manganese and iron, which are needed only in very small amounts. Although some shortages, for example of potash, show as a reddish-brown scorching of the leaf margin, it is usually necessary to obtain an analysis of the soil or of the leaf tissues to identify such deficiencies.

The absence of iron shows as a lack of green color until the leaves are very pale yellow or even white and this condition often results from excess of lime and is called lime-induced chlorosis'. This, however, is easily tested by one of the soil-testing field kits which will give the pH measurement of the soil showing its acidity or alkalinity. A quick remedy for iron shortage is now available by watering with the chelated compound called Sequestrene 138 Fe, which should be applied in January. A modern treatment for deficiencies is to use the spray called Foliar Feed. This is sprayed on to a crop and is said to supply all the necessary food elements through the leaves.

1

CHAPTER

FUNGAL AND BACTERIAL BIOCONTROL OF TOMATO

Robert P. Larkin and Deborah R. Fravel

Keywords: biological control, Cladorrhinum foecundissimum, Laetisaria arvalis, Lycopersicon esculentum, Stilbella aciculosa

INTRODUCTION

Fusarium wilt diseases, caused by pathogenic formae speciales of the soilinhabiting fungus Fusarium oxysporum Schlectend.:Fr., can cause severe losses in a wide variety of crop plants. On tomato, two symptomologically distinct forms of the pathogen can cause either a vascular wilt (F. oxysporum f. sp. lycopersici W. C. Snyder & H. N. Hans.) or a crown and root rot (F. oxysporum f. sp. radicis-lycopersici W. R. Jarvis & Shoemaker). Both of these pathogens occur throughout most tomato growing areas and either can devastate a crop. Although the use of Fusarium-resistant tomato cultivars can provide some degree of control of these diseases, the occurrence and development of new pathogenic races is a continuing problem, and currently there are no commercially acceptable cultivars with adequate resistance to F. oxysporum f. sp. radicis-lycopersici or race 3 of F. oxysporumfsp. lycopersici (13,14). These diseases are generally controlled in tomato (and several other crops) by preplant soil fumigation with methyl bromide (MBr). Tomatoes represent the largest single-crop use of MBr in the United States, accounting for 25% of the total MBr use for soil fumigation, or over 5,000 t/year (45). However, fumigation with MBr is expensive and not always effective. In addition to other potential health, safety, and environmental risks, MBr is classified as an ozone-depleting compound and, as required by the Clean Air Act, is scheduled to be removed

Citation: Robert P. Larkin and Deborah R. Fravel, Fungal and Bacterial Biocontrol of Tomato, http://dx.doi.org/10.1094/PDIS.1998.82.9.1022.

from the market in the United States by 2001. Therefore, alternative control measures are necessary and need to be made available as soon as possible. Biological control has potential for the management of these diseases. A variety of soil microorganisms have demonstrated activity in the control of various soilborne plant pathogens, including Fusarium wilt pathogens. Fusarium wilt-suppressive soils are known to occur in many regions of the world, and suppression has generally been shown to be biological in origin. Antagonists recovered from Fusarium wilt-suppressive soils, especially nonpathogenic F. oxysporum, have been used to reduce Fusarium wilt diseases of several different crops (1,18,35,37,39). Other biocontrol fungi, such as Trichoderma and Gliocladium spp., have been used to control a variety of fungal pathogens, including Rhizoctonia, Pythium, Sclerotinia, Sclerotium, and Fusarium spp. (11,26,27,30, 32,44), and may also be effective against Fusarium wilt diseases (5,34,43,49). In addition, several lesser-known groups of biocontrol fungi, including Laetisaria, Stilbella, Cladorrhinum, and Penicillium spp., have been used to control soilborne pathogens (6,25,28,29) and also may have activity against Fusarium diseases. Rhizobacterial strains of Pseudomonas, Burkholderia, and Bacillus spp. also have been used to reduce disease caused by a variety of soilborne pathogens (47,48), including Fusarium spp. (12,31,40,42,46). Although many different biocontrol strains have shown potential for some degree of control of Fusarium diseases, strains which can provide the best control of Fusarium wilt of tomato and have potential for effective implementation in commercial agriculture have not yet been identified. It has been suggested that microorganisms isolated from the root or rhizosphere of a specific crop may be better adapted to that crop and may provide better control of diseases than organisms originally isolated from other plant species (3). Such plantassociated microorganisms may make better biocontrol agents because they are already closely associated with and adapted to the plant or plant part as well as the particular environmental conditions in which they must function. The screening of such locally adapted strains has yielded improved biocontrol in some cases (3). The use of combinations of multiple antagonist organisms also may provide improved disease control over the use of single organisms. Multiple organisms may enhance the level and consistency of control by providing multiple mechanisms of action, a more stable rhizosphere community, and effectiveness over a wider range of environmental conditions. In particular, combinations of fungi and bacteria may provide protection at different times or under different conditions, and occupy different or complementary niches. Such combinations may overcome inconsistencies in the performance of individual isolates. Several researchers have observed improved disease control using various combinations of multiple compatible biocontrol organisms (7,8,38). Lemanceau and Alabouvette and others (19,22,36) have demonstrated enhanced biocontrol of Fusarium wilt by combining certain nonpathogenic strains of F. oxysporum with fluorescent strains of Pseudomonas. The objectives of this research were to examine existing biocontrol strains, including commercial formulations, with known activity against soilborne fungal pathogens for their efficacy in controlling Fusarium wilt of tomato, as well as to isolate potential new antagonists from the rhizosphere and roots of field-grown tomato plants and screen them for their biocontrol capabilities. In addition, preliminary investigations on the potential benefits of combining various antagonist organisms for the control of Fusarium wilt were performed. This research was conducted as a first step toward the development of effective biological control as an alternative strategy for the management of Fusarium wilt diseases. A preliminary report of aspects of this work has been published (16).

MATERIALS AND METHODS

Known biocontrol isolates.

Various fungi and bacteria with known biocontrol activity against soilborne fungal pathogens, including biocontrol isolates of Fusarium spp., Trichoderma spp., Gliocladium virens, Pseudomonas spp., Burkholderia cepacia, and others, were used in screening tests. Isolates of Fusarium spp. included several F. oxysporum and F. solani isolates collected from a Fusarium wilt-suppressive soil in Florida in previous studies (18) and F. oxysporum isolate Fo47 obtained from C. A. Alabouvette (Dijon, France; 1). Isolates of T. hamatum (TRI-4), T. harzianum (Th-87), T. viride (Tv-101), Laetisaria arvalis (Zh-1), Cladorrhinum foecundissimum (Cf-1), and Stilbella aciculosa (112-B2), as well as numerous isolates of G. virens (Gl-2 to Gl-21), were obtained from the collection of the United States Department of Agriculture-Agricultural Research Service (USDA-ARS), Biocontrol of Plant Diseases Laboratory (BPDL). Numerous isolates of Burkholderia cepacia, Pseudomonas fluorescens, and P. corrugata originally isolated from corn and soybean roots or soil were obtained from K. P. Hebbar, BPDL. Additional isolates of P. fluorescens (Pf-5 and C7) were obtained from J. E. Loper (Corvallis, OR) and C. Alabouvette, respectively. Fungal isolates were grown on potato dextrose agar (PDA) and bacterial isolates were grown on nutrient agar (NA). Commercially available formulations of G. virens strain Gl-21 (SoilGard, Thermo-Trilogy Corp., Columbia, MD) and T. harzianum strain T-22 (RootShield, Bio-Works, Inc., Geneva, NY) also were tested for efficacy in this system. Isolation of soil and root organisms.

Soil and roots were collected from established tomato field plots that had been planted with tomatoes for at least two consecutive years at the Beltsville Agricultural Research Center, Maryland. Tomato field soil was placed in flats in the greenhouse and planted with tomato seeds. After 2 weeks, seedlings were harvested and root systems with adhering rhizosphere soil were used to isolate rhizosphere and rhizoplane organisms as previously described (17). Segments of root systems collected directly from the field also were used. Root sections of approximately 0.2 g were added to 100 ml sterile water in flasks and shaken on a rotary shaker at 150 rpm for 30 min. Root segments, as well as a 10-fold dilution series of the resulting water suspensions, were plated on various general and selective media to recover root organisms. General bacteria were recovered using tryptic soy agar and NA, whereas Pseudomonas and Burkholderia spp. were preferentially recovered using King's medium B with antibiotics (41) and P. cepacia azaleic acid tryptamine (PCAT) medium (2), respectively. Trichoderma and Gliocladium spp. were recovered on a Trichoderma-selective medium (9), and Fusarium spp. were recovered using Komada's Fusarium-selective medium (KM; 15). Other fungi were recovered using PDA amended with 1 ml tergitol and 50 mg chlortetracycline per liter.

Screening trials

Tomato seeds of the cultivar "Bonny Best" were planted in soilless potting mix (Redi-Earth, ScottsSierra Inc., Marysville, OH), infested with the antagonist to be tested, and grown in seedling plug trays (plug size 3.4 by 3.4 by 5 cm, 98 plugs/tray). For initial screenings, isolates of nonpathogenic Fusarium spp. were grown for 7 to 10 days on PDA and 3 ml conidial suspensions of approximately 106 conidia/ml were added to each plug cell. Fungal antagonists Trichoderma and Gliocladium spp., L. arvalis, C. foecundissimum, and S. aciculosa were added to potting mix as wheat bran culture germlings (27) at the rate of 1% (wt/vol). Germlings were prepared by adding cell or conidial suspensions of approximately 106 /ml to autoclaved, moistened wheat bran (1:1 bran:water wt/vol) and incubated under ambient conditions (25°C) for 3 to 5 days. Bacterial antagonists were grown on NA for 3 to 5 days, then 3 ml of an aqueous suspension of approximately 108 bacterial cells/ml was added per plug of potting mix. Five tomato seeds were planted in each cell of the plug tray and maintained in the greenhouse. After 2 weeks, plugs containing the tomato plants were transplanted into 10-cm-diameter pots containing nonsterile field soil (Galestown gravelly loamy sand) infested with the Fusarium wilt pathogen. Pathogen inoculum consisted of a race 1 strain of F. oxysporum f. sp. lycopersici (strain IA-7 from R. W. Schneider, Louisiana State University) grown in a liquid culture of 1% ground soybean hull fiber (Dietfiber soy fiber, Lauhoff Grain Co., Danville, IL). Cultures were grown for 10 to 14 days on a rotary shaker at 150 rpm at 25°C, blended, and propagule counts determined on a hemacytometer. Inoculum consisted of a mixture of conidia and chlamydospores and was added to field soil at a rate of 104 CFU/g soil. Subsequent screening assays involving nonpathogenic Fusarium spp. used the liquid soybean hull fiber cultures to produce the antagonist as well as the pathogen inoculum as described above.

Commercial biocontrol preparations of G. virens and T. harzianum (SoilGard and RootShield granules, 1 × 106 and 1 × 107 CFU/g inoculum, respectively) were incorporated into Redi-Earth prior to planting tomato seeds at the rates of 0.05, 0.1, 0.2, and 0.5% (wt/vol). An incorporation rate of 0.5% corresponded to final propagule counts of approximately 7.5 × 104 and 1 × 105 CFU/g soilless mix for SoilGard and RootShield, respectively. Each treatment consisted of five replicate pots of four to five plants/pot. Disease was monitored for 4 to 6 weeks and assayed as the total percentage of seedlings showing any wilt symptoms due to the pathogen (yellowing and dropping of leaves, vascular discoloration, wilting, and so on). Stem sections of wilted seedlings were surface-disinfested in 0.5% sodium hypochlorite and plated on KM to confirm the presence of the wilt pathogen. Stem sections of asymptomatic plants were also plated at the conclusion of the experiment to evaluate potential pathogen infection. Additional assays on watermelon and muskmelon were conducted in the same manner using a race 1 isolate of F. oxysporum f. sp. niveum (strain CS93-8 from D. L. Hopkins, University of Florida) and watermelon cultivar "Crimson Sweet," and a race 1 isolate of F. oxysporum f. sp. melonis (strain 660K from E. Dutkey, University of Maryland) and melon cultivar "Goldstar." Biocontrol isolates that were most effective in individual assays were tested in various combinations using the same procedures as with individual isolates. Combinations tested included multiple Fusarium isolates; Fusarium isolates combined with bacteria, such as P. fluorescens or B. cepacia; and Fusarium isolates combined with G. virens. Treatments were prepared such that the total propagule counts for the combinations were the same as for the individual treatments.

Statistical analyses

All experiments were conducted at least twice. Statistical analyses were conducted using the general linear models procedures of the Statistical Analysis Systems version 6.08 (SAS Institute, Inc., Cary, NC). Experiments were analyzed using standard analysis of variance (ANOVA) with interactions. Significance was evaluated at $P < 0.05$ for all tests. Mean separation was accomplished using Fisher's protected least significant difference test. All data expressed as percentages were arcsine-transformed (sine–1 √ x) before analysis. Since repeated tests yielded similar results, data from a single representative experiment are presented.

Table 1. Summary of potential biocontrol organisms tested for their ability to reduce Fusarium wilt of tomato in field soil transplant tests.

		Reduction of disease[z]		
Organism group	**Total isolates tested**	**Effective isolates**	**Range of reduction**	**Average reduction**
Known biocontrol isolates				
Fusarium spp.	10	8	50–100%	66.6%
Trichoderma and *Gliocladium* spp.	14	4	37–75%	57.3%
Pseudomonas spp.	8	3	30–63%	46.1%
Burkholderia spp.	10	5	19–57%	36.4%
Other	4	2	35–61%	49.2%
Field isolates				
Bacteria	120	29	30–65%	43.2%
Fungi	60	14	40–66%	45.1%

[z] Values represent the number of isolates within each group that significantly ($P < 0.05$) reduced disease incidence (relative to pathogen-infested control soil), the range of percent reduction of disease for effective isolates, and the average percent reduction of disease for all effective isolates in that group.

Table 2. Isolates of nonpathogenic Fusarium spp. most effective in reducing Fusarium wilt of tomato in a greenhouse test

Isolate	% wilt	% reduction	Species	Source[y]
Pathogen-only	56.7 a[z]	0		
Fo47	28.0 b	50.6	*F. oxysporum*	France, Fusarium-suppressive soil
CS-21	26.5 b	53.3	*F. oxysporum*	Florida, wilt-suppressive soil
CS-10	17.2 bc	69.7	*F. oxysporum*	Florida, wilt-suppressive soil
CS-6	12.9 bc	77.3	*F. solani*	Florida, wilt-suppressive soil
CS-2	12.4 bc	78.1	*F. oxysporum*	Florida, wilt-suppressive soil
CS-24	8.6 c	84.9	*F. oxysporum*	Florida, wilt-suppressive soil
CS-20	6.7 c	88.2	*F. oxysporum*	Florida, wilt-suppressive soil
CS-1	0.0 c	100	*F. solani*	Florida , wilt-suppressive soil

[y] Sources: France (1) and Florida (17,18).

[z] Values followed by the same letter are not significantly different ($P < 0.05$) according to Fisher's protected least significant difference (LSD) test (LSD = 17.7).

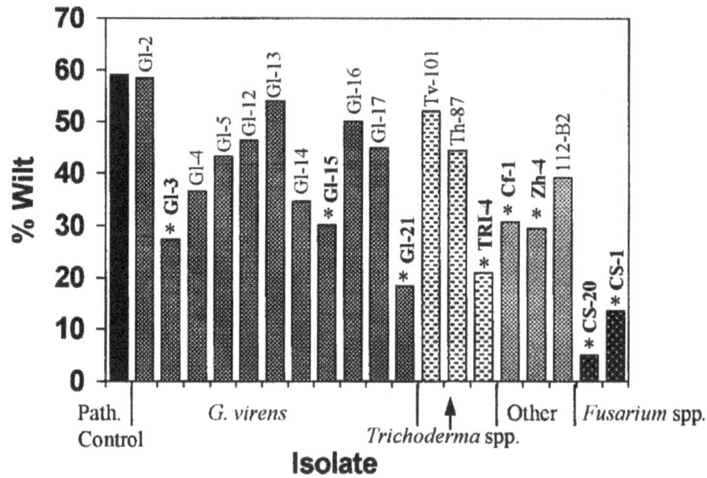

Figure 1. Development of Fusarium wilt in tomato seedlings as affected by treatments with various biocontrol fungi. Bars topped by an asterisk denote significant reduction of disease relative to the pathogen-infested control treatment according to Fisher's protected least significant difference (LSD) at P = 0.05 (LSD = 24). Data represent results of a single representative test.

RESULTS

Known biocontrol isolates

Individual isolates within all groups of the known biocontrol organisms, including isolates of F. oxysporum, T. hamatum, G. virens, P. fluorescens, and B. cepacia, significantly reduced Fusarium wilt of tomato in greenhouse tests (Table 1). Reductions in disease incidence ranged from 35 to 100% relative to the pathogen-infested controls. As a group, the most effective antagonists were the Fusarium isolates collected from wiltsuppressive soils, with 8 of 10 isolates tested reducing disease incidence by 50 to 100%. Isolate Fo47, which was collected from a wilt-suppressive soil in France (1) and is in the process of being developed as a biological control agent for the control of Fusarium diseases in Europe, reduced disease by 50% in a representative test (Table 2). Several isolates of Fusarium spp. collected from watermelon roots in a wiltsuppressive soil in Florida (18) reduced disease by up to 100% (Table 2). In general, the pathogen could not be isolated from stem sections of asymptomatic plants protected by the biocontrol treatments, indicating that the disease assay was effective in identifying all diseased plants, and that the asymptomatic plants were generally free of disease and did not merely represent attenuation of symptom development. These asymptomatic plants continued to remain healthy and developed normally even after the duration of these studies were concluded, suggesting that they would develop into productive fruiting plants. On the other hand, symptoms continued to worsen in diseased plants over time, eventually resulting in loss of all leaves and death of the plant. Several other fungal antagonists, including G. virens strain Gl-3, Gl-15, and Gl-21; T. hamatum strain TRI-4; L. arvalis strain Zh-4; and C. foecundissimum strain Cf-1; also significantly reduced Fusarium wilt in these screening tests (Fig. 1). The most effective of these other fungal antagonists were G. virens strain Gl-21, which reduced wilt incidence by 68%, and T. hamatum strain TRI-4, which reduced wilt incidence by 64%. However, even these isolates were not quite as effective as Fusarium isolates CS-20 and CS-1, which reduced wilt by 92 and 77%, respectively, in a corresponding test (Fig. 1). Several rhizobacterial isolates, including biocontrol strains of B. cepacia, P. fluorescens, and P. corrugata, also significantly reduced Fusarium wilt compared to the pathogenonly control (Fig. 2). However, although selected bacterial isolates reduced disease by 30 to 65% in individual tests, these isolates, in general, were not as effective or consistent in the reduction of disease as the nonpathogenic F. oxysporum isolates or some of the other fungal antagonists in repeated tests. The commercial biocontrol formulations, SoilGard and RootShield, each significantly reduced Fusarium wilt disease when granules were incorporated into the potting medium at the 0.2% level, but not at 0.1 or 0.05% (Table 3). At the 0.2% level, disease control (62 to 68% reduction) was comparable to that provided by Fusarium isolates CS-20 and CS-1 (73 to 78% reduction). SoilGard also was effective at the higher incorporation rate of 0.5%, whereas RootShield was not consistently effective at this higher rate in initial tests. However, in an additional test of SoilGard and RootShield at 0.2 and 0.5% incorporation rates, both formulations significantly reduced ($P < 0.05$) disease incidence compared to the pathogen-infested controls. Both formulations provided comparable disease reduction at both incorporation rates (rate effect and rate × treatment interaction not significant), averaging 27.8 and 38.5% disease incidence for the SoilGard and RootShield treatments, respectively, compared to 83.4% in the control treatment, representing reductions in disease incidence of 66.7 and 53.8% by SoilGard and RootShield, respectively. In this same test, disease incidence in CS-20- treated plants was 15.6%, for a disease reduction of 81.3% relative to the pathogen-infested controls

Field isolates

A total of 180 isolates, including 120 bacteria and 60 fungi, were collected from tomato roots and rhizosphere soil and tested for their ability to control Fusarium wilt of tomato in greenhouse screening tests. Of the 60 fungi tested, 18 were isolates of Fusarium spp., 24 were Trichoderma- or Gliocladium-type fungi, and 18 were unidentified. Bacterial isolates consisted of various unidentified rhizobacteria, including numerous Pseudomonas and Burkholderia-type isolates. Numerous isolates of these fungi and bacteria also significantly reduced Fusarium wilt of tomato by 30 to 65% relative to pathogen-infested controls (Table 1). Approximately 23% of the fungal isolates tested, including 7 of 18 Fusarium (39%), 5 of 24 Gliocladium or Trichoderma (21%), and 2 of 18 other fungi (11%), as well as 24% of the bacterial isolates tested, significantly reduced Fusarium wilt. However, since the level of disease control provided by these field isolates was generally not as good as and certainly no better than that provided by the nonpathogenic Fusarium spp. and several other of the known biocontrol isolates, no further tests were performed using these field isolates. Additional screening tests. Because the nonpathogenic Fusarium isolates were the most effective antagonists in all screening tests, additional tests concentrated primarily on these isolates. Subsequent testing on three different hosts of Fusarium wilt (tomato, watermelon, and muskmelon) determined that these Fusarium isolates were equally effective in reducing Fusarium wilt disease (50 to 80% disease reduction) on all three host crops (Table 4). Disease reduction by each isolate was consistent across all three hosts (host × treatment interaction was not significant), indicating no detectable host-specific effect on disease reduction. As in previous tests on tomato alone, isolates CS-1 and CS-20 were consistently the most effective isolates on all hosts, resulting in >80% overall reduction of disease incidence. In preliminary tests using combinations of antagonists, including multiple Fusarium strains, bacteria with Fusarium, and other fungi (G. virens) with Fusarium, all combinations tested significantly reduced Fusarium wilt of tomato relative to pathogen controls (Fig. 3). However, no combination of antagonists, including combinations of Fusarium and fluorescent pseudomonads, provided consistent or significant improvement on the disease control provided by nonpathogenic Fusarium strains alone. In fact, most combinations were not quite as effective in reducing disease incidence as were the most effective isolates of Fusarium alone, although there were no statistically significant (P > 0.05) differences among all the biocontrol treatments.

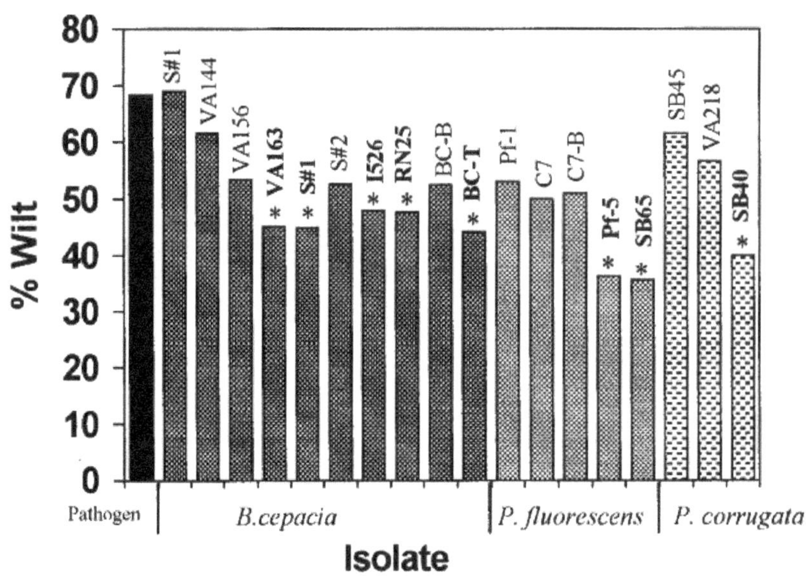

Figure 2. Development of Fusarium wilt in tomato seedlings as affected by treatments with various biocontrol bacteria. Bars topped by an asterisk denote significant reduction of disease relative to the pathogen-infested control treatment according to Fisher's protected least significant difference (LSD) at P = 0.05 (LSD = 18). Data represent results of a single representative test.

Table 3. Development of Fusarium wilt of tomato as affected by biocontrol treatments with two commercially available formulations at four application rates and with antagonistic Fusarium isolates

Treatment	Rate[y]	% wilt	% reduction
Pathogen-only		73.4	0
SoilGard	0.05	64.4	12
	0.1	47.6	39
	0.2	23.2*[z]	68
	0.5	21.4*	71
RootShield	0.05	57.0	22
	0.1	55.0	25
	0.2	28.0*	62
	0.5	43.0	41
CS-20 (*F. oxysporum*)		20.0*	73
CS-1 (*F. solani*)		16.0*	78
LSD ($P = 0.05$)		32.0	

[y] Rate = % incorporation (wt/vol).
[z] Values followed by an asterisk are significantly lower ($P < 0.05$) than the pathogen-only control treatment according to Fisher's protected least significant difference (LSD) test.

DISCUSSION

Several isolates of nonpathogenic Fusarium spp. collected from a Fusarium wilt-suppressive soil previously shown to reduce Fusarium wilt of watermelon (18) also were found to be highly effective in the control of Fusarium wilt of tomato, as well as other crops. Isolates CS-20 and CS- 1 were the most effective antagonists, consistently providing from 50 to 80% reduction of disease relative to pathogen-treated control plants over all tests and crops, and as high as 90 to 100% control in selected individual tests. Isolates of nonpathogenic Fusarium spp. originally isolated from watermelon roots were just as effective in controlling Fusarium wilt on tomato and muskmelon as on watermelon. Preliminary data on basil and spinach (unpublished) also indicated that control is provided for Fusarium wilt diseases on a variety of different crops. This suggests that the mechanism of action is not host-specific. Most of these isolates (7 of 8) came from a unique wilt-suppressive soil in Florida. This suppressive soil (CSS) developed as a result of prolonged monoculture to a particular cultivar of watermelon, which promoted suppressiveness (17,18).

Fusarium isolates collected from this soil were more effective in reducing Fusarium wilt disease, and effective isolates were more abundant in this soil than in a similar nonsuppressive soil (18). Other previously described biocontrol organisms, including G. virens, T. hamatum, P. fluorescens, and B. cepacia, also demonstrated varying degrees of efficacy in reducing Fusarium wilt of tomato. The most effective of these isolates, G. virens strain Gl-21, previously has been used to control damping-off diseases of various crops caused by Pythium ultimum, P. arrhenomanes, Rhizoctonia solani, and others (26,27,32), as well as Sclerotium rolfsii (30). However, this isolate had not been previously shown to be effective against Fusarium wilt diseases. Zhang et al. (49), using different G. virens isolates (G-4 and G-6), demonstrated significant reduction of Fusarium wilt of cotton and colonization of roots by F. oxysporum. Trichoderma spp., particularly T. harzianum, have been effective against diseases caused by F. oxysporum in studies by other researchers (5,34,43). In our tests, only the TRI-4 isolate of T. hamatum was effective in reducing disease, whereas isolates Th-87 and Tv-101 of T. harzianum and T. viride, respectively, were not consistently effective. This research also demonstrated for the first time that two other lesser known biocontrol agents, L. arvalis and C. foecundissimum, which had previously been shown to control R. solani (25,28) also have biocontrol activity against Fusarium wilt diseases. Commercial biocontrol formulations of G. virens strain Gl-21 (SoilGard) and T. harzianum strain T-22 (RootShield) were also effective at reducing disease when incorporated at rates of 0.2% or higher. The recommended rates for control of root rot and damping-off pathogens for these products are 0.6 to 0.9 kg/m3 or 0.06 to 0.09% (wt/vol).

However, in our tests, neither product was effective against Fusarium wilt at incorporation levels of 0.1% or below. Thus, higher levels of these antagonists apparently are necessary for adequate control of Fusarium wilt of tomato. In our assays, 0.1% incorporation corresponded to approximately 1 to 2 × 104 CFU/g soilless mix or less, whereas 0.2% incorporation or higher corresponded to propagule counts of approximately 3 to 4 × 104 CFU/g mix and higher for both products. Nonetheless, an incorporation rate of 0.2% still involves a relatively small amount of formulated product (2 kg/m^3) and should not affect the practical use of these products for greenhouse and transplant applications. Thus, this research has also demonstrated that commercial biocontrol products already available on the market were effective in controlling Fusarium wilt of tomato. The active ingredient of RootShield had also previously been shown to reduce Fusarium crown and root rot of tomato in the field (5). Several of the biocontrol bacterial strains of B. cepacia and Pseudomonas fluorescens provided significant reductions of disease incidence, but generally did not provide control levels greater than approximately 30 to 40% reduction, which

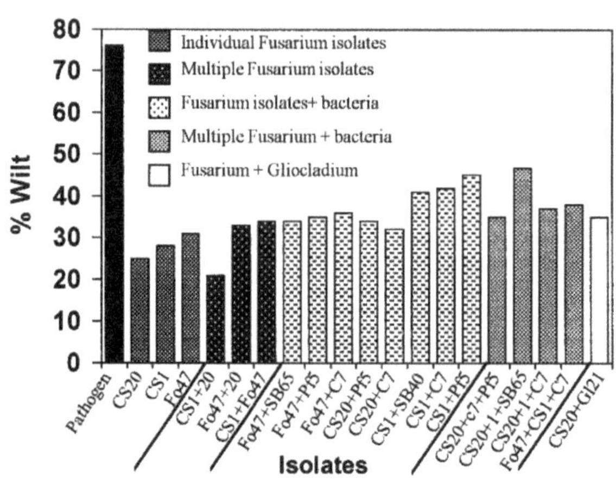

Figure 3. Development of Fusarium wilt in tomato seedlings as affected by treatments with various combinations of biocontrol organisms. All antagonist treatments significantly reduced disease relative to the pathogen-infested control treatment according to Fisher's protected least significant difference (LSD) at P = 0.05 (LSD = 22), but there were no differences among antago-

nists. Other organisms tested included Cladorrhinum foecundissimum (Cf-1), Laetisaria arvalis (Zh-4), and Stilbella aciculosa (112-B2). Data represent results of a single representative test.

Table 4. Effectiveness of selected nonpathogenic strains of Fusarium spp. in controlling Fusarium wilt diseases of tomato, watermelon, and muskmelon

	Overall[y]		Tomato		Watermelon		Muskmelon	
Isolate	% wilt	% reduction	% wilt	% reduction	% wilt	% reduction	% wilt	% reduction
Pathogen only	58.7	0.0	54	0	56	0	66	0
CS-2	29.3*[z]	50.3*	28	47	20*	64*	40	39
CS-10	24.1*	57.8*	26	51	27*	53*	20*	70*
CS-6	23.1*	60.2*	25*	54*	20*	64*	25*	63*
Fo47	20.4*	64.7*	24*	55*	20*	64*	17*	74*
CS-21	18.8*	67.4*	22*	59*	19*	67*	16*	76*
CS-24	17.3*	69.6*	25*	53*	12*	79*	16*	76*
CS-20	10.7*	81.5*	4*	92*	20*	64*	8*	88*
CS-1	10.6*	81.6*	11*	79*	10*	82*	11*	84*

[y] Overall results represent combined data over all three hosts (isolate × host interactions were not significant). Disease incidence (% wilt) is represented as the percentage of wilted plants over a 4-week period. Disease reduction (% reduction) represents the percent reduction of disease relative to the pathogen-only control.

[z] Values followed by an asterisk represent a significant reduction of disease relative to the pathogen-only control according to Fisher's protected least significant difference test ($P < 0.05$).

under severe disease pressure resulted in wilt incidences of approximately 40 to 50%. This level of control is comparable to those reported by others using rhizobacterial strains for the control of Fusarium wilt diseases, including those using bacterial isolates that induce resistance within the host plant (20,31,40,46). However, for development as biocontrol agents, this level of disease reduction is generally not sufficient for effective and consistent disease control. Attempts to find superior biocontrol microorganisms through isolation of tomato root-colonizing organisms were not successful in this study. Although several isolates significantly reduced disease, no tomato root isolate tested resulted in biocontrol performance as good or better than that produced by existing biocontrol strains. However, this does not suggest that this approach does not have merit. Ideally, organisms should be isolated from plants grown in fields where both tomato plants and the pathogen have been co-existing for several years (3).

Such fields were not available for the present study. Fusarium wilt was not observed to be present in the fields we sampled; therefore, effective antagonist populations probably were not present or were present at such low numbers that our relatively small sample did not recover them. Approximately 20 to 25% of the isolates tested (both fungal and bacterial) showed some activity to suppress Fusarium wilt. These numbers are comparable to the percentage of antagonists isolated from soils in other studies, such as in a nonsuppressive soil in Florida (18). The primary mechanisms of action associated with biocontrol by nonpathogenic Fusarium strains include competition, either for nutrients in the rhizosphere or infection sites on the root, and induced systemic resistance (1,18,33). Isolate Fo47 has previously been shown to involve saprophytic competition as a primary mechanism of action (1,4,21). Recent evidence also suggests that protection by Fo47 also involves at least some degree of induced resistance (10,16).

Some isolates from the Florida CSS wilt-suppressive soil, including CS-6, were previously shown to induce resistance in watermelon, as determined by split-root tests (18). Preliminary results of split-root tests using isolates CS-20 and CS-1 on tomato and watermelon indicated that these isolates also function primarily through induced resistance rather than through competition (16). In this study, combinations of antagonist organisms, including multiple Fusarium isolates, Fusarium and bacterial strains, and Fusarium and Gliocladium strains, did not improve the level of control observed using the Fusarium isolates alone. Lemanceau and Alabouvette (22) reported improved biocontrol performance when Fusarium isolate Fo47 was used in combination with a fluorescent Pseudomonas strain (C7). Other researchers (19,36) have also reported a similar enhanced response with combinations of Fusarium and Pseudomonas spp.

The mechanism involves competition for carbon provided by F. oxysporum and competition for iron provided by the bacteria. Evidence suggests that the presence of the siderophore produced by P. fluorescens increases the intensity of the antagonism provided by Fo47, making the pathogen more sensitive to glucose competition by the antagonistic Fusarium (23,24). However, in our tests, we did not observe this response, even in combinations using strain Fo47. This may be due to differences in the application or assay methods, or the environmental or soil conditions present. Isolates which function primarily through induced resistance instead of competition also may not show any enhanced control in combination with fluorescent pseudomonads. Even though no improvement in disease control over that produced by F. oxysporum alone could be detected with the use of combinations in this study, combinations may still be advantageous for other reasons, including control of multiple pathogens, more consistent control, or control over a wider range of environmental conditions, which were not evaluated in this study. Further research on the potential advantages of using combinations of these effective antagonists under varying stress conditions is needed. Although only a limited number of potential biocontrol isolates could be tested in this study, representative isolates of a variety of biocontrol species were tested. Several of the Fusarium isolates, particularly CS-20 and CS-1, have potential for further development as biological control agents for the control of Fusarium wilt diseases.

These isolates were highly effective and consistently more effective in controlling Fusarium wilt than all other isolates tested. The level of disease control observed was also comparable or better than that reported in other studies using nonpathogenic strains of Fusarium spp. (1,35,37,39), and generally better than the control reported in studies using other biocontrol organisms, such as rhizobacteria (20,31,40,46,49) or other fungal antagonists (5,6,43,49). Identification of effective antagonist strains represents only the first step toward the development of effective biological control. In order for biocontrol to be implemented on a practical level, the antagonists must be ecologically fit to survive, become established, and function within the particular conditions of the ecosystem. To attain this, much more information regarding the mechanisms of action, ecological characteristics, and interactions with the soil and rhizosphere microbial communities is needed. Thus, now that we have identified several organisms as potential antagonists, we are continuing to work with these strains to determine the specific mechanisms, interactions, conditions, and requirements responsible for effective biological control. Through an understanding of these characteristics, we can establish the limitations as well as the full potential for biocontrol within this pathosystem, and develop strategies for its implementation and management.

ACKNOWLEDGMENTS

We thank C. Tate for valuable technical assistance; C. Alabouvette, K. P. Hebbar, and J. E. Loper for providing biocontrol isolates; E. Dutkey, D. L. Hopkins, and R. W. Schneider for providing pathogen isolates; and Thermo-Trilogy Corp. and Bio-Works Inc. for providing samples of their biocontrol products.

REFERENCES

1. Alabouvette, C., Lemanceau, P., and Steinberg, C. 1993. Recent advances in the biological control of Fusarium wilts. Pestic. Sci. 37:365-373.
2. Burbage, D. A., Sasser, M., and Lumsden, R. D. 1982. A medium selective for Pseudomonas cepacia. (Abstr.) Phytopathology 72:706.
3. Cook, R. J. 1993. Making greater use of introduced microorganisms for biological control of plant pathogens. Annu. Rev. Phytopathol. 31:53-80.
4. Couteaudier, Y. 1992. Competition for carbon in soil and rhizosphere; a mechanism involved in biological control of Fusarium wilts. Pages 99-104 in: Biological Control of Plant Diseases: Progress and Challenges for the Future. E. C. Tjamos, G. C. Papavizas, and R. J. Cook, eds. Plenum Press, New York.
5. Datnoff, L. E., Nemec, S., and Pernezny, K. 1995. Biological control of Fusarium crown and root rot of tomato in Florida using Trichoderma harzianum and Glomus intraradices. Biol. Control 5:427-431.
6. De Cal, A., Pascual, S., Larena, I, and Melgarejo, P. 1995. Biological control of Fusarium oxysporum f. sp. lycopersici. Plant Pathol. 44:909-917.
7. Duffy, B. K., Simon, A., and Weller, D. M. 1996. Combination of Trichoderma koningii with fluorescent pseudomonads for control of take-all on wheat. Phytopathology 86:188- 194.
8. Duffy, B. K., and Weller, D. M. 1995. Use of Gaemannomyces graminis var. graminis alone and in combination with fluorescent Pseudomonas spp. to suppress take-all of wheat. Plant Dis. 79:907-911.
9. Elad, Y., Chet, I., and Henis, Y. 1981. A selective medium for improving quantitative isolation of Trichoderma spp. from soil. Phytoparasitica 9:59-67.
10. Fuchs, J. -G., Moënne-Loccoz, Y., and Défago, G. 1997. Nonpathogenic Fusarium oxysporum strain Fo47 induces resistance to Fusarium wilt of tomato. Plant Dis. 81:492- 496.
11. Harman, G. E. 1991. Seed treatments for biological control of plant disease. Crop Prot. 10:166-171.
12. Hebbar, K. P., Atkinson, D., Tucker, W., and Dart, P. J. 1992. Suppression of Fusarium moniliforme by maize root-associated Pseudomonas cepacia. Soil Biol. Biochem. 24:1009-1020.
13. Jarvis, W. R. 1988. Fusarium crown and root rot of tomatoes. Phytoprotection 69:49-64.
14. Jones, J. B., Jones, J. P., Stall, R. E., and Zitter, T. A., eds. 1991. Compendium of

Tomato Diseases. American Phytopathological Society, St. Paul, MN.

15. Komada, H. 1975. Development of a selective medium for quantitative isolation of Fusarium oxysporum from natural soil. Rev. Plant Prot. Res. 8:115-125.

16. Larkin, R. P., and Fravel, D. R. 1996. Efficacy of various biocontrol organisms in the control of Fusarium wilt of tomato. Phytopathology 86:S83.

17. Larkin, R. P., Hopkins, D. L., and Martin, F. N. 1993. Ecology of Fusarium oxysporum f. sp. niveum in soils suppressive and conducive to Fusarium wilt of watermelon. Phytopathology 83:1105-1116.

18. Larkin, R. P., Hopkins, D. L., and Martin, F. N. 1996. Suppression of Fusarium wilt of watermelon by nonpathogenic Fusarium oxysporum and other microorganisms recovered from a disease-suppressive soil. Phytopathology 86:812-819.

19. Leeman, M., Den Ouden, F. M., van Pelt, J. A., Cornelissen, Matamala-Garros, A., Bakker, P. A. H. M., and Schippers, B. 1996. Suppression of Fusarium wilt of radish by co-inoculation of fluorescent Pseudomonas spp. and root-colonizing fungi. Eur. J. Plant Pathol. 102:21-31.

20. Leeman, M., Den Ouden, F. M., van Pelt, J. A., Hendrickx, M. J., Scheffer, R., Bakker, P. A. H. M., and Schippers, B. 1995. Biocontrol of Fusarium wilt of radish in commercial greenhouse trials by seed treatment with Pseudomonas fluorescens WCS374. Phytopathology 85:1301-1305.

21. Lemanceau, P. 1989. Role of competition for carbon and iron in mechanisms of soil suppressiveness to Fusarium wilts. Pages 385- 395 in: Vascular Wilt Diseases of Plants. E. C. Tjamos and C. Beckman, eds. Springer-Verlag, New York.

22. Lemanceau, P., and Alabouvette, C. 1991. Biological control of fusarium diseases by fluorescent Pseudomonas and non-pathogenic Fusarium. Crop Prot. 10:279-286.

23. Lemanceau, P., Bakker, P. A. H. M., de Kogel, W. J., and Alabouvette, C. 1992. Effect of pseudobactin 358 production by Pseudomonas putida WCS358 on suppression of Fusarium wilt of carnations by nonpathogenic Fusarium oxysporum Fo47. Appl. Environ. Microbiol. 58:2978-2982.

24. Lemanceau, P., Bakker, P. A. H. M., De Kogel, W. J., Alabouvette, C., and Schippers, B. 1993. Antagonistic effect on nonpathogenic Fusarium oxysporum strain Fo47 and pseudobactin 358 upon pathogenic Fusarium oxysporum f. sp. dianthi. Appl. Environ. Microbiol. 59:74-82.

25. Lewis, J. A., Fravel, D. R., and Papavizas, G. C. 1995. Cladorrhinum foecundissimum: a potential biological control agent for the reduction of Rhizoctonia solani. Soil Biol. Biochem. 27:863-869.

26. Lewis, J. A., Lumsden, R. D., and Locke, J. C. 1996. Biocontrol of damping-off diseases caused by Rhizoctonia solani and Pythium ultimum with alginate prills of Gliocladium virens, Trichoderma hamatum, and various food bases. Biocontrol Sci. Technol. 6:163- 173.

27. Lewis, J. A., and Papavizas, G. C. 1985. Effect of mycelial preparations of Trichoderma and Gliocladium on populations of Rhizoctonia solani and the

incidence of damping-off. Phytopathology 75:812-817.

28. Lewis, J. A., and Papavizas, G. C. 1992. Potential of Laetisaria arvalis for the biocontrol of Rhizoctonia solani. Soil Biol. Biochem. 24:1075-1079.

29. Lewis, J. A., and Papavizas, G. C. 1993. Stilbella aciculosa: a potential biocontrol fungus against Rhizoctonia solani. Biocontrol Sci. Technol. 3:3-11.

30. Lewis, J. A., Papavizas, G. C., and Hollenbeck, M. D. 1993. Biological control of damping-off of snapbeans caused by Sclerotium rolfsii in the greenhouse and field with formulations of Gliocladium virens. Biol. Control 3:109-115.

31. Liu, L., Kloepper, J. W., and Tuzun, S. 1995. Induction of systemic resistance in cucumber against Fusarium wilt by plant growth-promoting rhizobacteria. Phytopathology 85:695- 698.

32. Lumsden, R. L., and Locke, J. C. 1989. Biological control of damping-off caused by Pythium ultimum and Rhizoctonia solani with Gliocladium virens in soilless mix. Phytopathology 79:361-366.

33. Mandeel, Q., and Baker, R. 1991. Mechanisms involved in biological control of Fusarium wilt of cucumber with strains of nonpathogenic Fusarium oxysporum. Phytopathology 81:462-469.

34. Marois, J. J., Mitchell, D. J., and Sonoda, R. M. 1981. Biological control of Fusarium crown and root rot of tomato under field conditions. Phytopathology 71:1257-1260.

35. Minuto, A., Migheli, Q., and Garabaldi, A. 1995. Evaluation of antagonistic strains of Fusarium spp. in the biological and integrated control of Fusarium wilt of cyclamen. Crop Prot. 14:221-226.

36. Park, C. S., Paulitz, T. C., and Baker, R. 1988. Biocontrol of Fusarium wilt of cucumber resulting from interactions between Pseudomonas putida and nonpathogenic isolates of Fusarium oxysporum. Phytopathology 78:190-194.

37. Paulitz, T. C., Park, C. S., and Baker, R. 1987. Biological control of Fusarium wilt of cucumber with nonpathogenic isolates of Fusarium oxysporum. Can. J. Microbiol. 33:349-353.

38. Pierson, E. A., and Weller, D. M. 1994. Use of mixtures of fluorescent pseudomonads to suppress take-all and improve the growth of wheat. Phytopathology 84:940-947.

39. Postma, J., and Rattink, H. 1992. Biological control of Fusarium wilt of carnation with a non-pathogenic isolate of Fusarium oxysporum. Can. J. Bot. 70:1199-1205.

40. Raaijmakers, J. M., Leeman, M., van Oorschot, M. M. P., van der Sluis, I., Schippers, B., and Bakker, P. A. H. M. 1995. Dose-response relationships in biological control of Fusarium wilt of radish by Pseudomonas spp. Phytopathology 85:1075-1081.

41. Sands, D. C., and Rovira, A. R. 1970. Isolation of fluorescent pseudomonads with a selective medium. Appl. Microbiol. 20:513-514.

42. Scher, F. M., and Baker, R. 1982. Effect of Pseudomonas putida and a synthetic iron chelator on induction of suppressiveness to Fusarium wilt pathogens.

Phytopathology 72:1567-1573.

43. Sivan, A., and Chet, I. 1993. Integrated control of fusarium crown and root rot of tomato with Trichoderma harzianum in combination with methyl bromide or soil solarization. Crop Prot. 12:380-386.

44. Taylor, A. G., Harman, G. E., and Nielsen, P. A. 1994. Biological seed treatments using Trichoderma harzianum for horticultural crops. Hortic. Technol. 4:105-108.

45. UNEP Methyl Bromide Technical Options Committee. 1994. Montreal protocol on substances that deplete the ozone layer: 1994 report of the MBTOC. EPA-430/K94/029.

46. Van Peer, R., Niemann, G. J., and Schippers, B. 1991. Induced resistance and phytoalexin accumulation in biological control of Fusarium wilt of carnation by Pseudomonas sp. strain WCS417r. Phytopathology 81:1508- 1512.

47. Weller, D. M. 1988. Biological control of soilborne plant pathogens in the rhizosphere with bacteria. Annu. Rev. Phytopathol. 26:379-407.

48. Weller, D. M., and Cook, R. J. 1983. Suppression of take-all of wheat by seed treatments with fluorescent pseudomonads. Phytopathology 73:463-469.

49. Zhang, J., Howell, C. R., and Starr, J. L. 1996. Suppression of Fusarium colonization of cotton roots and Fusarium wilt by seed treatments with Gliocladium virens and Bacillus subtilis. Biocontrol Sci. Technol. 6:175-187.

2
CHAPTER

DEVELOPMENT OF FUNGICIDE CROSS RESISTANCE

M. G. Cunha and D. M. Rizzo,

Keywords: benzimidazole, potato tuber disease, silver scurf

INTRODUCTION

Helminthosporium solani, causal agent of silver scurf of potato tubers, has assumed great importance with the development of washing and prepacking of potato tubers because it often adversely affects the appearance and skin color (20,22). Soon after tuber initiation, H. solani can infect the progeny tubers. Lesions spread slowly on the surface of progeny tubers when in the soil; more rapid spread takes place during storage (16,22,29). The application of benzimidazole fungicides was the primary means of disease until appearance of fungicide resistance among field isolates of H. solani in the mid-1990s (28,32). Benzimidazole resistance has developed independently in a number of fungal species and it has always been associated with point mutations in the β-tubulin gene (18,33). This gene has been mapped, cloned, and sequenced from a variety of ascomycete fungi (7,19,34,35). A single base-pair point mutation in the β-tubulin gene leading to substitution of glutamic acid for alanine or glutamine was detected at codon 198 of thiabendazole-resistant isolates of H. solani (27). Mutations leading to substitutions at amino acids 198 and 200 conferred benomyl resistance in field isolates of Botrytis cinerea (35) and mutations mapped to the first position of codon 6 in Septoria nodorum produced a benomyl-resistant phenotype (7). In addition, it recently was found that sensitive isolates of Rhynchosporium secalis can become

Citation: Cunha, M. G., and Rizzo, D. M. 2003. Models for Predicting Potential Yield Loss of Wheat Caused by Stripe Rust. 87:798-803.

benzimidazole-resistant by transformation with a -tubulin gene carrying a point mutation (4). Methods for detection of benzimidazole resistance in plant pathogens recently have been improved. In addition to the classical methods that used fungicide-amended medium, molecular approaches have been developed. Wheeler et al. (34) accurately detected carbendazim-resistant isolates of R. secalis by combining polymerase chain reaction (PCR) amplification of partial β- tubulin gene sequences with hybridization of allele-specific probes. McKay and Cooke (27) identified thiabendazole resistance in H. solani by PCR amplification of a portion of its β-tubulin gene followed by restriction digest of the PCR product. Molecular approaches are very efficient and rapid, which makes them suitable to evaluate the extent of fungicide resistance in fungal field populations. We recently reported the first occurrence of silver scurf in California and observed its progress in several commercial potato sheds in the Tulelake area (8), which is the major potato-growing area in California for the fresh table market. In this region, growers usually store potato tubers for 7 months, during which time the incidence of silver scurf increases sharply, affecting up to 95% of the tubers (8). Using seven isolates of H. solani isolated from tubers collected from potato sheds and packaging plants in Tulelake, a preliminary fungicide resistance survey revealed the presence of thiophanate-methyl resistance among H. solani isolates from California, but no inference could be made about the significance of this fungicide resistance due to the small number of isolates (8). Resistance of H. solani to thiabendazole or benomyl fungicides has been associated with outbreaks of silver scurf in several potato-growing regions (5,15,21,28). It has been shown that seed treatments or postharvest application of these fungicides will not control silver scurf during storage once the majority of H. solani isolates have developed fungicide resistance (15,17,21). It often has been observed in other plantpathogenic fungi that isolates resistant to one benzimidazole also have reduced sensitivity to the other benzimidazoles (24,25,31); however, cross-resistance among benzimidazole fungicides has not been reported in H. solani. This study was conducted to determine the frequency of resistance to thiophanate-methyl and benomyl in H. solani populations from Tulelake, CA, frequently treated with thiophanate-methyl, a benzimidazole-related fungicide, but not benomyl. Thiophanate-methyl is not a true benzimidazole (11); therefore, molecular changes in the β-tubulin gene due to recurrent application of this fungicide were examined and compared with those changes associated with thiabendazole resistance. Finally, the molecular approach developed by McKay and Cooke (27) for identifying thiabendazole resistance was assessed for detecting thiophanate-methyl and benomyl resistance in H. solani populations in California.

MATERIALS AND METHODS

Collection of H. solani isolates.

In 1998, three geographically isolated potato fields located in Tulelake, CA, were sampled 1 week before harvest. Potato was not grown in these fields in the previous 3 years, but silver scurf had been reported previously on potato crops in these fields. Before planting, certified seed potato plants were treated with thiophanatemethyl (Tops

2.5D, 2.5% a.i.; Haco, Inc., Madison, WI) as recommended by its manufacturer. Thirty-six plants were selected arbitrarily and manually harvested from each potato field (leases 8240-82-41, 8310, and SC-92-2 in the Tulelake irrigation district map). These fields are referred to as fields 1, 2, and 3, respectively, throughout the remainder of the article. All tubers from each single plant represented a sample. Tubers were stored for approximately 15 days at 5°C until processed. The tubers were removed from storage, washed in water, disinfested in a 2.5% NaOCl solution for 3 min, dried, and incubated at 16°C in the dark at high relative humidity (>90%) for 15 to 21 days. Using a dissecting microscope, H. solani conidia produced on individual lesions of infected tubers were removed with a sterile needle and placed into 0.5 ml of antibiotic solution (0.25% streptomycin sulfate) on 2% water agar in a petri dish. This spore suspension then was manually agitated to disperse H. solani conidia over the medium surface. Conidia were incubated at 25°C in the dark for 4 to 6 days. Using a dissecting microscope, single germinating conidia or hyphal tips of monoconidial colonies were transferred to petri dishes with clarified V8 juice agar (V8A) (12) amended with a 0.25% streptomycin sulfate solution at 20 ml/liter and incubated as above for 30 days. One to five isolates were collected from each plant, but each isolate originated from a different progeny tuber. Isolates were identified as H. solani based on morphological features (2,13,14). In addition to isolates from California, isolate HSME27 from Maine, which is resistant to thiabendazole (G. Secor, North Dakota State University, personal communication), and isolate 1-SS3 from New York, which is thiabendazole sensitive (28), were included in all assays as positive and negative controls. All isolates were maintained on V8A at 5°C.

Field fungicide resistance survey

Commercial formulations of thiophanatemethyl (Tops 2.5D , 2.5% a.i.; Haco, Inc.) and benomyl (Benlate, 50% a.i.; E.I. DuPont de Nemours Co., Inc., Wilmington, DE) were employed in all trials. To construct a representative fungicide dose-response curve, an assay was conducted using the isolate HSCA06, arbitrarily chosen from among resistant isolates from California detected in the preliminary fungicide resistance survey. Either benomyl or thiophanate-methyl was added to V8A before autoclaving to achieve a suspension of 0, 1.0, 3.0, 10, 30, 100, 300, or 1,000 mg a.i./liter. Plugs 0.5 cm in diameter were taken from the edges of 30-dayold cultures and placed onto six replicate plates. After incubating for 30 days in the dark at room temperature, two perpendicular colony diameters were measured on the bottom of each plate. The diameter of the agar plug was subtracted from each measurement. The average of the two colony diameters was determined for each plate. Thiophanate-methyl or benomyl dose-response curves were constructed for isolate HSCA06, fitting a quadratic function to the relative radial growth data (Excel 97; Microsoft Corporation, Redmond, WA). The equations for the fitted curves were used to calculate the effective concentrations that reduced growth by 50% (EC50) for each isolate. We collected 121, 74, and 41 isolates of H. solani from potato fields 1, 2, and 3, respectively; all were tested for resistance to either thiophanate-methyl or benomyl. Agar plugs from the edge of actively growing colonies were transferred to four replicate petri dishes containing clarified V8A amended with either benomyl at 35 mg a.i./liter or thiophanate-methyl at 400 mg a.i./liter. The control petri dishes had no fungicide added to the clarified V8A medium. All cultures were incubated and evaluated as described above. Isolates were considered either resistant if growth was observed or sensitive if no growth occurred. A 2 analysis was performed to test the hypothesis that fungicide resistance is uniformly spread among potato fields in Tulelake.

DNA extraction

Isolates of H. solani were grown on V8A covered with a sterile cellophane sheet. After 2 weeks, a small amount of mycelium was removed from the edge of colonies for DNA extraction. The rapid DNA extraction protocol described by Cenis (6) was followed for extracting the fungal total genomic DNA. The DNA concentration was adjusted to 10 ng/μl with Tris-EDTA (TE) buffer. All DNA extractions were maintained at –20°C.

PCR amplification. DNA amplifications were performed in 50-μl total volume reactions containing 60 ng of DNA-template, 120 μM of each deoxyribonucleotide triphosphate, dNTPs (dATP, dGTP, dCTP, dTTP), 75 ng of SS-forward primer (5- AGCATAGGCTGATGCTCGT), 75 ng of SS-reverse primer (5-GACGATGAGTCCTGAGTAA) specific for amplifying the -tubulin gene of H. solani (27), 1.5 U of taq polymerase (Promega Corp., Madison, WI), 5.0 mM Tris-HCl, 10 mM NaCl, 0.01 mM EDTA, 0.1 mM dithiothreitol, 5% glycerol, 0.1% Triton X-100, 2.5 mM MgC12, and 3 μl of 0.01% Blotto milk suspension (10). As a negative control, 6 μl of purified water (milliQ) was used instead of fungal DNA. All reactions were overlaid with approximately 50 μl of mineral oil. The amplification was performed on a Gradient 96 Robocycler (Stratagene Cloning Systems, La Jolla, CA). DNA was denatured for 3 min at 94°C, followed by 40 cycles of denaturation at 94°C for 45 s, annealing at 58°C for 1 min, extension at 72°C for 1.5 min, and a final extension at 72°C for 5 min. After amplification, PCR products were electrophoresed on 1% agarose gel in 1× Trisborate EDTA (TBE) buffer at constant power, 60 volts, for approximately 2.5 h, stained in ethidium bromide (0.5 μg/ml) and visualized and recorded with a UVP ImageStore 7500 white/UV Transilluminator (UVP Inc., Upland, CA).

DNA sequencing

To study molecular changes in the β-tubulin gene of H. solani associated with recurrent application of thiophanate-methyl on potato seed tubers, total DNA was extracted from two isolates showing either resistance or sensitivity to thiophanate-methyl from each potato field. A section of the -tubulin gene from each isolate was amplified as outlined above. The PCR-amplified β-tubulin gene was initially cloned using the original TA cloning kit (Invitrogen, Carlsbad, CA). Plasmid extraction was carried out with QIAprep miniprep kit (QIAGEN, Inc., Valencia, CA). Purified plasmid concentration was adjusted to 150 to 200 ng/μl with EB buffer (10 mM Tris-Cl, ph 8.5). Both DNA strands of either two or three β-tubulin clones from each isolate of H. solani were sequenced with an ABI PRISM 377 DNA Sequencer (PerkinElmer/ABI, Forest City, CA) at the University of California DNA sequencing facility. M13F and M13R primers (Invitrogen, Carlsbad, CA) were used for sequencing. For determining the entire sequence of the amplified β-tubulin gene fragment, all sequences, including forward and reverse sequences of the same H. solani isolate, were aligned with the PRETTY program of GCG Sequence Analysis Software Package (Genetic Computer Group, Madison, WI). Aligned sequences were edited using a word processor (Microsoft Word 97) and adjustments were performed manually on the consensus sequences for resolving the β-tubulin gene fragment delimited by the SS-forward primer and the SS-reverse primer. I

Identification of thiophanate-methyl resistance by PCR-based restriction fragment length polymorphism assay

Molecular analysis of the -tubulin gene fractions of H. solani isolates from California revealed a similar mutation pattern to thiabendazole resistance; therefore, the PCR-based restriction fragment length polymorphism (RFLP) technique used to identify thiabendazole resistance developed by McKay and Cooke (27) was evaluated for detecting thiophanate-methyl and benomyl resistance in H. solani populations from California. Ten isolates identified as resistant or sensitive to both thiophanate-methyl and benomyl by assay on the fungicide-amended medium were selected from each potato field. Total genomic DNA was extracted, and a section of the -tubulin gene was amplified as described previously. The amplified -tubulin gene was digested with BsaI (New England Biolabs, Beverly, MA). The digest was performed in approximately 45 μl of total volume that contained 40 μl of PCR product and 5 μl (25 units) of BsaI. After incubating at 50°C for 2 h, electrophoresis for resolving β- tubulin gene fragments was carried out with 15 μl of the restriction digest on a 2.0% Seakem LE agarose gel (BMA, Rockland, ME) and 1× TBE buffer at constant power, 60 volts, for approximately 6.5 h. The gel was stained in ethidium bromide, visualized, and recorded as outlined above.

RESULTS

The relative growth of isolate HSCA06 challenged with increasing concentrations of either thiophanate-methyl or benomyl fungicides were fitted on concentrationresponse curves (Fig. 1). Quadratic equations that best described these concentration response curves were y = –0.0639x2 – 0.0371x + 1.0277 (R2 = 0.94) for thiophanate-methyl and y = –0.1115x2 – 0.1223x + 0.9455 (R2 = 0.94) for benomyl (y = relative colony diameter and x = log10 fungicide concentration, mg/liter). These equations were used for calculating fungicide EC50 values. The EC_{50} values were approximately 400 mg/liter for thiophanatemethyl and 35 mg/liter for benomyl. These EC_{50} concentrations then were used to assess three H. solani populations for thiophanate-methyl and benomyl resistance, respectively. In all studied populations, the majority of H. solani isolates exhibited cross-resistance to both fungicides. Of 238 tested isolates, 182 were resistant to both thiophanate-methyl and benomyl, while 56 isolates were sensitive to both of them.

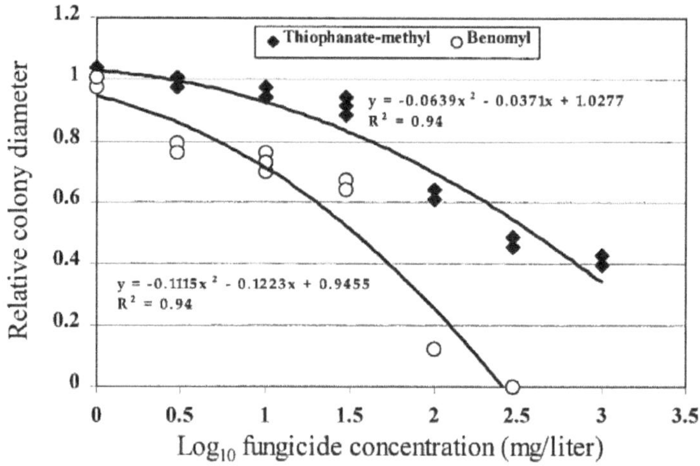

Figure 1. Thiophanate-methyl and benomyl concentration response curves for isolate HSCA06 (Helminthosporium solani) from California. After subtracting the diameter of the original plug, relative colony diameters were calculated by dividing the diameter on fungicide-amended medium by the diameter on nonamended medium.

None of the isolates exhibited a differential reaction to the two fungicides (i.e., each isolate behaved as resistant to both fungicides or as sensitive to both). Despite the fact that potato fields yielded different numbers of isolates, the proportion of resistance varied significantly among H. solani populations (2 = 36.7, P = 0.05). In field 1, 73 out of 121 isolates (60%) were resistant, while in field 2, 74 out of 77 isolates (96%) exhibited resistance to both fungicides. Likewise, 35 out of 40 isolates (87%) were resistant in field 3. Isolates 1- SS3 and HSME27 were always sensitive or resistant, respectively, to both fungicides in all assays in which they were included as negative and positive controls. Examining the potato plants that produced four to five H. solani isolates, a regular pattern of fungicide resistance isolates spreading on adjacent progeny tubers was not observed. Of 34 plants studied, 4 and 19 plants yielded only sensitive and resistant isolates, respectively. A mixture of resistant and sensitive isolates was observed on the other 11 plants. Plants yielding either benzimidazole-resistant or benzimidazole-sensitive isolates, or both phenotypes, were found in all three potato fields.

Molecular analysis of β-tubulin gene.

An 871-bp fragment of the β-tubulin gene containing the "benzimidazole box" (codons 198 and 200) from three resistant and three sensitive H. solani isolates was amplified and sequenced. All sequenced fragments were deposited in the GenBank (AF461127, AF461128, AF461129, AF461130, AF461131, and AF461132). The molecular differences in the β-tubulin gene between fungicide-resistant and fungicide-sensitive H. solani isolates are shown in Figure 2. There were only two point mutations observed in the β-tubulin gene from H. solani isolates from California compared with the sequence of the β- tubulin gene of H. solani sensitive to thiabendazole deposited in GenBank (Y10670). These mutations were at the 455-nucleotide and 836-nucleotide positions; however, only the mutation at the 455-nucleotide position was strongly associated with thiophanate-methyl resistance. In all isolates showing thiophanate-methyl and benomyl resistance, the adenine at the 455-nucleotide position was converted to cytosine, which led to a change from glutamic acid to alanine at codon 198. An additional mutation at the 836 site was detected in all isolates from California. This mutation changed adenine to cytosine, but it did not change the amino acid assigned by its codon (codon 309), which was still argenine.

```
         421                                                                490
HSCA1021 ---------- ---------- ---------- ----C----- ---------- ---------- ----------
HSCA2051 ---------- ---------- ---------- ----C----- ---------- ---------- ----------
HSCA3051 ---------- ---------- ---------- ----C----- ---------- ---------- ----------
HSCA1031 ---------- ---------- ---------- ----A----- ---------- ---------- ----------
HSCA2061 ---------- ---------- ---------- ----A----- ---------- ---------- ----------
HSCA3171 ---------- ---------- ---------- ----A----- ---------- ---------- ----------
HSY10670 CTTTCCGTTC ACCAGCTTGT CGAGAACTCG GACGAGACCT TCTGCATTGA CAACGAGGTA CGTAATATAC

         771                                                                840
HSCA1021 ---------- ---------- ---------- ---------- ---------- ---------- -----C----
HSCA2051 ---------- ---------- ---------- ---------- ---------- ---------- -----C----
HSCA3051 ---------- ---------- ---------- ---------- ---------- ---------- -----C----
HSCA1031 ---------- ---------- ---------- ---------- ---------- ---------- -----C----
HSCA2061 ---------- ---------- ---------- ---------- ---------- ---------- -----C----
HSCA3171 ---------- ---------- ---------- ---------- ---------- ---------- -----C----
HSY10670 GAGCTCACAC AGCAAATGTT CGACCCTAAG AACATGATGG CTGCCTCTGA CTTCCGTAAT GGCCGATACC
```

Figure 2. Two segments of the sequence alignment of the -tubulin gene fragment from selected Helminthosporium solani isolates obtained from California and a thiabendazole-sensitive isolate from GenBank (Y10670). Alignments were performed using the PRETTY program of the GCG Sequence Analysis Software Package. Gap creation and extension penalties were 5 and 1, respectively. A dash (–) indicates nucleotide identity and numbers on the top line represent first and last nucleotides, respectively. Codons 198, 200, and 309 are underlined.

Identification of thiophanate-methyl resistance by PCR-based RFLP assay

From 10 isolates identified as resistant or sensitive to both thiophanate-methyl and benomyl, the total genomic DNA was extracted, and a section of the β-tubulin gene was amplified and digested with Bsa I (Table 1). The RFLP patterns of β-tubulin fragments produced by resistant (lanes 2 to 5) and sensitive (lanes 6 to 9) isolates are shown in Figure 3. As expected, all resistant H. solani isolates exhibited two DNA fragments of 390 and 482 bp (lanes 2 to 5), whereas sensitive isolates showed three fragments of 62, 390, and 420 bp. The smallest fragment most likely ran off the agarose gel so that it was not present on the gels. There were no differences between methods for detecting thiophanate-methyl or benomyl resistance so that every isolate identified either as resistant or sensitive by the PCR-based RFLP assay correlated 100% with the results of the assay with the fungicide-amended medium in which either thiophanate-methyl or benomyl was employed (Table 1).

Table 1. Identification of thiophanate-methyl and benomyl resistance by fungicide-amended medium and polymerase chain reaction (PCR)-based restriction fragment length polymorphism (RFLP) assays in Helminthosporium solani isolates from Tulelake, California in 1999

				Fungicide-amended medium[a]		
Location[b]	Field	Plant	Tuber	Thiophanate-methyl	Benomyl	Genotypes[c]
CA	1	3	1	S	S	S
CA	1	5	1	S	S	S
CA	1	6	1	S	S	S
CA	1	8	1	S	S	S
CA	1	10	1	S	S	S
CA	1	1	1	R	R	R
CA	1	2	1	R	R	R
CA	1	4	1	R	R	R
CA	1	7	1	R	R	R
CA	1	9	1	R	R	R
CA	2	6	1	S	S	S
CA	2	13	2	S	S	S
CA	2	28	2	S	S	S
CA	2	1	1	R	R	R
CA	2	2	1	R	R	R
CA	2	3	1	R	R	R
CA	2	4	1	R	R	R
CA	2	5	1	R	R	R
CA	2	7	1	R	R	R
CA	2	8	1	R	R	R
CA	3	17	1	S	S	S
CA	3	20	1	S	S	S
CA	3	4	4	S	S	S
CA	3	15	4	S	S	S
CA	3	20	4	S	S	S
CA	3	1	1	R	R	R
CA	3	2	1	R	R	R
CA	3	3	1	R	R	R
CA	3	4	1	R	R	R
CA	3	5	1	R	R	R
ME	...	...	...	R	R	R
NY	...	...	...	S	S	S

[a] Fungicide sensitivity (S) and resistance (R). Isolates were considered resistant if growth was observed on medium amended with thiophanate-methyl at 400 mg a.i./liter or benomyl at 35 mg a.i./liter. If no growth occurred, it was considered fungicide sensitive.
[b] State where isolate originally was collected: CA = California, ME = Maine, and NY = New York.
[c] PCR-based RFLP genotypes.

DISCUSSION

In this work, we found that thiophanatemethyl resistance was common and widespread in populations of H. solani from potato fields in Tulelake area, CA. These populations also exhibited the same resistance patterns when exposed to benomyl, which had not been used in this potatogrowing area. Positive cross-resistance is very common among benzimidazole fungicides (23–26). Other researchers have reported development of thiabendazole resistance in H. solani isolates from several states in the United States (3,17,23,28); however, this is the first report of thiophanate-methyl resistance in H. solani populations in the United States. Relative toxicity and effectiveness for disease control vary among benzimidazole fungicides, but all of them are site-specific inhibitors (11). In natural fungal populations, any individual that has a mutated β-tubulin gene at this specific site is very likely to be selected by continual or repeated exposure to benzimidazole fungicides. Therefore, considering the regular application of thiophanate-methyl as a potato seed treatment and the fact that relatively stable benzimidazole resistance has arisen in other plant-pathogenic fungal populations around the world, it is not surprising that resistance to both fungicides is so prevalent among the analyzed H. solani populations. This high frequency of resistant isolates could also account for the recent outbreaks of silver scurf in the Tulelake area. Furthermore, once thiophanate-methyl resistance is established in H. solani populations, replacement with another benzimidazole would likely be useless. Even though thiophanate-methyl and benomyl are very closely related and have the same mode of action, the EC50 values were reasonably distinct. In resistant H. solani isolates from California, the EC50 values were approximately 35 mg/liter for benomyl and 400 mg/liter for thiophanatemethyl. Despite inhibitory differences, the EC50 values for thiophanate-methyl and benomyl were consistent with other reports. For example, thiophanate-methyl EC50 was greater than 500 mg/liter among isolates of H. solani from 11 different locations in Canada (23) and B. cinerea populations from commercial greenhouses growing ornamental crops in South Carolina (36), whereas EC50 values for benomyl ranged from 30 to 40 mg/liter in isolates of Didymella bryoniae from South Carolina and New York (24) and Cochliobolus heterostrophus (19). Conversion of thiophanate-methyl into carbendazim, which is the active form of benzimidazole fungicides, is slower than conversion of benomyl to this active ingredient. Moreover, analysis of benzimidazole and its degraded molecule in plants showed that benomyl and carbendazim have similar fungicide toxicity while intact thiophanate-methyl is much less toxic than carbendazim (11). These properties might explain the higher EC50 values of thiophanate-methyl in contrast to benomyl. It is possible that autoclaving might somehow affect the activity of the fungicides. However, the 100% correlation between the results of the fungicide-amended medium assay and the PCR-based RFLP assay clearly support the idea that both benomyl and thiophanate-methyl were still effective against H. solani even after being autoclaved. Further investigations should be done to demonstrate that autoclaving has neither positive nor negative impact on benomyl or thiophanate-methyl fungicide activity. It also was observed that neither fungicide dissolved completely during the preparation of the fungicide-amended medium at concentrations over 100 mg a.i./liter. This resulted in a fungicide suspension rather than a true solution. In addition, thiophanate-methyl EC_{50} was unable to be determined in a preliminary fungicide-amended medium assay in which 100 mg a.i./liter was the highest concentration (8). Because of the solubility limits of thiophanate-methyl, these results strongly indicate that the true thiophanate methyl EC_{50} is probably greater than 100 mg a.i./liter, but less than 400 mg a.i./liter. It was not the objective of our study to determine the exact fungicide EC_{50} but, rather, a threshold for screening H. solani populations for fungicide resistance; therefore, we consider that our results accomplished this goals and were corroborated by the results of the PCR-based RFLP assay.

H. solani is an asexual fungus disseminated mainly by infected seed tubers (30,32). These features, coupled with the approach of collecting isolates from different infected tubers in the same plant, allow for the possibility of clonal relationships among all isolates from the same plant or even within a potato field. Among multiple isolates from tubers on the same plant, both resistant and sensitive genotypes were detected in approximately 32% of the sampled plants. This mixture of genotypes also was observed among isolates within potato fields, especially in field 1. This suggests that H. solani populations are not uniform and diverse genotypes may exist in close proximity on diseased progeny tubers. Studying thiabendazole sensitivity in multiple isolates of the same tuber lot, Merida and Loria (28) found predominantly thiabendazole-resistant genotypes within the same lot. Analyzing fewer samples than our study might have misled them to observing almost uniform populations. On the other hand, the H. solani populations in New York could have been under longer and more intensive pressure for selection of benzimidazole resistance than H. solani populations from California. Several mutations have been detected on the β-tubulin gene of benzimidazole-resistant fungi but mutation occurs most frequently at codon 6, 198, or 200 in plantpathogenic fungi (1,7,9). The present work revealed that thiophanate-methyl resistance always was associated with a mutation in the -tubulin gene of H. solani in the same manner as other benzimidazole resistance in other plant-pathogenic fungi. In H.

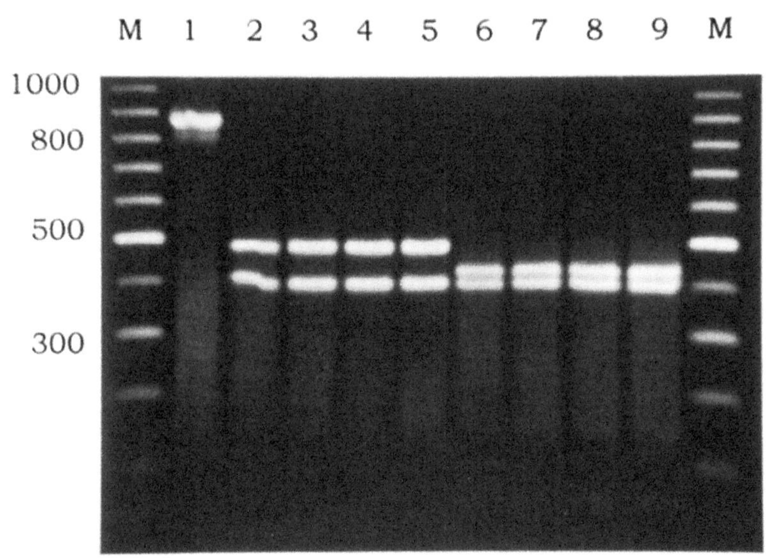

Figure 3. DNA fingerprinting patterns of -tubulin gene fractions of thiophanate-methyl-resistant and thiophanate-methyl-sensitive Helminthosporium solani isolates, which were polymerase chain reaction amplified by species-specific primers and Bsa-I digested. Lane 1 contains undigest-

ed - tubulin gene, lanes 2 to 5 represent digested -tubulin gene of resistant isolates (HSCA1021, HSCA2051, HSCA3051, and HSME27), and lanes 6 to 9 represent sensitive isolates (HSCA1031 HSCA2061, HSCA3117, and 1-SS3). Lanes marked with M contain DNA size markers (100-bp DNA ladder).

Solani populations from California, thiophanate-methyl resistance was consistently related to a point mutation at codon 198 of the β-tubulin gene (Fig. 2). A single transversion mutation was consistently associated with an amino acid change at codon 198 from GAG (glutamic acid) to GCG (alanine). Similarly, McKay and Cooke (27) described a single base changed at the same codon linked to thiabendazole resistance; however, they observed two kinds of mutations in which GAG sequence was converted to either GCG (alanine) or CAG (glutamine) sequences. In addition, their PCR-based RFLP technique was more reliable for detecting thiabendazole resistance in isolates that carried GCG conversion than the other mutation. The predominance of the GCG conversion also was observed in the present study, in which 30 randomly selected isolates were successfully screened. Thus, it is likely that the second mutation (CAG) has not occurred in H. solani populations from California or that it occurs at very low frequency. The mutation at codon 198 of the β- tubulin gene plays an important role in the development of benzimidazole resistance in H. solani. However, it does not rule out other mutations in this gene being involved with benzimidazole resistance because only partial molecular analysis of the β- tubulin gene was performed. Molecular analysis of the partial β-tubulin gene from Tapesia yallundae and T. acuformis isolates exhibiting field resistance to benzimidazole showed that every resistant isolate carries a point mutation at codons 198, 200, or 240 of the β-tubulin gene (1). This latter mutation was observed in field strains with low benzimidazole resistance. In the present work, we did not quantify thiophanate-methyl resistance in H. solani isolates beyond two categories (resistant and sensitive). Expanding the number of categories of resistance, followed by molecular analysis within categories, may reveal whether other minor mutations occur in the β-tubulin gene of H. solani carrying benzimidazole resistance. Moreover, transformation of a sensitive H. solani isolate with a β-tubulin gene modified only at codon 198 will provide a better insight into the mechanisms of H. solani resistance to benzimidazole fungicides. The occurrence of benomyl-resistant isolates that are as fit as sensitive isolates has been observed with most genera of fungi in which benzimidazole resistance has been documented. Merida and Loria (28) studied several aspects of H. solani isolates from New York, such as optimal temperature for radial growth, spore production, and virulence, but they found no significant differences between sensitive and resistant isolates. Hide and Hall (21) observed similar disease severity on healthy micropropagated plants inoculated with either sensitive or resistant H. solani isolates, which indicated that both types of isolates were equally virulent. The similarity in pathogenic fitness of sensitive and resistant isolates of H. solani indicates that mutation in the β-tubulin gene is not associated with a fitness cost and may allow the fungus to survive for long periods even in the absence of fungicide selection pressure. For instance, Romero and Sutton (31) detected a high frequency of benomyl resistance in Mycosphaerella fijiensis even after 5 years without spraying this fungicide. This situation makes disease management in the Tulelake area rather challenging because potato growers will have to rely on chemicals outside of the benzimidazole group or modify some of the cropping and storage practices to minimize losses due to silver scurf. Likewise, growers should monitor the sensitivity of H. solani populations to benzimidazole before considering any use of fungicides in this group, regardless of the number of years since previous use of these products. This study not only showed the degree of thiophanate-methyl resistance in field populations of H. solani but also demonstrated that its molecular basis resembles resistance observed in other pathogens that have developed resistance to fungicides in the benzimidazole group. Although the occurrence of thiophanate-methyl resistance among H. solani populations and its molecular basis has been documented, the persistence of this resistance in field populations of H. solani remains to be established. Molecular detection of benzimidazole resistance is more convenient than assays based on fungicide-amended medium because it can be performed in a very short time and is as accurate as the latter assay. Using the PCR-based method, benzimidazole resistance can be detected within 2 to 3 days compared with 4 weeks by conventional assay procedures.

ACKNOWLEDGMENTS

We thank the Brazilian Ministry of Education (CAPES) for sponsoring the Ph.D. studies of M. G. Cunha, R. M. Davis and T. Gordon for their suggestions and assistance with the manuscript, D. Kirby for potato storage and sampling, and G. Secor and C. Olivier for providing the nonCalifornia isolates of H. solani.

REFERENCES

1. Albertini, C., Gredt, M., and Leroux, P. 1999. Mutations of the beta-tubulin gene associated with different phenotypes of benzimidazole resistance in the cereal eyespot fungi Tapesia yallundae and Tapesia acuformis. Pestic. Biochem. Physiol. 64:17-31.
2. Alcorn, J. L. 1988. The taxonomy of "Helminthosporium" species. Annu. Rev. Phytopathol. 26:37-56.
3. Bains, P. S., Bisht, V. S., and Benard, D. A. 1996. Soil survival and thiabendazole sensitivity of Helminthosporium solani isolates from Alberta, Canada. Potato Res. 39:23-30.
4. Butters, J. A., and Hollomon, D. W. 1999. Resistance to benzimidazole can be caused by changes in beta-tubulin isoforms. Pestic. Sci. 55:501-503.
5. Carnegie, S. F., Cameron, A. M., Hide, G. A., and Hall, S. M. 1994. The occurrence of thiabendazole-resistant isolates of Polyscytalum pustulans and Helminthosporium solani on seed potato tubers in relation to fungicide treatment and storage. Plant Pathol. 43:961- 971.
6. Cenis, J. L. 1992. Rapid extraction of fungal DNA for PCR amplification. Nucleic Acids Res. 20:2380.
7. Cooley, R. N., and Caten, C. E. 1993. Molecular analysis of the Septoria nodorum beta-tubulin gene and characterization of a benomyl-resistance mutation. Mol. Gen. Genet. 237:58-64.

8. Cunha, M. G. 2001. Epidemiology of silver scurf and population biology of Helminthosporium solani Dur. & Mont. (1849) in California. Ph.D., Department of Plant Pathology, University of California, Davis.

9. Davidse, L. C., and Ishii, H. 1995. Biochemical and molecular aspects of the mechanisms of action of benzimidazole, N-phenylcarbamates and N-phenylformamidoximes and the mechanisms of resistance to these compounds. Pages 304-322 in: Modern Selective Fungicides—Properties, Applications, Mechanisms of Action. H. Lyr, ed. Gustav Fischer Verlag, New York.

10. De Boer, S. H., Ward, L. J., Li, X., and Chittarajan, S. 1995. Attenuation of PCR inhibition in the presence of plant compounds by addition of BLOTTO. Nucleic Acids Res. 23:2567-2568.

11. Delp, C. J. 1995. Benzimidazole and related fungicides. Pages 291-303 in: Modern Selective Fungicides—Properties, Applications, Mechanisms of Action. H. Lyr, ed. Gustav Fischer Verlag, New York.

12. Dhingra, O. D., and Sinclair, J. B. 1995. Basic plant pathology methods. 2nd ed. CRC, Boca Raton, FL.

13. Ellis, M. B. 1968. Helminthosporium solani. Descriptions of Pathogenic Fungi and Bacteria, Number 166. Commonwealth Mycological Institute, The Eastern Press, London.

14. Ellis, M. B. 1976. Dematiaceous Hyphomycetes. Commonwealth Mycological Institute, Kew, England.

15. Errampalli, D., Saunders, J. M., and Holley, J. D. 2001. Emergence of silver scurf (Helminthosporium solani) as an economically important disease of potato. Plant Pathol. 50:141- 153.

16. Firman, D. M., and Allen, E. J. 1995. Transmission of Helminthosporium solani from potato seed tubers and effects of soil conditions, seed inoculum and seed physiology on silver scurf disease. J. Agric. Sci. 124:219-234.

17. Frazier, M. J., Shetty, K. K., Kleinkopf, G. E., and Nolte, P. 1998. Management of silver scurf (Helminthosporium solani) with fungicide seed treatments and storage practices. Am. J. Potato Res. 75:129-135.

18. Free, S. J., Holtz, B. A., and Michailides, T. J. 1996. Mating behavior in field populations of Monilinia fructicola. Mycologia 88:208-211.

19. Gafur, A., Tanaka, C., Shimizu, K., Ouchi, S., and Tsuda, M. 1998. Molecular analysis and characterization of the Cochliobolus heterostrophus beta-tubulin gene and its possible role in conferring resistance to benomyl. J. Gen. Appl. Microbiol. 44:217-223.

20. Heiny, D. K., and McIntyre, G. A. 1983. Helminthosporium solani Dur. & Mont. development on potato periderm. Am. Potato J. 60:773-789.

21. Hide, G. A., and Hall, S. M. 1993. Development of resistance to thiabendazole in Helminthosporium solani (silver scurf) as a result of potato seed tuber treatment. Plant Pathol. 42:707-714.

22. Jellis, G. J., and Taylor, G. S. 1977. The development of silver scurf

(Helminthosporium solani) disease of potato. Ann. Appl. Biol. 86:19-28.

23. Kawchuk, L. M., Holley, J. D., Lynch, D. R., and Clear, R. M. 1994. Resistance to thiabendazole and thiophanate-methyl in Canadian isolates of Fusarium sambucinum and Helminthosporium solani. Am. Potato J. 71:185-192.
24. Keinath, A. P., and Zitter, T. A. 1998. Resistance to benomyl and thiophanate-methyl in Didymella bryoniae from South Carolina and New York. Plant Dis. 82:479-484.
25. Lamondia, J. A., and Douglas, S. M. 1997. Sensitivity of Botrytis cinerea from Connecticut greenhouses to benzimidazole and dicarboximide fungicides. Plant Dis. 81:729-732.
26. Leroux, P., Chapeland, F., Desbrosses, D., and Gredt, M. 1999. Patterns of cross-resistance to fungicides in Botryotinia fuckeliana (Botrytis cinerea) isolates from French vineyards. Crop Prot. 18:687-697.
27. McKay, G. J., and Cooke, L. R. 1997. A PCRbased method to characterise and identify benzimidazole resistance in Helminthosporium solani. FEMS Microbiol. Lett. 152:371-378.
28. Merida, C. L., and Loria, R. 1994. Comparison of thiabendazole-sensitive and -resistant Helminthosporium solani isolates from New York. Plant Dis. 78:187-192.
29. Merida, C. L., and Loria, R. 1994. Survival of Helminthosporium solani in soil and in vitro colonization of senescent plant tissue. Am. Potato J. 71:591-598.
30. Olivier, C. 1998. Molecular phylogenetic analysis of Helminthosporium solani and alternative disease management strategies for silver scurf. Ph.D., Plant Pathology, Cornell University, Ithaca, NY.
31. Romero, R. A., and Sutton, T. B. 1998. Characterization of benomyl resistance in Mycosphaerella fijiensis, cause of black Sigatoka of banana, in Costa Rica. Plant Dis. 82:931- 934.
32. Secor, G. A., and Gudmestad, N. C. 1999. Managing fungal diseases of potato. Can. J. Plant Pathol. 21:213-221.
33. Stanis, V. F., and Jones, A. L. 1984. Genetics of benomyl resistance in Venturia inaequalis from North and South America, Europe and New Zealand. Can. J. Plant Pathol. 6:283-290.
34. Wheeler, I. E., Kendall, S. J., Butters, J., Hollomon, D. W., and Hall, L. 1995. Using allelespecific oligonucleotide probes to characterize benzimidazole resistance in Rhynchosporium secalis. Pestic. Sci. 43:201-209.
35. Yarden, O., and Katan, T. 1993. Mutations leading to substitutions at amino acids 198 and 200 of beta-tubulin that correlate with benomyl-resistance phenotypes of field strains of Botrytis cinerea. Phytopathology 83:1478- 1483.
36. Yourman, L. F., and Jeffers, S. N. 1999. Resistance to benzimidazole and dicarboximide fungicides in greenhouse isolates of Botrytis cinerea. Plant Dis. 83:569-575.

3
CHAPTER

MODELS FOR PREDICTING POTENTIAL YIELD LOSS OF WHEAT CAUSED BY STRIPE RUST

D. Sharma-Poudyal and X. M. Chen

Keywords: Triticum aestivum, yellow rust.

INTRODUCTION

Stripe rust of wheat (Triticum aestivum L.) is caused by Puccinia striiformis Westend. f. sp. tritici Eriks. The disease can cause severe yield loss in many wheat growing areas of the world including the United States (3–5,7,23,32). Stripe rust epidemics are more frequent in the western United States, especially the Pacific Northwest (PNW), which includes Washington (WA), Oregon (OR), and Idaho (ID). Yield loss caused by stripe rust depends on several factors, including the degree of cultivar susceptibility, infection time, rate of disease development, and duration of disease (3). These crop and disease factors are influenced by environmental factors, among which temperature and moisture are the most important in determining disease severity and yield loss (4,17,28,35). Because temperature and moisture conditions vary significantly from year to year in most wheat producing regions, they are the major limiting factors for the development of stripe rust epidemics and therefore have been used to develop descriptive and forecasting models for the disease. Forecasting models have been developed and used in management of many plant diseases (2,12–14,18,19,27,29,30, 34,36) including stripe rust (10,11,23). Because temperature has the most profound effect on the P. striiformis f. sp. tritici life cycle, influencing pathogen survival, infection, latent period, and sporulation, it has been used to develop forecasting models for stripe rust (9–11,23). Coakley (8)

Citation: Sharma-Poudyal, D., and Chen, X. M. 2011. Models for predicting potential yield loss of wheat caused by stripe rust 101:544-554. doi:10.1094 / PHYTO-08-10-0215

initially studied the relationship between climatic variables and stripe rust epidemics in Pendleton, Oregon and found epidemics were favored by winter temperatures that are higher than normal (the mean winter temperature from 1961 to 1975).

In a subsequent study, Coakley et al. (9) quantified the relationships between temperature and stripe rust epidemics on winter wheat using disease data from Pullman, WA from 1963 to 1979. Negative and positive degree days were used to determine the effect of cumulative temperatures on disease development using 7°C as a base temperature. This base temperature was selected because it was found to be the most optimum for urediniospore germination and infection (31). Coakley et al. (9) found that the mean temperature for January was positively correlated with stripe rust intensity recorded on susceptible cultivars. However, the cumulative negative (1 December to 31 January) and positive degree days (1 April to 30 June) had higher negative correlations to disease intensity than the January mean temperature. Therefore, simple linear regression models for predicting stripe rust severity were developed using the positive and negative degree days as descriptors. Later, Coakley and Line (10) extended their stripe rust forecasting models developed for Pullman to other locations in the PNW using standardized negative degree days accumulated during December and January. Additional predictive models were developed by Coakley et al. (11) using disease index, temperature, and other meteorological factors such as the amount and frequency of precipitation from 1968 to 1986 at Pullman, WA using multiple regression analysis. However, the simpler models based on negative and positive degree days have been mostly used in forecasting for stripe rust management in the PNW (3,23). The previous models can generally predict disease intensity in a 0 to 9 scale at the flowering stage of wheat growth (9–11).

Though these models have been used successfully to predict stripe rust in most years, they could not accurately predict the epidemic levels in some years when the weather conditions did not follow the normal pattern. It is possible that the weather pattern in the PNW may have changed after the period used for developing the previous forecast models, which might make the models less suitable for stripe rust forecasting in the recent years. The major limitation of these models using negative and positive degree days is the lumping of the temperature during December and January, which usually are coldest months. If temperatures increase dramatically in this period, they negate the effect of the colder period. For example, the mean monthly temperature in December 2005 was significantly lower than normal and that of January 2006 was much higher than normal.

Using these models, stripe rust was predicted in 2006 to be as severe as in 2005, the most severe year of stripe rust since the early 1980s (4), however, the actual damage in 2006 was much less than in 2005 (6). Furthermore, the highest stripe rust intensity index (9 in the 0 to 9 scale) or severity (>95% of the foliage infected) at the flowering stage, which was used in the previous models (10), occurred in the experimental fields at the Pullman location in 2004, 2005, and 2006, but yield loss varied considerably with 40% in 2004, 74% in 2005, and 30% in 2006 for susceptible winter wheat (X. M. Chen, unpublished data). The difference in yield loss was caused by variation in duration of epidemic. The previous models do not consider the duration factor and thus, cannot give a fine scale estimation of epidemic severity or yield loss when disease reaches the highest severity level by the flowering stage. This highlights the need for prediction models that provide information on potential yield loss rather than disease severity. Because the previous models require experts' help on translation of predicted disease index to yield loss, growers have relied on experts to do forecasting and interpretation. Prediction models requiring simple calculations may be directly used by growers, extension agents, and farm consultants. Therefore, the major objective of this study was to identify weather variables correlated to yield loss and to develop a series of models for early prediction of potential yield loss caused by stripe rust in the PNW.

MATERIALS AND METHODS

Meteorological data.

Data of daily minimum and maximum temperature and precipitation for Pullman, WA from 1975 to 2007 were obtained from the National Oceanic and Atmospheric Administration (NOAA) National Data Centers (NNDC) Climatic Data Online service (http://cdo.ncdc.noaa.gov/CDO/dataproduct). The Pullman meteorological station is at the 46°45′N latitude, 117°12′W longitude, and 775 m elevation. A year was divided into four seasons based on P. striiformis f. sp. tritici biology and wheat crop seasons applicable to the Pullman situation: winter (1 November to 28 February), spring (1 March to 30 May), summer (1 June to 31 August), and fall (1 September to 31 October). The relatively long winter season was used to insure that all cold periods were included. Daily mean temperatures were calculated based on the daily maximum and minimum temperatures. If the maximum, minimum, and average temperatures of 1 day were missing, they were calculated by averaging the previous and following day temperatures. Any year with two or more continuous days of missing daily temperature was excluded from analysis. The minimum, maximum, and mean temperatures for a period were calculated based on daily minimum, maximum, and mean temperatures in that period, respectively. In addition, minimum and maximum moving average temperatures (MAT) were calculated based on daily minimum and maximum mean temperatures for 10-, 20-, 30-, and 60-day periods of 1, 2, 3, and 4 months in each season wherever applicable.

The sum of accumulated positive degree days (APDD) and the sum of accumulated negative degree days (ANDD) were calculated using a previously described method (10). Instead of the sum of degree days, the sum of temperatures (ST) was calculated due to the easiness of computation. The amounts of rainfall and snowfall, rainfall days, snow depth, and snow ground cover days (>2.54 cm) were also used. Thus, a total of 1,376 weather variables were tested in this study. Disease data. Historical data of stripe rust yield loss for winter and spring wheat were from the fungicide and cultivar yield loss tests conducted in Pullman, WA over the years in our program, some of which were published in Fungicide and Nematicide Tests Reports and Plant Disease Management Reports (http://www. plantmanagementnetwork.org/pub/trial/pdmr/). In the fungicide tests, the standard susceptible wheat genotypes PS 279 and Lemhi were used for winter wheat and spring wheat, respectively. A completed randomized block design was used with four replications for each of nonsprayed check and various fungicide-sprayed treatments. Percentage yield loss caused by stripe rust was obtained using the following formula: $Y(\%) = (Y1 - Y2)/Y1 \cdot 100\%$, where Y1 was the mean grain yield of a fungicide treatment providing the complete or best control of stripe rust and Y2 was the mean grain yield of the nonsprayed check treatment.

In the cultivar tests, PS 279 or Lemhi was tested together with 23 winter or 15 spring wheat cultivars grown in that year using a randomized split-block design with four replications for each of sprayed and nonsprayed treatments of each cultivar. A fungicide, the most popularly used in any particular period (e.g., Tilt at 4 fl oz/A used from 2000 to 2007), was sprayed before stripe rust reached 5% severity and when plants were usually at boot to heading stages. Percentage of yield loss was calculated using the above formula with Y1 as the mean grain yield of the sprayed treatment and Y2 as the nonsprayed treatment of PS 279 or Lemhi. In both fungicide and cultivar tests, the length (about 5 m) and width (about 1.5 m) were measured for each plot. The tests were conducted under natural infection, without artificial inoculation, of stripe rust and fields were managed using the standard cultural practice of commercial fields in the region. Stripe rust severity for each plot was recorded 1 day before or on the same day of fungicide application and two to four times after application. Stripe rust yield loss data confounded by other diseases (in most cases, leaf rust) >5% were excluded from analysis. Yield loss data were available for winter and spring wheat from 1993 to 2007 and 1995 to 2007, respectively.

Data analyses

The variables included in this study were yield loss percentage of winter wheat or spring wheat due to stripe rust (Y) and temperature- and precipitation-based variables (X). Pearson's correlation coefficients were calculated between yield loss percentage and temperature-based variables (minimum, maximum, mean, minimum MAT, maximum MAT, ANDD, APDD, and ST) as well as precipitation variables (total rainfall, total rainfall days [RFD], snow depth [SD], and snow ground cover days [SCD]) for different periods. Separate but similar analyses were performed to develop yield loss models for winter and spring wheat. Simple linear regression models were developed for each variable that was significantly correlated with yield loss. Models for different months within a season were selected based on the coefficient of determination (R^2) and prediction sum of squares (PRESS) value (22). Stepwise regression was used to select the explanatory variables for different winter periods: 1, 2, 3, and 4 months. Selected independent variables were subjected again for stepwise regression to select the descriptors for the entire winter season. Stepwise regression identified variables that explained the maximum variation in yield loss. Explanatory variables identified by stepwise regression were subjected for the best subset regression algorithm for model selection (22). Mallows' Cp, which is an indication of goodness of fit of the regression equations with different numbers of independent variables, was used to select models. In addition, the adjusted coefficient of multiple determination (R2 adj), which accounts for the number of independent variables, was used to identify the total variation in a response variable associated with the multiple descriptors (12,22). A multicolinearity test was performed as indicated by variance inflation factors (VIFs) (15,22,26). After the stepwise regression analysis, all explanatory variables from each season were used for the yearly model development. Descriptors for yearly models were identified by stepwise regression analysis. Correlated independent variables with a small regression sum of square were excluded from further model development in multiple regressions. The best subset regression algorithm was used to identify different models in the seasonal and yearly multiple regressions. Principal component analysis was also used to select explanatory variables for the most information (1,20,21). Principal component analysis addresses the problem of multicolinearity transforming the variables into a set of orthogonal or uncorrelated variables (15,16).

Model validation

Model validation was done based on the PRESS value, which is equivalent to the leave-one out cross validation method (25,33). The best fitting models should have the lowest PRESS value. Models with small PRESS values fit well because these models have small prediction errors (15,22, 26). Additionally, prediction accuracy percentage for selected models was calculated using contingency tables (11).

Table 1. Correlation coefficients obtained between weather variables and yield loss percentage of winter wheat (1993 to 2007) and spring wheat (1995 to 2007) caused by stripe rust in Pullman, Washington

Period	Variable no.	Weather variable[a]	Winter wheat		Spring wheat	
			r^b	P^c	r^b	P^c
Winter						
Temperature						
One-month						
Dec	X_1	Dec10D-MaxMAT(Mean)	0.27	0.45	**0.68**	**0.02**
Dec	X_2	DecANDD(Max)	**−0.72**	**0.02**	−0.41	0.21
Dec	X_3	DecANDD10D-MAT(Max)	**−0.64**	**0.04**	−0.29	0.37
Dec	X_4	DecANDD20D-MAT(Max)	**−0.65**	**0.04**	−0.33	0.32
Dec	X_5	DecAPDD(Min)	**0.70**	**0.02**	**0.65**	**0.03**
Dec	X_6	DecAPDD10D-MAT(Max)	0.26	0.46	**0.63**	**0.04**
Dec	X_7	DecST(Max)	**0.71**	**0.02**	0.45	0.16
Dec	X_8	DecST10D-MAT(Max)	**0.65**	**0.03**	0.32	0.32
Jan	X_9	JanAPDD10D-MAT(Max)	**0.66**	**0.03**	**0.60**	**0.03**
Feb	X_{10}	Feb10D-MinMAT(Max)	0.35	0.31	**0.73**	**0.01**
Feb	X_{11}	FebANDD(Max)	**−0.84**	**0.002**	**−0.63**	**0.04**
Feb	X_{12}	FebANDD(Mean)	**−0.78**	**0.008**	−0.43	0.18
Feb	X_{13}	FebANDD10D-MAT(Max)	**−0.80**	**0.005**	−0.45	0.15
Feb	X_{14}	FebANDD20D-MAT(Max)	**−0.76**	**0.01**	−0.32	0.32
Feb	X_{15}	FebAPDD(Max)	**0.82**	**0.004**	0.15	0.65
Feb	X_{16}	FebAPDD(Mean)	**0.71**	**0.02**	0.23	0.48
Feb	X_{17}	FebAPDD20D-MAT(Max)	0.51	0.13	**0.70**	**0.02**
Feb	X_{18}	FebST(Max)	**0.94**	**0.0001**	0.42	0.19
Feb	X_{19}	FebST(Mean)	**0.79**	**0.007**	0.42	0.19
Feb	X_{20}	FebST10D-MAT(Max)	**0.79**	**0.007**	−0.20	0.54
Feb	X_{21}	FebST20D-MAT(Max)	**0.81**	**0.005**	−0.21	0.53
Two-month						
Nov–Dec	X_{22}	NovDecANDD(Max)	**−0.73**	**0.01**	0.31	0.32
Nov–Dec	X_{23}	NovDecAPDD10D-MAT(Max)	0.03	0.92	**0.64**	**0.03**
Dec–Jan	X_{24}	DecJanAPDD(Min)	**0.70**	**0.02**	−0.36	0.26
Jan–Feb	X_{25}	JanFeb10D-MaxMAT(Max)	**0.71**	**0.02**	0.10	0.70
Jan–Feb	X_{26}	JanFeb10D-MaxMAT(Mean)	**0.74**	**0.01**	0.05	0.86
Jan–Feb	X_{27}	JanFeb20D-MaxMAT(Max)	**0.85**	**0.002**	0.44	0.17
Jan–Feb	X_{28}	JanFeb20D-MaxMAT(Mean)	**0.82**	**0.003**	0.44	0.17
Jan–Feb	X_{29}	JanFeb30D-MaxMAT(Max)	**0.80**	**0.005**	0.37	0.34
Jan–Feb	X_{30}	JanFeb30D-MaxMAT(Mean)	**0.66**	**0.03**	0.31	0.34
Jan–Feb	X_{31}	JanFebAPDD(Max)	**0.77**	**0.009**	0.31	0.34
Jan–Feb	X_{32}	JanFebAPDD(Mean)	**0.67**	**0.03**	0.20	0.57
Jan–Feb	X_{33}	JanFebAPDD20D-MAT(Max)	0.25	0.28	**0.68**	**0.02**
Jan–Feb	X_{34}	JanFebAPDD30D-MAT(Max)	0.13	0.31	**0.63**	**0.04**

(Continued on following page)

[a] D = days, Max = maximum, MAT = moving average temperature, (Min) = based on daily minimum temperature, (Mean) = based on daily mean temperature, (Max) = based on daily maximum temperature, ST = sum of temperatures, ANDD = accumulated negative degree days, APDD = accumulated positive degree days, SD = snow depth, and RFD = rain fall days.

[b] *r* = correlation coefficient. Significant *r* values are highlighted in bold.

[c] *P* = probability. Significant *P* values (< 0.05) are highlighted in bold.

In addition to data used for model development, yield loss data from 2008 and 2009 of winter and spring wheat, Pullman, WA were included to compute prediction accuracy percentage, chi-squared test, and predicted R^2 . Because stripe rust was able to cause yield loss ranging from 3 to 86% during the period of 1975 to 2007 on susceptible cultivars in the PNW, yield loss of 20% or less was generally considered light, more than 20 to 50% moderate, and more than 50% severe. The probability of chi-squared test was used to compare the observed and expected frequency of yield loss categories for different selected models. Furthermore, predicted R2 values were determined to find out the predictive ability of the selected models (11).

RESULTS

Winter temperature variables correlated with winter wheat yield loss

None of the November based variables were significantly correlated with yield loss percentage. Out of seven significantly correlated December variables, accumulated negative degree days based on daily maximum [DecANDD(Max), X2] had the highest correlation coefficient ($r = -0.72$, $P = 0.02$). In January, only one variable, accumulated positive degree days based on 10 days MAT using daily maximum temperature [JanAPDD10D-MAT(Max), X9], had a significant correlation coefficient ($r = 0.66$, $P = 0.03$) (Table 1). Among the monthly winter temperature variables, February variables had relatively higher correlation coefficient than the others. February sum of temperatures based on daily maximum temperature [(FebST(Max), X18] had the highest correlation coefficient ($r = 0.94$, $P = 0.0001$). Among significantly correlated bi-monthly variables, January to February 20 days maximum MAT based on daily maximum temperature [(JanFeb20D-MaxMAT(Max), X27] had the highest correlation coefficient ($r = 0.85$, $P = 0.002$). However, all correlated bi-monthly variables had lower correlation coefficients than the single month variable, [FebST(Max), X18]. None of the 3-month period temperature-based variables from November to January were significantly correlated to yield loss of winter wheat. All significant temperature variables were from December to February periods. None of the significantly correlated tri-monthly variables had a higher correlation coefficient than the 1 month [FebST(Max), X18] and bi-monthly variable [(JanFeb20D-MaxMAT(Max), X27]. Only two variables based on entire winter season (November to February) were significantly correlated to yield loss of winter wheat, but these variables had a lower correlation coefficient than most of the significant variables for 1-, 2-, and 3-month periods (Table 1). All variables were positively correlated to yield loss of winter wheat, except for ANDD variables that were negatively correlated. These results indicated that higher temperatures during the winter season were likely to result in greater yield loss due to stripe rust.

Winter precipitation variables correlated with winter wheat yield loss

Among different precipitation variables, total rainfall days (RFD) of different winter periods were significantly correlated to yield loss of winter wheat (Table 1). January RFD ($r = -0.76$, $P = 0.01$) had a higher correlation coefficient than the December RFD ($r = -0.64$, $P = 0.04$), however, the highest correlation coefficient ($r = -0.92$, $P = 0.001$)

was associated with the three-month precipitation variable, December to February RFD (DecFeb-RFD, X56). Except for monthly precipitation variables, correlation coefficients obtained for 2-, 3-, and 4-month winter periods had higher values than the temperature variables for the same periods of winter. The snow depth and snow cover days were not significantly correlated to yield loss. All of the RFD variables were negatively correlated with yield loss of winter wheat. These results indicate that a higher number of rainfall days in the winter season is associated with less yield loss caused by stripe rust.

Spring and summer weather variables correlated with winter wheat yield loss

Only three and four temperature variables of the spring and summer seasons were significantly correlated with winter and spring wheat yield loss, respectively (Table 1). None of the significantly correlated variables had higher correlation coefficients than the best winter temperature and precipitation variables. Precipitation variables of both seasons were not significantly correlated with yield loss.

Table 1. (Continued from previous page)

			Winter wheat		Spring wheat	
Period	Variable no.	Weather variable[a]	r^b	P^c	r^b	P^c
Three-month						
Dec–Feb	X_{35}	DecFeb10D-MaxMAT(Max)	**0.70**	**0.02**	0.05	0.86
Dec–Feb	X_{36}	DecFeb20D-MaxMAT(Max)	**0.80**	**0.005**	0.47	0.14
Dec–Feb	X_{37}	DecFeb30D-MaxMAT(Max)	**0.76**	**0.01**	0.35	0.28
Dec–Feb	X_{38}	DecFeb60D-MaxMAT(Mean)	**0.69**	**0.02**	0.33	0.31
Dec–Feb	X_{39}	DecFebAPDD(Max)	**0.73**	**0.01**	0.37	0.26
Dec–Feb	X_{40}	DecFebAPDD(Min)	**0.70**	**0.02**	–0.09	0.79
Dec–Feb	X_{41}	DecFebAPDD10D-MAT(Max)	**0.74**	**0.01**	0.42	0.19
Dec–Feb	X_{42}	DecFebAPDD20D-MAT(Max)	**0.75**	**0.01**	**0.67**	**0.02**
Dec–Feb	X_{43}	DecFebAPDD30D-MAT(Max)	**0.73**	**0.01**	0.62	0.04
Dec–Feb	X_{44}	DecFebST(Max)	**0.70**	**0.02**	0.45	0.17
Dec–Feb	X_{45}	DecFebST10D-MAT(Max)	**0.67**	**0.03**	0.41	0.21
Four-month						
Nov–Feb	X_{46}	NovFebAPDD(Max)	**0.68**	**0.03**	0.42	0.19
Nov–Feb	X_{47}	NovFebST(Max)	**0.68**	**0.03**	0.52	0.10
Nov–Feb	X_{48}	NovFebAPPD30D-MAT(Max)	0.48	0.15	**0.63**	**0.04**
Precipitation						
One-month						
Dec	X_{49}	Dec-RFD	**–0.64**	**0.04**	–0.50	0.11
Jan	X_{50}	Jan-RFD	**–0.76**	**0.01**	–0.42	0.20
Two-month						
Dec–Jan	X_{51}	DecJan-RFD	**–0.90**	**0.001**	**–0.64**	**0.03**
Dec–Jan	X_{52}	DecJan-SD	–0.43	0.21	**–0.72**	**0.01**
Jan–Feb	X_{53}	JanFeb-RFD	**–0.82**	**0.004**	–0.31	0.35
Three-month						
Nov–Jan	X_{54}	NovJan-RFD	**–0.87**	**0.001**	–0.48	0.15
Nov–Jan	X_{55}	NovJan-SD	–0.38	0.28	**–0.71**	**0.01**
Dec–Feb	X_{56}	DecFeb-RFD	**–0.92**	**0.001**	–0.51	0.11
Dec–Feb	X_{57}	DecFeb-SD	–0.51	0.12	**–0.64**	**0.03**
Four-month						
Nov–Feb	X_{58}	NovFeb-RFD	**–0.82**	**0.004**	–0.38	0.14
Nov–Feb	X_{59}	NovFeb-SD	–0.46	0.17	**–0.64**	**0.03**
Spring						
Mar	X_{60}	Mar20D-MaxMAT(Max)	**0.67**	**0.04**	0.21	0.53
Apr	X_{61}	Apr10D-MaxMAT(Mean)	**0.74**	**0.02**	0.01	0.97
May	X_{62}	MayANDD(Max)	–0.27	0.48	**–0.60**	**0.05**
Mar–Apr	X_{63}	MarApr10D-MaxMAT(Mean)	**0.74**	**0.02**	–0.03	0.93
Summer						
Jul	X_{64}	Jul10D-MinMAT(Min)	**0.70**	**0.02**	0.006	0.99
Jul	X_{65}	Jul10D-MinMAT(Mean)	**0.64**	**0.04**	0.002	0.99
Jul–Aug	X_{66}	JulAug30D-MaxMAT(Max)	**0.70**	**0.02**	0.33	0.31
Jun–Aug	X_{67}	JunAug30D-MaxMAT(Max)	**0.70**	**0.02**	0.18	0.58

Simple linear models developed for winter wheat yield loss

Simple linear regression models developed based on temperature variables explained different amounts of variation in yield loss. Because not all of the models were equally important, models with relatively high R2 values, and low PRESS values were selected from different periods within a season (Table 2). Different prediction models were developed using the different temperature variables during the winter period (Models 1 to 5, Table 2). Yield loss could be predicted using the temperature data as early as December, but the total variation explained was relatively low (Model 1, Table 2). Among the simple linear models, a model with 1-month explanatory variable, Model 2 (Table 2), described the highest yield loss variation. Model 3 (Table 2), developed with the 2-month variables, explained more variation in yield loss than the best early single month model (Model 1, Table 2). Model 8 had the highest R2 among the precipitation-based models; however, it explained less variation than the temperature-based Model 2. Models 9 to 11, based on spring and summer variables, had lower R2 values and higher PRESS values than did most of the simple linear models based on winter weather variables. Therefore, models based on winter weather variables are more predictive than the other seasons. Since none of the bi-monthly, trimonthly, and winter season models explained more variation in yield loss than the monthly Model 2, the latter was considered the best for predicting yield loss due to stripe rust.

Multiple linear models developed for winter wheat yield loss

All of the 54 significantly correlated variables were subjected to a stepwise regression for selection of variables that could explain most of the variation in yield loss. After stepwise regression, five independent variables were identified (Model 12, Table 2). All selected explanatory variables were winter temperature variables. Multicolinearity as indicated by the VIF values was found among weather variables. Therefore, explanatory variables with a high VIF value and a low regression sum of squares were excluded for model refinement. Another set of yield loss prediction models (Models 13 to 17, Table 2) was developed for different numbers of descriptors through the best subset algorithm. These multiple regression models explained more yield loss variation (6 to 11.4% more) than the best simple linear model (Model 2, Table 2). Multiple regression models (Models 18 and 19, Table 2) based on spring and summer variables had lower adjusted R2 and higher PRESS values than did Models 13 to 17. Therefore, multiple regression models based on winter weather variables explained the most variation in yield loss among all multiple-variable models. Model 12 with the highest predicted R2 was the best predicted model followed by Model 14 among the multiple regression models (Table 2).

Weather variables correlated with spring wheat yield loss

Twelve temperature and six precipitation variables were significantly correlated with yield loss (Table 1). Among the 12 winter temperature variables, X10 [Feb10D-MinMAT(Max)] had the highest correlation coefficient ($r = 0.73$, $P = 0.01$). Of the earlier variables, the best correlation was with Dec10D-MaxMAT(Mean), X1 ($r = 0.68$, $P = 0.02$). Among the precipitation variables, December to January snow depth (DecJan-SD, X52) had the highest correlation coefficient ($r = -0.72$, $P = 0.01$), which is the second highest coefficient after Feb10D-MinMAT(Max), X10. Only one spring variable, [MayANDD(Max), X62], was significantly correlated with yield loss (r =

–0.60, P = 0.01). None of the summer weather variables were correlated with spring wheat yield loss.

Prediction models for yield loss of spring wheat

Five models were developed using temperature and precipitation variables (Models 1 to 5, Table 3). Winter temperature (Models 1 and 2) and precipitation (Models 3 and 4) explained more variation in yield loss than a spring temperature-based model (Model 5). In general, models developed for predicting yield loss of spring wheat, including multiple regression models, were not as good as those for winter wheat. A simple linear model (Model 1, Table 3) with Dec10D-MaxMAT(Mean) (X1) as a descriptor allows for prediction of spring yield loss in early January. However, the temperature-based Model 2 [Feb10D-MinMAT(Max)] and the precipitation-based Model 3 (DecJan-SD) explained the highest amounts of variation of yield loss and can predict the yield loss in the end of the winter season. Four multiple regression models were developed with one temperature- and three precipitationbased variables (Models 6 to 9, Table 3). Due to presence of multicolinearity among the precipitation variables, other multiple regression models with different numbers of descriptors (Models 7 to 9) were developed through the best subset of algorithm.

Table 2. Single-variable and multiple-variable models obtained through stepwise and the best subset regression analyses for predicting winter wheat yield loss caused by stripe rust

No.	Models[a]	P^b	R^2/R^2_{adj} (%)[c]	PRESS[d]	Mallows' C_p	VIF[e]
	Single-variable models					
1	$Y = 84.3 - 0.323 X_2$	0.001	51.1	2,255.8		
2	$Y = -6.22 + 0.312 X_{18}$	0.0001	88.1	605.6		
3	$Y = 89.1 - 0.274 X_{22}$	0.01	53.3	2,139.7		
4	$Y = -17.2 + 8.34 X_{27}$	0.002	73.0	1,147.5		
5	$Y = -16.0 + 7.96 X_{36}$	0.005	64.4	1,533.7		
6	$Y = 84.7 - 3.64 X_{49}$	0.04	41.5	2,510.1		
7	$Y = 112 - 2.56 X_{51}$	0.0001	80.0	732.3		
8	$Y = 116 - 2.01 X_{56}$	0.0001	85.4	1,343.7		
9	$Y = -139 + 17.3 X_{61}$	0.023	54.8	1,985.6		
10	$Y = -139 + 17.3 X_{63}$	0.023	54.8	1,985.6		
11	$Y = -19.1 + 6.74 X_{64}$	0.025	48.5	1,934.3		
	Multiple-variable models					
	Based on yearly/winter weather variables					
12	$Y = 68.6 + 0.448 X_{18} - 0.0502 X_{47} - 0.390 X_3 - 0.368 X_7 + 0.265 X_{43}$	0.0001	99.4	119.5	6.0	5.7 X_{18}, 5.2 X_{47}, 38 X_3, 55.6 X_7, 3.4 X_{43}
13	$Y = 17.3 + 0.444 X_{18} - 0.0640 X_{47} - 0.0840 X_3 - 0.021 X_{43}$	0.0001	96.8	7,316.8	25.4	5.7 X_{18}, 4.2 X_{47}, 1.6 X_3, 2.3 X_{43}
14	$Y = 52.2 + 0.469 X_{18} - 0.0569 X_{47} - 0.304 X_3 - 0.255 X_7$	0.0001	98.7	110.7	10.1	4.8 X_{18}, 4.4 X_{47}, 27.9 X_3, 37.1 X_7
15	$Y = 17.3 + 0.441 X_{18} - 0.0637 X_{47} - 0.0831 X_3$	0.0001	97.3	125.7	23.4	4.2 X_{18}, 4.0 X_{47}, 1.5 X_3
16	$Y = 4.00 + 0.438 X_{18} - 0.0649 X_{47} + 0.0769 X_7$	0.0001	96.0	197.4	35.6	4.4 X_{18}, 4.1 X_{47}, 2 X_7
17	$Y = 3.16 + 0.467 X_{18} - 0.0592 X_{47}$	0.001	94.0	238.3	59.4	4 X_{18}, 4 X_{47}
	Based on spring weather variables					
18	$Y = -148 + 3.57 X_{60} + 14.2 X_{61}$	0.008	73.4	1,206.9	3.0	1.1 X_{60}, 1.1 X_{61}
	Based on summer weather variables					
19	$Y = -132 + 4.36 X_{64} + 4.29 X_{66}$	0.03	52.7	1,974.4	3.0	1.4 X_{64}, 1.4 X_{66}

[a] X_n, explanatory variables as described in Table 1.
[b] P = probability.
[c] R^2 = coefficient of determination for single-variable variable models, R^2_{adj} = adjusted coefficient of multiple determination for multiple-variable models.
[d] PRESS = prediction sum of squares.
[e] VIF = variance inflation factor.

Model validation

Six models each for winter and spring wheat were chosen for model validation. In addition to PRESS values, validations were done by the chi-squared test using yield loss categories (Table 4) and by regressing prediction and actual values of yield loss for selected models (Figs. 1 and 2). For winter wheat, all selected multiple regression models had a prediction accuracy (based on categories) of 92% and were superior to the two single-variable models (Table 4). The same pattern was evident with regression analysis of the percent yield loss; multiplevariable models had higher R2 values than single-variable models, although in this case the multiple-variable models had different R2 values, with Model 12 (Table 2) having the highest. Predictions were generally better with both validation approaches for winter wheat than for spring wheat. In the case of the latter, there was less of a clear superiority of multiple-variable models although one model (Model 9, Table 3) had a prediction accuracy of 92%, which was the highest. However, the R2 of Model 9 was slightly lower than that of Model 6 (58.9 and 64%, respectively) and overall, R2 values were lower for spring wheat (34.9 to 64%) than for winter wheat (55.9 to 87.6%) (Figs. 1 and 2). A partial comparison of a new model and an older model was made for winter wheat. Model 12 (Table 2) predicted yield loss similar to that observed in all 3 years (Fig. 3), while the older model based on disease index predicted the actual index accurately in only 2 of the 3 years.

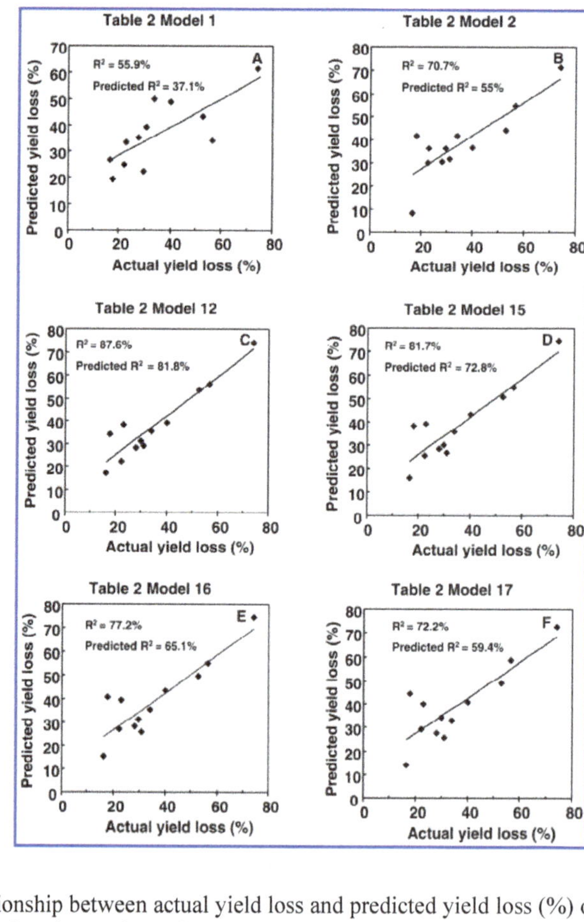

Figure 1. Relationship between actual yield loss and predicted yield loss (%) obtained using six models selected for winter wheat; data represent trials from 1993 to 2009 at Pullman, Washington.

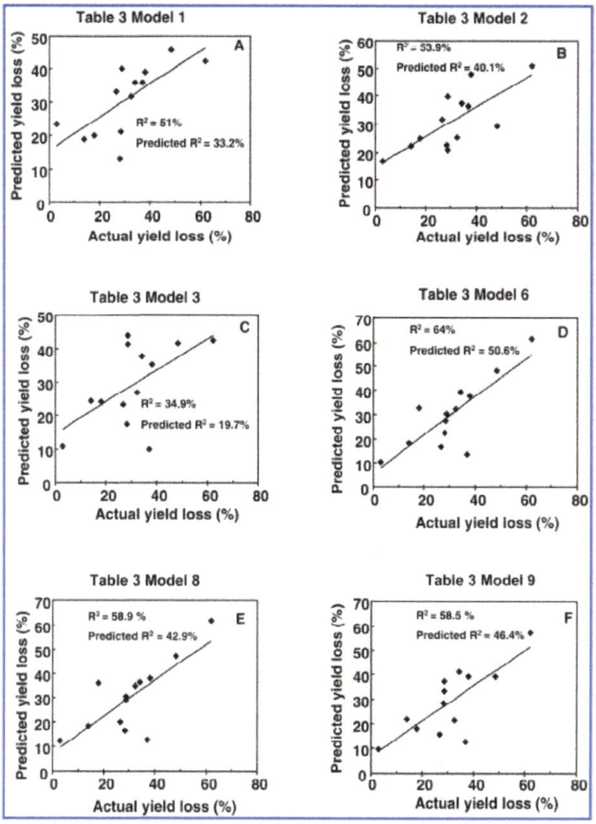

Figure 2. Relationship between actual yield loss and predicted yield loss (%) obtained through selected models on spring wheat; data represent trials from 1995 to 2009 at Pullman, Washington.

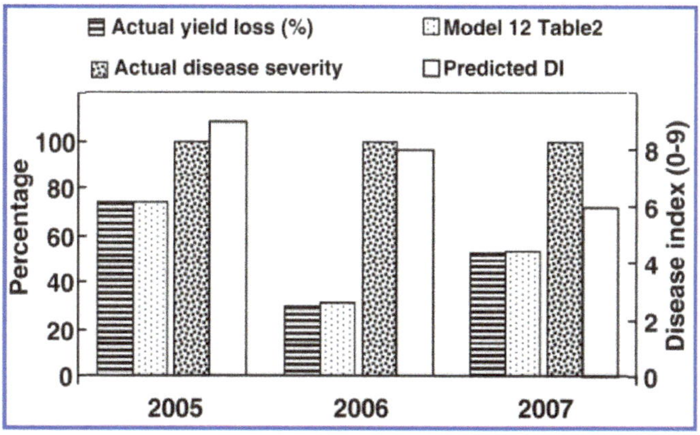

Figure 3. Comparison of actual and predicted yield loss percentages, and disease severity indices for 2005 to 2007 for Pullman, Washington.

Table 3. Single-variable and multiple-variable models obtained through stepwise and the best subset regression analyses for predicting spring wheat yield loss caused by stripe rust

No.	Models[a]	P[b]	R^2/R^2_{adj} (%)[c]	PRESS[d]	Mallows' C_p	VIF[e]
	Single-variable models					
1	$Y = 13.9 + 6.35\ X_1$	0.02	46.8	2,156.0		
2	$Y = 15.0 + 5.60\ X_{10}$	0.01	52.7	1,845.9		
3	$Y = 50.3 - 0.357\ X_{52}$	0.01	52.7	1,843.6		
4	$Y = 50.4 - 0.306\ X_{55}$	0.01	50.5	1,935.1		
5	$Y = 35.2 - 15.0\ X_{62}$	0.04	36.4	2,256.7		
	Multiple-variable models					
	Based on winter weather variables					
6	$Y = 56.9 + 2.82\ X_{17} - 0.731\ X_{51} - 0.905\ X_{52} + 0.679\ X_{57}$	0.004	83.5	1,251.4	4.3	$3.3\ X_{17}$, $2.8\ X_{51}$, $13.9\ X_{52}$, $15.0\ X_{57}$
7	$Y = 68.0 + 0.37\ X_{10} + 1.28\ X_{17} - 0.902\ X_{51} - 0.268\ X_{52}$	0.03	62.1	1,985.1	11.1	$3.5\ X_{10}$, $4.3\ X_{17}$, $2.7\ X_{51}$, $1.7\ X_{52}$
8	$Y = 34.3 + 4.42\ X_{17} - 0.925\ X_{52} + 0.712\ X_{57}$	0.002	82.1	645.7	3.7	$1.3\ X_{17}$, $13.8\ X_{17}$, $14.9\ X_{57}$
9	$Y = 83.8 -- 1.33\ X_{51} - 0.299\ X_{52}$	0.003	69.7	1,092.1	7.9	$1.1\ X_{51}$, $1.1\ X_{52}$

[a] X_n, explanatory variables as described in Table 1.
[b] P = probability.
[c] R^2 = coefficient of determination for single-variable models. R^2_{adj} = adjusted coefficient of multiple determination for multiple-variable models.
[d] PRESS = prediction sum of squares.
[e] VIF = variance inflation factor.

Table 4. Prediction accuracies and probabilities of $\chi 2$ tests for selected stripe rust yield loss models for winter and spring wheat based on predicted and actual yield loss data of winter wheat (1993 to 2009) and spring wheat (1995 to 2009) at Pullman, Washington

Model for winter wheat	Prediction accuracy (%)[a]	χ^2 test P[b]	Model for spring wheat	Prediction accuracy (%)[a]	χ^2 test P[b]
Model 1, Table 2	75	0.21	Model 1, Table 3	83	0.57
Model 2, Table 2	83	0.49	Model 2, Table 3	85	0.41
Model 12, Table 2	92	0.49	Model 3, Table 3	69	0.57
Model 15, Table 2	92	0.49	Model 6, Table 3	85	0.80
Model 16, Table 2	92	0.49	Model 8, Table 3	84	0.41
Model 17, Table 2	92	0.49	Model 9, Table 3	92	0.80

[a] The accuracy values were calculated as frequency (%) of predicted and actual values in the same yield loss categories.
[b] P = chi-squared test probability.

DISCUSSION

From a total of 1,376 weather variables, 54 (40 temperature and 14 precipitation variables) were identified to be correlated to yield loss of winter wheat and 18 (12 temperature and 6 precipitation variables) to yield loss of spring wheat caused by stripe rust (Table 1). Using these correlated variables, 19 models with single or multiple variables were developed for predicting yield loss of winter wheat and nine for spring wheat (Tables 2 and 3). Based on their greater practicality, six models each (Table 4) were selected for both winter and spring wheat crops to be validated, and eventually to be used in disease management. Use of the series of the selected models will allow prediction of yield loss as early as just after December and also more accurate predictions as the season progresses and more weather data become available. Compared with previous models (9–12), the new system of prediction using the series of models has several advantages. First, the new models directly predict yield loss rather than disease severity at the flowering stage, which can make predictions more meaningful. Second, the new model system makes prediction 1 month earlier than the old models (December versus January). Growers in the PNW would benefit from knowing the potential level of a stripe rust epidemic in the following crop season as early as possible to select spring cultivars to plant and prepare for early management of winter wheat crops. Third, the new system uses a series of models, which allows refinement of prediction as weather condition changes and the crop growth progresses. Fourth, several selected models, which use simple weather factor values without complicated calculation and conversion of negative or positive degree days, should be easy to understand by growers. Finally, the new models, which were developed using weather and disease data from 1993 to 2007, should be more suitable than the old models, which were based on data from 1963 to 1979 (9) for reasons discussed below. In the present study, two different periods of weather data were analyzed: 1975 to 2007 and 1993 to 2007 for winter wheat and 1979 to 2007 and 1995 to 2007 for spring wheat. In addition to the data in the 1960s and 1970s used in previous studies (9,10), the yield loss data since 1975 were used in this study. The variables that had the highest correlation to stripe rust yield loss were not the same for the periods 1975 to 2007 and 1993 to 2007.

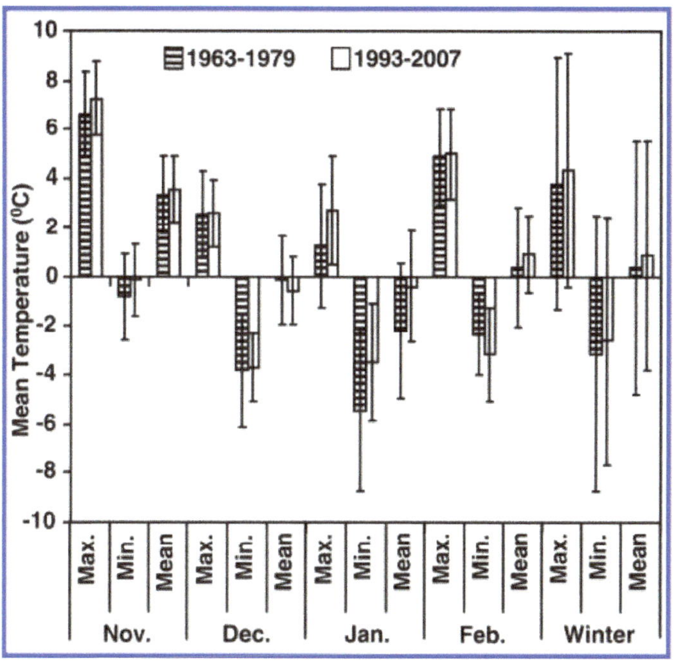

Figure 4. Comparison of monthly winter normal (long-term mean) temperatures for the periods of 1963 to 1979 and 1993 to 2007 for Pullman, Washington.

A number of weather variables from 1993 to 2007 had higher correlation coefficients with yield loss percentage than the weather variables from 1975 to 2007 (data not shown). For example, the highest correlation coefficient obtained for the 1975 to 2007 period was for the entire winter season, NovFebST(Max), ($r = 0.71$, $P = 0.0001$), while the highest correlation coefficient obtained for variables of the 1993 to 2007 period was with FebST(Max) ($r = 0.94$, $P = 0.0001$). Since the weather variables of 1993 to 2007 had higher correlation coefficients than those of 1975 to 2007, the new models were developed using weather variables of 1993 to 2007. Within the winter season, December temperatures were found more highly correlated with winter and spring yield loss than the January temperatures in the present study. However, January temperatures were previously thought to be a more limiting factor (9). This difference may indicate changes in weather patterns between the periods. Comparison between the normal winter temperatures for the period of 1963 to 1979 used in a previous study (9) and that of 1993 to 2007 revealed increases and pattern changes in winter temperature during the winter period (November to February) in the PNW except for December mean and February minimum temperatures (Fig. 4).

The coldest month for the period of 1963 to 1979 was January, while December was the coldest month for the period of 1993 to 2007, although overall winter temperatures have risen. Therefore, the more recent winter temperature probably had less limiting effect on the pathogen winter survival and thus in stripe rust development compared with the previous period, 1963 to 1979 (Fig. 4). The monthly mean minimum temperature for April was significantly correlated with stripe rust intensity during the period of 1963 to 1979 (9). In the present study, none of the spring weather variables based on minimum temperature correlated with yield loss. These results indicate that spring temperature in the period of 1993 to 2007 was not as critical for stripe rust development as in 1963 to 1979. For example, April and spring season mean maximum, minimum, and mean temperatures were 13.74, 1.94, and 7.84°C; and 14.00, 2.53, and 8.26°C, respectively, for the period of 1993 to 2007. In contrast, during the period of 1963 to 1979, these temperatures were 8.03, –1.03, and 3.5°C; and 12.66, 1.68, and 7.17°C, respectively. Normal mean temperatures of April and spring season of 1993 to 2007 were closer to the basal temperature, 7°C, for P. striiformis f. sp. tritici than during the 1963 to 1969 period. The 1993 to 2007 April temperatures were more favorable for stripe rust development and therefore the spring-season temperatures were more favorable to stripe rust and might no longer be a critical limiting factor for stripe rust development in the PNW. The changes in winter and spring temperatures in the periods compared here demonstrate the necessity of periodically updating forecasting models based on these weather variables. The disease-favorable temperature changes in the spring, together with the changes in the winter season toward conditions less detrimental for pathogen survival, could partially explain the more frequent severe epidemics that have occurred in the region in the recent years (3–7). A number of approaches are available to develop crop loss models (19,24,28,29). We relied on the commonly used least squares regression analysis to develop yield loss prediction models using weather variables (24). Absolute yield loss may provide direct information for wheat growers; however, relative yield loss can be expressed in different forms (24). Relative yield loss should be interpreted cautiously in accordance with wheat productivity of a locality. For example, 10% yield loss represents different actual yield losses in Whitman and Benton counties (WA) because of different productivities, 4.7 and 2.4 t ha–1, respectively. However, yield loss percentage gives a good indication of absolute loss for a particular cultivar in a given region. Our program annually conducts yield loss experiments for major wheat cultivars grown in the PNW to estimate percentage and absolute value of yield loss for each cultivar and uses this information to guide disease management on the basis of individual cultivars (X. M. Chen, unpublished data). Simple linear models developed using an independent variable of different months/periods can be used for long-term, mid-term, and short-term yield loss predictions. Prediction models such as Model 1 (Table 2) based on December temperature would be applicable for very early forecasting of yield loss. Model 2 (Table 2) with FebST(Max) as a descriptor is the best simple linear model because this model has the least PRESS and the highest predicted R2 and prediction accuracy percentage among the simple regression models. Model 7 (Table 2) with DecJan-RFD can predict yield loss in early February, earlier than Model 2 (Table 2). Without measuring other weather parameters, just counting raining days during December to January, this model can be used to predict potential yield loss. Relatively few weather variables were significantly correlated with spring wheat yield loss compared with winter wheat yield loss. Among the simple linear models developed for spring wheat, Model 1 (Table 3) can be used for yield loss prediction in early January. Models 2 and 3 (Table 3) can be used for yield loss prediction in early February and March, respectively. Better prediction can be obtained with multiple regression Model 6 (Table 3) or Model 8 (Table 3). Since the winter wheat and spring wheat yield loss due to stripe rust is highly correlated ($r = 0.91$, $P = 0.001$), yield loss prediction models developed for winter wheat are practically useful for managing stripe rust of spring wheat.

In the PNW, winter wheat is generally planted in September and October and harvested in July and August; and spring wheat is planted in March and April and harvested in August and September. Urediniospores produced on winter crops provide inoculum for spring wheat. Any control measures for stripe rust management of winter wheat, such as growing resistant cultivars and timely use of fungicides should reduce the disease pressure for spring wheat. However, it is also important to grow resistant spring wheat cultivars because stripe rust only varies in the inoculum level from year to year, but never disappears in the PNW, and also because the summer weather conditions allow stripe rust to develop on spring wheat. Although the selected models were developed with yield loss and weather data at Pullman, WA, they can be useful for other areas in the PNW with the exceptions of western Oregon, western Washington, and possibly southern Idaho. However, the stripe rust epidemic levels in the Pullman region (southeastern Washington and north-central Idaho) give indications of severe, moderate, and light damage levels for the whole PNW even though damage can vary among different areas in the region. The prediction models can be directly used for stripe rust management in the PNW. Based on our observations, if stripe rust is predicted to be light (1 to 20% yield loss) on susceptible cultivars, fungicide application is not necessary for most of the currently grown cultivars that are resistant to moderately susceptible. Moderately susceptible to susceptible cultivars should be sprayed and moderately resistant to resistant cultivars may not need to be sprayed with fungicides if stripe rust is predicted to be moderate (>20 to 50% yield loss). If stripe rust is predicted to be severe (>50% yield loss), all cultivars except those highly resistant should be sprayed with fungicides one or two times.

ACKNOWLEDGMENTS

This research was supported by the U.S. Department of Agriculture, Agricultural Research Service (Project 5348-22000-014-00D) and Washington State University (Project 11W-3061-7824 and 13C-3061- 3923). PPNS 0541, Department of Plant Pathology, College of Agricultural, Human, and Natural Resource Sciences, Agricultural

Research Center, Project Number WNP00823, Washington State University, Pullman, WA 99164-6430. We thank D. A. Johnson and T. D. Murray for critical review of the manuscript and M. Evans for statistical consultation.

REFERENCES

1. Al-Kandari, N. M., and Jolliffe, I. T. 2005. Variable selection and interpretation in correlation principal components. Environmetrics 16:659-672.
2. Audsley, E., Milne, A., and Paveley, N. 2005. A foliar disease model for use in wheat disease management decision support systems. Ann. Appl. Biol. 147:161-172.
3. Chen, X. M. 2005. Epidemiology and control of stripe rust [Puccinia striiformis f. sp. tritici] on wheat. Can. J. Plant Pathol. 27:314-337.
4. Chen, X. M. 2007. Challenges and solutions for stripe rust control in the United States. Aust. J. Agric. Res. 58:648-655.
5. Chen, X. M., Moore, M. K., Milus, E. A., Long, D. L., Line, R. F., Marshall, D., and Jackson, L. 2002. Wheat stripe rust epidemics and races of Puccinia striiformis f. sp. tritici in the United States in 2000. Plant Dis. 86:39-46.
6. Chen, X. M., Moore, M. K., and Wood, D. A. 2003. Epidemics and control of stripe rust on spring wheat in the Pacific Northwest in 2002. Phytopathology 93(suppl.):S16.
7. Chen, X. M., Penman, L., Wan, A. M., and Cheng, P. 2010. Virulence races of Puccinia striiformis f. sp. tritici in 2006 and 2007 and development of wheat stripe rust and distributions, dynamics, and evolutionary relationships of races from 2000 to 2007 in the United States. Can. J. Plant Pathol. 32:315-323.
8. Coakley, S. M. 1978. The Effect of climate variability on stripe rust of wheat in the Pacific North-west. Phytopathology 68:207-212.
9. Coakley, S. M., Boyd, W. S., and Line, R. F. 1982. Statistical models for predicting stripe rust on winter wheat in the Pacific Northwest. Phytopathology 72:1539-1542.
10. Coakley, S. M., and Line, R. F. 1981. Quantitative relationships between climatic variables and stripe rust epidemics on winter wheat. Phytopathology 71:461-467.
11. Coakley, S. M., Line, R. F., and McDaniel, L. R. 1988. Predicting stripe rust severity on winter wheat using an improved method for analyzing meteorological and rust data. Phytopathology 78:543-550.
12. Coakley, S. M., McDaniel, L. R., and Shaner, G. 1985. Model for predicting severity of Septoria tritici blotch on winter wheat. Phytopathology 75:1245-1251.
13. De Wolf, E. D., and Isard, S. A. 2007. Disease cycle approach to plant disease prediction. Annu. Rev. Phytopathol. 45:203-220.
14. De Wolf, E. D., Madden, L. V., and Lipps, P. E. 2003. Risk assessment models for wheat Fusarium head blight epidemics based on within-season weather data. Phytopathology 93:428-435.

15. Draper, N. R., and Smith, H. 1981. Pages 419-422, 709 in: Applied Regression Analysis. John Wiley & Sons, New York.

16. Fekedulegn, B. D., Colbert, J. J., Hicks, R. R., Jr., and Schuckers, M. E. 2002. Coping with multicollinearity: An example on application of principal components regression in dendroecology. Pages 1-40 in: Research Paper NE-721, USDA Forest Service, Northeastern Research Station, Newton Square, PA.

17. Gladders, P., Langton, S. D., Barrie, I. A., Hardwick, N. V., Taylor, M. C., and Paveley, N. D. 2007. The importance of weather and agronomic factors for the overwinter survival of yellow rust (Puccinia striiformis) and subsequent disease risk in commercial wheat crops in England. Ann. Appl. Biol. 150:371-382.

18. Hardwick, N. V. 1998. Disease forecasting. Pages 207-230 in: The Epidemiology of Plant Diseases. D. G. Jones ed. Kluwer Academic, Boston, MA.

19. James, W. C. 1974. Assessment of plant disease and losses. Annu. Rev. Phytopathol. 12:27-48.

20. Jolliffe, I. T. 1972. Discarding variables in a principal component analysis. I: Artificial data. Appl. Statist. 21:160-173.

21. King, J. R., and Jackson D. A. 1999. Variable selection in large environmental data sets using principal components analysis. Environmetrics 10:67-77.

22. Kutner, M. H., Christopher, J. N., John, N., and William, L. 2005. Pages 134-137; 256-293; 343-382 in: Applied Linear Statistical Models. McGraw-Hill, New York.

23. Line, R. F. 2002. Stripe rust of wheat and barley in North America: A retrospective historical review. Annu. Rev. Phytopathol. 40:75-118.

24. Madden, L. V., Hughes, G., and van den Bosch, F. 2007. Pages 40; 360 in: The Study of Plant Disease Epidemics. The American Phytopathological Society, St. Paul, MN.

25. Michaelsen, J. 1987. Cross-validation in statistical climate forecast models. J. Climate Appl. Meteorol. 26:1589-1600.

26. Montgomery, D. C., and Peck, E. A. 1982. Pages 424-429 in: Introduction to Linear Regression Analysis. John Wiley & Sons, New York.

27. Moschini, R. C., and Perez, B. A. 1999. Predicting wheat leaf rust severity using planting date, genetic resistance, and weather variables. Plant Dis. 83:381-384.

28. Park, R. F. 1990. The role of temperature and rainfall in the epidemiology of Puccinia striiformis f. sp. tritici in the summer rainfall area of eastern Australia. Plant Pathol. 39:416-423.

29. Savary S., Teng, P. S., Willocquet, L., and Nutter, F. W. 2006. Quantification and modeling of crop losses: A review of purposes. Annu. Rev. Phytopathol. 44:89-112.

30. Savary, S., Willocquet, L., Elazegui, F. A., Castilla, N. P., and Teng, P. S. 2000. Rice pest constraints in tropical Asia: quantification of yield losses due to rice pests in a range of production situations. Plant Dis. 84:357- 369.

31. Sharp, E. L. 1965. Prepenetration and postpenetration environment and development of Puccinia striiformis on wheat. Phytopathology 55:198- 203.

32. Stubbs, R. W. 1985. Stripe rust. Page 61-101 in: The Cereal Rusts II. Diseases, Distribution, Epidemiology and Control. A. P. Roelfs and W. R. Bushnell, eds. Academic Press, New York.

33. Weisberg, S. 1985. Page 324 in: Applied Linear Regression. John Wiley, New York.

34. Xu, X. M. 1999. Modeling and forecasting epidemics of apple powdery mildew (Podosphaera leucotricha). Plant Pathol. 48:462-471.

35. Zadoks, J. C. 1961. Yellow rust on wheat: Studies in epidemiology and physiologic specialization. Neth. J. Plant Pathol. 67:69-259.

36. Zadoks, J. C. 1984. A quarter century of disease warning, 1958–1983. Plant Dis. 68:352-355.

4
CHAPTER

PHYTOCHEMICAL DETERMINATION AND ANTIBACTERIAL ACTIVITY OF PUNICA GRANATUM PEEL EXTRACTS AGAINST PLANT PATHOGENIC BACTERIA

Ayad Ismael Khaleel, Kamaruzaman Sijam, Tavga Sulaiman Rashid, Khairulmazmi Bin Ahmad

Keywords: Punica granatum, Plant Extraction, Pathogenic Bacteria, Phytochemical Screening

INTRODUCTION

For several years now, plants and plant materials have been used as a source of medicinal agent and numerous natural products acquired from medicinal plants either as a crude extract or as purified products have been employed in disease control. Medicinal plant parts have been extensively used to extract raw drugs owing to possession of various medicinal properties. They constitute credible sources for a huge number of modern drugs, several of which are usually based on their traditional folk medicine. The World Health Organization (WHO) has stated that medicinal plants are the best source for obtaining a variety of therapeutic agents, and several medicinal plants have been employed as a source of medicine in daily life for treatment of various types of ailments globally [1] . Clinical microbiologists have greatly used medicinal plants for the screening of new therapeutic agents [2] . A great range of biotic molecules referred to as secondary metabolites are produced by plants [3] , thereby making them a rich source for diverse forms of medicine. Additionally, the primary advantage of using these naturally derived products include safety for human consumption, possess

Citation: Ayad Ismael Khaleel, Kamaruzaman Sijam, Tavga Sulaiman Rashid, Khairulmazmi Bin Ahmad, Phytochemical Determination and Antibacterial Activity of Punica granatum Peel Extracts against Plant Pathogenic Bacteria, . http://dx.doi.org/10.1016/j.jep.2012.07.004.

no harmful effects on the environment, and low cost is incurred in treating microbial infections when they are used [4] . In this study, Punica granatum were chosen as the antibacterial agent against plant pathogenic bacteria because these plants contain many biological activity potentials of being anti-diarrheal, antimicrobial, antimalarial, hepatoprotective effects, wound healing and many more. Punica granatum L. back to the family Punicaceae usually known as Pomegranate. It has just two species P. protopunica Balf and P. granatum Linn. Punica granatum is a plant or small tree with some upright, prickly stems, the leaves are roughly 2 × 1 inches, elliptic, flowers white or red, double-flowered races being also known. P. granatum is reported to be used for many disease conditions in traditional medicine. Some of its reported uses include enhancement of semen formation, gastrointestinal problems, memory activation, boosting of hemoglobin. Various parts of the plant are active in the management of many diseases such as dyspepsia, leprosy, bronchitis and hypertension. The plant has also been used as an antispasmodic and anthelmintic. In Hausa land, the flowers are used as vermifuge. The fruit and bark have also been used in tanning in ancient times. The plant is reported to contain over 28% of Gallotannic acid and the alkaloid pelletierine, methypelletierine, isopelletierine, gallic acid, psuedopelletierine, calcium oxalate, tannic acid, sugar [5] . In addition, the main advantage of using these naturally derived products are that they are safe to human health, do not leave any harmful effects on the environment and incur little cost in counteracting microbial infections [4] . R. Solanacearum, X. gardneri, and E. caratovora sub.sp. caratovora are believed to be the most important plant pathogens that are becoming destructive when affecting particular crops that strengthen the agriculturist trust that these plant pathogenic bacteria cannot be left to spread widely amongst the valuable crops that will lead to severe crop and economic losses. All of these pathogenic bacteria have treated as early as possible before the whole crops are damaged. These pathogens started to be resilient towards the antibiotics, and chemical pesticide used, so the use of chemical pesticides and antibiotics in controlling these destructive plant pathogens seem to be less effective. To face this problem, the use of P. granatum L. peel extracts could be another successful method in controlling severe plant diseases. For that reason, the mean objective of this study was to determine the antibacterial activity of P. granatum L. peel extracts against plant pathogenic bacteria.

MATERIALS AND METHODS

Bacteria Isolate

All plant pathogenic bacteria used in this study were obtained from University Protection, the Microbiology Laboratory. They are X. gardneri, R. Solanacearum, E. caratovora sub.sp. caratovora. The cultures were maintained on nutrient agar (NA) for continued viability.

Plant Materials

Fresh fruit of pomegranate (P. granatum L.) was collected from the local market. The fruit was washed with running tap water to remove dirt and insects from the peel surfaces and then separate the seeds from the peel and then dried in an oven at 50°C for three days. Then crushed into the fine peel powder using a cross beater mill (SK100, Retsch) machine with sieving size 0.50 mm (SK100, Retsch). Three hundred gm. of P. granatum peel powder placed into three separate (2 L) volume conical flask labeled with n-hexane, methanol, and ethyl acetate. Tow liters each of n-hexane, methanol, and ethyl acetate were poured into different labeled flask. The flask was shaken gently to mix well the organic solvents with the peel powder and was left under fume chamber for two days with occasional shaking. The extracts were filtered through Whatman No.1 filter paper with an aid of aspirator (A-3S, Eyela) into three different clean flasks. The filtrates were then lyophilized using rotavapor. The dry extracts were then collected into a small beaker and labeled based on the organic solvents used. The extracts then were kept in chillers (Chill-300, Protech) at 4°C for further use slight modification from [6] and [7] .

Antibacterial Activity of P. granatum Peels Extracts

The effects of P. granatum peel extracts on bacterial growth were measured by in vitro agar well diffusion assay according to the technique described by [8] with slight modification. 150 µl of standardized bacterial suspension with O.D. = 0.1 was spread over 20 mm thick Mueller Hinton Agar (MHA) with an L-shaped glass rod and left for 5 minutes to dry. Six wells with 0.6 mm diameter were. Each well was filled with 50 µl of extracts different concentration of 200, 100, 50 and 25 mg/ml. The other two empty wells were filled with 50 µl of Streptomycin sulfate (30 µg/ml) as a positive control and 50 µl of absolute methanol as a negative control. All plates were then incubated at 30°C for 24 h and were done in three replicates. Measured with a ruler in millimeter (mm) the zones around the wells were will record as inhibition zone for extracts.

Statistical Analysis

Data analysis, all variables were subjected to normality test, and the results showed that all variables were normally distributed. To compare different concentration (treatment) data were analyzed by analysis of variance (ANOVA) based on CRD (completely randomized design) using SAS version 9.1. The mean comparison was done by Duncan Multiple Range Test (DMRT). The significant differences were considered significant at $P < 0.05$ [9] . The results of IC_{50} and IC_{90} values were calculated using probit analysis using Polo Plus Ver 2.

Minimal Inhibitory Concentration (MIC)

Minimal inhibitory concentration (MIC) determination, the lowest concentration (25 mg/ml) of the antibacterial activity that was still able to inhibit the bacteria was taken as the starting concentration for making dilution. Eleven test tubes labeled (T1-T11) were each filled with 1 ml Mueller-Hinton Broth (MHB) for each bacterial isolate. Two-fold serial dilutions (1:1) were made by filling in 1 ml of the extracts to the first test tube and vortexed to mix the solution well. After that take 1 ml of the solution from T1 was withdrawn and diluted into (T2)a second test tube and repeated the procedure until test tubes 10 (T10). 100 µl of the standardized bacterial suspension was then inserted into each test tube except for T10 that served as positive control (MHB + plant extracts only). The last test tube (T11) served as t h e negative control (MHB + bacteria suspension). Finally, 50 μl of 2,3,5-Triphenyltetrazolium chloride (TTC) was added to each test tube. All the test tubes were then incubated for 24 h at 30°C. after that any test tube that did not change into red color was logged as the MIC.

Minimum Bactericidal Concentration (MBC)

The minimum bactericidal concentration value was determined by a subculture the test tubes from MIC essay that did not change into a red color, onto the sterile MHA plate. Then incubated overnight at 28°C. The least concentration that presented no detectable growth on the agar plates was considered as the MBC value [10] .

Protocol for Phytochemical Screening

The phytochemical screening of P. granatum L. ethyl acetate peel extracts was done to find the presence of active chemical constituents such as Flavonoids, Alkaloids, Saponins, Tannins, Phenolic, and terpenoids. The phytochemical screening was done based on previous procedures published somewhere else [11] -[18] .

RESULTS

In-vitro Antibacterial Activity of Punica granatum Peel Extracts

As the result shown, there was no antibacterial activity of Punica granatum methanol and n-hexane crude extracts detected against all species of plant pathogenic bacteria. The antibacterial activity of P. granatum L. ethyl acetate peel extracts was successfully evaluated in-vitro against three plant pathogenic bacteria. They are X. gardneri, R. Solanacearum, E. caratovora sub.sp. caratovora. These bacteria are known to cause a harmful effect on valued food crops. Table 1 and Figure 1 show inhibition zone diameter ranging from (8.50 ± 0.58 mm) to (22.75 ± 0.48 mm). As the positive control Streptomycin sulfate (30 μg/ml) showed inhibition towards three bacterial isolates ranging from (24.75 ± 0.48 mm) to (26.75 ± 0.48 mm). There was no inhibition recorded for all plant pathogenic bacteria when tested with absolute ethyl acetate as the negative control.

At the lowest concentration (25 mg/ml) of extracts used, the highest mean of inhibition zone was recorded for R. Solanacearum (13.0 ± 0.58^{e} mm), followed by then E. caratovora sub.sp. caratovora (9.5 ± 0.29^{e}) and X. gardneri (each 8.5 ± 0.58 mm). At a concentration of 50 mg/ml, the lowest mean diameter inhibition zones were recorded for X. gardneri (12.5 ± 0.29 mm) followed by E. caratovora sub.sp. caratovora (13.50 ± 0.29 mm)

Table 1. Means (n = 4) of inhibition zone diameter (mm) of P. granatum L. ethyl acetate peel extract against plant pathogenic bacteria. Within rows, values with different letters differ significantly.

Concentration	R. solanacearum	X. gardneri	E. carotovora
25	13.0 ± 0.58[e]	8.5 ± 0.29[e]	9.5 ± 0.29[e]
50	15.75 ± 0.48[d]	12.5 ± 0.29[d]	13.50 ± 0.29[d]
100	19.25 ± 0.48[c]	15.5 ± 0.29[c]	18.50 ± 0.29[c]
200	22.75 ± 0.48[b]	19.5 ± 0.29[b]	21.75 ± 0.25[b]
Positive Control	26.75 ± 0.48[a]	24.75 ± 0.48[a]	25.0 ± 0.41[a]
Negative Control	0[f]	0[f]	0[f]

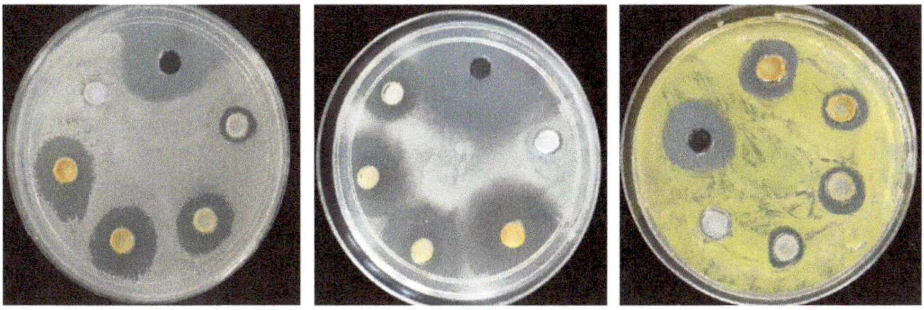
(a) (b) (c)

Figure 1. Inhibition zone diameter of Punica granatum L. ethyl acetate peel extracts against plant pathogenic bacteria. (a) E. caratovora sub.sp. caratovora; (b) R. Solanacearum; (c) X. gardneri.

while R. Solanacearum showed diameter inhibition zone of (15.75 ± 0.48). While a concentration of 100 mg/ml, the highest mean diameter inhibition zones was recorded for R. Solanacearum (19.25 ± 0.48), E. caratovora sub.sp. caratovora showed diameter inhibition zone of (18.50 ± 0.29 mm) while X. gardneri (15.5 ± 0.29 mm). At the highest extract concentration of 200 mg/ml, the highest mean diameter was recorded for R. Solanacearum (22.75 ± 0.48 mm), followed by E. caratovora sub.sp. caratovora (21.75 ± 0.25 mm) and X. gardneri showed diameter inhibition zone of (19.5 ± 0.29 mm).

Minimum Inhibitory Concentration (MIC)

The (MIC) all the tested bacterial isolates, the lowest concentration that still inhibited the plant pathogenic bacteria was at 3.125 mg/ml for R. Solanacearum followed by E. caratovora sub.sp. caratovora with the lowest concentration recorded at 3.125 mg/ml while X. gardneri with the lowest concentration recorded at 6.25 mg/ml. For MBC, The lowest concentration that showed no visible growth on the agar plates was considered the MBC value 25 mg/ml recorded for X. gardneri while concentration 12.5 mg/ml for R. Solanacearum and E. caratovora sub.sp. caratovora as shown inTable 2 and Figure 2.

The results of IC_{50} and IC_{90} values were calculated using probit analysis using Polo plus Ver 2 and results indicated that the chi square test for all these three bacteria has significant heterogeneity in the test population (Table 3). The highest IC_{50} was observed for X. gardneri followed by E. carotovorum and R. solancearum. This results indicated the X. gardneri had the highest IC_{90}.

Table 2. Minimal inhibitory concentration (MIC) and minimum bactericidal Concentration (MBC) of P. granatum L. ethyl acetate peel extracts against plant pathogenic bacteria by macro-broth dilution methods.

Bacteria species	MIC (mg / ml)	MBC (mg / ml)
Ralstonia solanacearum	3.125	12.5
Erwinia caratovora sub.sp. caratovora	3.125	12.5
Xanthomonas gardneri	6.25	25

Table 3. Probit regression line parameter and inhibition concentration (IC).

Bacteria	Regression equation	Chi-square (df)	IC_{50} with fiducial limits	IC_{90} with fiducial limits
R. solancearum	Y = 1.086 - 1.644 X	0.156 (2)	32.55 (5.8 - 55.2)	492.25 (197.9 - 39033.0)
X. gardneri	Y = 1.207 - 2.181 X	0.050 (2)	63.99 (32.8 - 111.6)	737.11 (276.6 - 39355.0)
P. carotovorum	Y = 1.570 - 2.625 X	0.063 (1)	46.971 (25.8 - 68.4)	307.64 (167.7 - 1622.8)

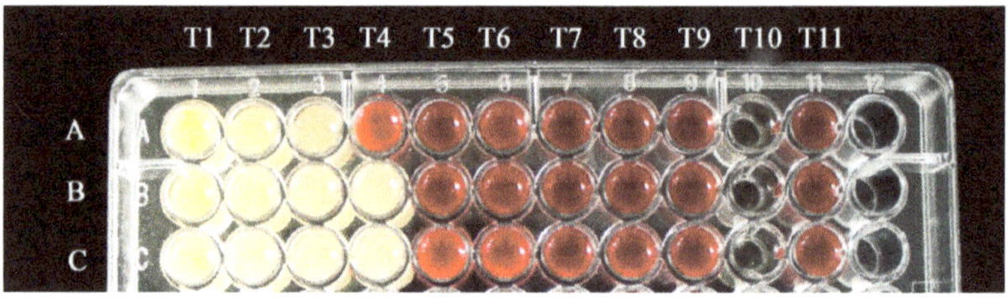

Figure 2. Minimum inhibitory concentration (MIC) of Punica granatum L. ethyl acetate peel extract against X. gardneri (A), R. solancearum (B), E. carotovorum (C). (T1 = 25 mg/ml, T2 = 12.5 mg/ml, T3 = 6.25 mg/ml, T4 = 3.125 mg/ml, T5 = 1.563 mg/ml, T6 = 0.781 mg/ml, T7 = 0.391 mg/ml, T8 = 0.195 mg/ml, T9 = 0.098 mg/ml, T10 = positive control (MHB + plant extract), and T11 = Negative control (MHB + bacteria suspension). The results obtained from macro broth dilution technique using test tubes were transferred into microliter plate for the photography purpose.

Phytochemical Screening of P. granatum L. Ethyl Acetate Peel Extracts

The phytochemical screening of P. granatum L. ethyl acetate peel extracts the presence some major compound known as secondary metabolites determine, the results obtained were shown in Table 4. It showed the positive results for the presence of alkaloid, saponin, phenol, tannin, flavonoid, terpenoid. As determined by the color change.

DISCUSSION

P. granatum L. against plant pathogenic bacteria were effective against all plant pathogenic bacteria tested. This can probably replace the use of synthetic antibiotics as they can become resistance to the antibiotics when used. Most of the previous research P. granatum L. is reported to be used for many disease conditions in folktale medicine. Previous studies showed that some researchers like [19] -[21] stated that Punica granatum peel extracts in different concentrations were effective against streptococci strains (streptococci mutans; streptococci aureus; streptococci salivarius; streptococci sanguinis and streptococci epidermidis). It was confirmed that this antibacterial activity may be connected to the presence of polyphenolics and hydrolysable tannins in the pomegranate extract specifically gallagic acid and punicalagin [22] [23] . It means that the antimicrobial influence of tannins was linked to its toxicity and structure of molecular. Tannins may be a performance on the cell wall and across the cell membrane because they can precipitate proteins [23] [24] . They may also suppress many enzymes such as glycosyltransferases [21] and [20] demonstrated that gallic acid (a tannic acid) has the highest antibacterial effect against tested sensitive strains even at low concentrations. Hence, the antibacterial activity of Punica granatum may be related to polyphenol structures because polyphenols may affect the bacterial cell wall, inhibit enzymes by oxidized agents, interact with proteins and disturb coaggregation of microorganisms [20] [21] . This study proved that P. granatum Ethyl Acetate peel extracts can be industrialized as next antibacterial agents to control plant pathogenic bacteria as it was able to inhibit the growth of plant pathogenic bacteria. All bacteria sp. isolates gave good inhibition zones when tested with all range of concentration. Another study has reported that pomegranate extract inhibited Pseudomonas aeroginosa growth and had an interactive effect against bacteria resistant to the known antibiotics. Synergistic effects of methanolic pomegranate extract and chloramphenicol, gentamicin, ampicillin, tetracycline and oxacillin against Staphylococcus aureus is also reported by [25] . [26] reported that the antibacterial activity of peel extracts of pomegranate against both Gram positive and negative bacteria strains and mention MIC values ranging from 0.25 to 4.0 mg/ml against the tested bacteria. Also, he reported a two-fold MIC value against a Gram-positive bacterium (S. aureus) than against a Gram-negative bacterium (E. coli). The phytochemical screening of Punica granatum L ethyl acetate peel extracts, the results obtained were shown the positive results for the presence of flavonoid, alkaloid, saponin, terpenoid, phenol, tannin. Study for [27] reported that the pomegranate peel indicated the presence of Alkaloid, Tannins, Flavonoids, Saponins, Glycosides, Sterols, Resins, Volatile oils, Carbohydrates, Balsams, Terpenes, and Free-Reducing sugar. The presence of these metabolites suggests great possibilities for the plant as a source of useful phytomedicines. For instance, some alkaloids are known to be used as antimalarial agents [28] . Tannins too could show that it is a severe, help in wound healing and anti-parasitic. The presence of resins and flavonoids might be responsible its use as anti-inflammatory properties [29] . The presence of terpenes proposes it possible use as an antiviral and antitumor [30] .

Table 4. Phytochemical screening of Punica granatum L ethyl acetate peel.

Chemical Constituents	Positive Results	Ethyl Acetate extracts
Saponin	Small bubbles (foam)	+
Alkaloid	Creamish precipitate	+
Phenol	Dark blue	+
Tannin	Blue-green color	+
Flavonoid	Dark yellow	+
Terpenoid	Reddish brown coloration	+

CONCLUSION

P. granatum Ethyl Acetate peel extracts gave significant antibacterial activity against X. gardneri, R. Solana- cearum, E. caratovora sub.sp. caratovora, which indicate that these extracts have the potential being the next natural biological control agent for plant pathogenic bacteria. The main compounds found in this study can be used to produce new and useful natural chemical product as an alternative to the use of synthetic antibiotics and chemicals for the control of plant pathogenic bacteria.

CITATION

Ayad Ismael Khaleel, Kamaruzaman Sijam, Tavga Sulaiman Rashid, Khairulmazmi Bin Ahmad (2016) Phytochemical Determination and Antibacterial Activity of *Punica granatum* Peel Extracts against Plant Pathogenic Bacteria. *American Journal of Plant Sciences*,**07**,159-166. doi:10.4236/ajps.2016.71017

REFERENCES

1. Alo, M.N., Anyim, C., Igwe, J.C., Elom, M. and Uchenna, D.S. (2012) Antibacterial Activity of Water, Ethanol and Methanol Extracts of Ocimum gratissimum, Vernonia amygdalina and Aframomum melegueta. Adv. Appl. Sci. Res., 3, 844-848.
2. Ashokkumar, P., Rajkumar and Kanimozhi, M. (2010) Phytochemical Screening and Antimicrobial Activity from Five Indian Medicinal Plants against Human Pathogens. Middle East J. Sci. Res., 5, 157-162.
3. Croteau, R., Kutchan, T.M. and Lewis, N.G. (2000) Secondary Metabolites. Biochem. Mol. Biol. Plants, 7, 1250-1318.
4. Al-zubaydi, S.R., Al-hmdany, M.A. and Raesan, S. (2009) Antibactrial Effect of Some Medicinal Plant Extracts against Some Pathogenic Bacterial Strains. Journal of Duhok University, 12, 244-249. [Paper reference:2]
5. Ismail, T., Sestili, P. and Akhtar, S. (2012) Pomegranate Peel and Fruit Extracts: A Review of Potential Anti-Inflammatory and Anti-Infective Effects. Journal of Ethnopharmacology, 143, 397-405. http://dx.doi.org/10.1016/j.jep.2012.07.004

6. Zaidan, M.R.S., Rain, N., Badrul, A.R., Adlin, A., Norazah, A. and Zakiah, I. (2005) In Vitro Screening of Five Local Medicinal Plants for Antibacterial Activity Using Disc Diffusion Method. Tropical Biomedicine, 22, 165-170.

7. Singh, S., Srivastava, R. and Choudhary, S. (2010) Antifungal and HPLC Analysis of the Crude Extracts of Acorus calamus, Tinospora cordifolia and Celestrus paniculatus. Journal of Agricultural Technology, 6, 149-158.

8. Patel, R., Patel, A., Vaghasiya, D. and Nagee, A. (2012) Antimicrobial Evaluation of Hibiscus rosasinensis Plant Extracts against Some Pathogenic Bacteria. Bulletin of Environmental and Scientific Research, 1, 14-17.

9. Obeidat, M. (2011) Antimicrobial Activity of Some Medicinal Plants against Multidrug Resistant Skin Pathogens. Journal of Medicinal Plants Research, 5, 3856-3860.

10. Taweechaisupapong, S., Singhara, S., Lertsatitthanakorn, P. and Khunkitti, W. (2010) Antimicrobial Effects of Boesenbergia pandurata and Piper sarmentosum Leaf Extracts on Planktonic Cells and Biofilm of Oral Pathogens. Pakistan Journal of Pharmaceutical Sciences, 23, 224-231.

11. Obadoni, B.O. and Ochuko, P.O. (2002) Phytochemical Studies and Comparative Efficacy of the Crude Extracts of Some Haemostatic Plants in Edo and Delta States of Nigeria. Global Journal of Pure and Applied Sciences, 8, 203-208. http://dx.doi.org/10.4314/gjpas.v8i2.16033

12. Alabri, T.H.A., Al Musalami, A.H.S., Hossain, M.A., Weli, A.M. and Al-Riyami, Q. (2014) Comparative Study of Phytochemical Screening, Antioxidant and Antimicrobial Capacities of Fresh and Dry Leaves Crude Plant Extracts of Datura metel L. Journal of King Saud University—Science, 26, 237-243. http://dx.doi.org/10.1016/j.jksus.2013.07.002

13. Khanam, Z., Wen, C.S. and Bhat, I.U.H. (2014) Phytochemical Screening and Antimicrobial Activity of Root and Stem Extracts of wild Eurycoma longifolia Jack (Tongkat Ali). Journal of King Saud University—Science, 27, 23-30.http://dx.doi.org/10.1016/j.jksus.2014.04.006

14. Sharma, A.K., Gangwar, M., Tilak, R., Nath, G., Sinha, A.S.K., Tripathi, Y.B. and Kumar, D. (2012) Comparative in Vitro Antimicrobial and Phytochemical Evaluation of Methanolic Extract of Root, Stem and Leaf of Jatropha curcas Linn. Pharmacognosy Journal, 4, 34-40. http://dx.doi.org/10.5530/pj.2012.30.7

15. Maskovic, P., Manojlovic, N., Mandic, A., Misan, A., Milovanovic, I., Radojkovic, M., Cvijovic, M. and Solujic, S. (2012) Phytochemical Screening and Biological Activity of Extracts of Plant Species Halacsya sendtneri (Boiss.) D&oumlrfl. Chemical Industry, 66, 43-51. http://dx.doi.org/10.2298/HEMIND110828068M

16. Sheikh, N., Kumar, Y., Mishra, A.K. and Pfoze, L. (2013) Phytochemical Screening to Validate the Ethnobotanical Importance of Root Tubers of Dioscorea Species of Meghalaya, North East India. Journal of Medicinal Plants Studies, 1, 62-69.

17. Krasteva, I., Nikolova, I., Danchev, N. and Nikolov, S. (2004) Phytochemical Analysis of Ethyl Acetate Extract from Astragalus corniculatus Bieb. and Brain

Antihypoxic Activity. Acta Pharmaceutica, 54, 151-156.

18. Swamy, K.M., Pokharen, N. and Dahal, S. (2011) Phytochemical and Antimicrobial Studies of Leaf Extract of Euphorbia neriifolia. Journal of Medicinal Plant Research, 5, 5785-5788.
19. Singh, R.P., Chidambara Murthy, K.N. and Jayaprakasha, G.K. (2002) Studies on the Antioxidant Activity of Pomegranate (Punica granatum) Peel and Seed Extracts Using in Vitro Models. Journal of Agricultural and Food Chemistry, 50, 81-86. http://dx.doi.org/10.1021/jf010865b
20. Naziri, Z., Rajaian, H. and Firouzi, R. (2012) Antibacterial Effects of Iranian Native Sour and Sweet Pomegranate (Punica granatum) Peel Extracts against Various Pathogenic Bacteria. Iranian Journal of Veterinary Research, 13, 282-288. [Paper reference:2]
21. Vasconcelos, L.C.D.S., Sampaio, M.C.C., Sampaio, F.C. and Higino, J.S. (2003) Use of Punica granatum as an Antifungal Agent against Candidosis Associated with Denture Stomatitis Verwendung von Punica granatum als Antimykotikum gegen Candidose in Verbindung mit Zahnprothesen-Stomatitis. Mycoses, 46, 192-196. [Paper reference:3]
22. Reddy, M.K., Gupta, S.K., Jacob, M.R., Khan, S.I. and Ferreira, D. (2007) Antioxidant, Antimalarial and Antimicrobial Activities of Tannin-Rich Fractions, Ellagitannins and Phenolic Acids from Punica granatum L. Planta Medica, 73, 461-467. http://dx.doi.org/10.1055/s-2007-967167
23. Vasconcelos, L.C.D.S., Sampaio, F.C., Sampaio, M.C.C., Pereira, M.D.S.V., Higino, J.S. and Peixoto, M.H.P. (2006) Minimum Inhibitory Concentration of Adherence of Punica granatum Linn (Pomegranate) Gel against S. mutans, S. mitis and C. albicans. Brazilian Dental Journal, 17, 223-227. http://dx.doi.org/10.1590/S0103-64402006000300009
24. Braga, L.C., Shupp, J.W., Cummings, C., Jett, M., Takahashi, J.A., Carmo, L.S., Chartone-Souza, E. and Nascimento, A.M.A. (2005) Pomegranate Extract Inhibits Staphylococcus aureus Growth and Subsequent Enterotoxin Production. Journal of Ethnopharmacology, 96, 335-339. http://dx.doi.org/10.1016/j.jep.2004.08.034
25. Braga, L.C., Leite, A.A.M., Xavier, K.G.S., Takahashi, J.A., Bemquerer, M.P., Chartone-Souza, E. and Nascimento, A.M.A. (2005) Synergic Interaction between Pomegranate Extract and Antibiotics against Staphylococcus aureus. Canadian Journal of Microbiology, 51, 541-547. http://dx.doi.org/10.1139/w05-022
26. Al-Zoreky, N.S. (2009) Antimicrobial Activity of Pomegranate (Punica granatum L.) Fruit Peels. International Journal of Food Microbiology, 134, 244-248. http://dx.doi.org/10.1139/w05-022
27. Omoregie, H., Nkiruka, O.P., Sabo, M., Koma, S. and Ibumeh, J. (2010) Phytochemical Analysis and Antimicrobial Activity of Punica granatum L. (Fruit Bark and Leaves). New York Science Journal, 3, 91-98.
28. Batista, R., De Jesus Silva Jr., A. and De Oliveira, A.B. (2009) Plant-Derived Antimalarial Agents: New Leads and Efficient Phytomedicines. Part II. Non-Alkaloidal Natural Products. Molecules, 14, 3037-3072. http://dx.doi.

org/10.3390/molecules14083037

29. Egharevba, H.O. and Kunle, F.O. (2010) Preliminary Phytochemical and Proximate Analysis of the Leaves of Piliostigma thonningii (Schumach.) Milne-Redhead. http://www.ethnoleaflets.com/leaflets/eghokufo
30. Ehiabhi, S., Omachonu, M. and Adeola, I. (2010) Phytochemical and Pharmacognostic Investigation of Antidiabetic Scoparia dulcis Linn Scrophulariaceae Whole Plant Grown in Nigeria. Researcher, 2, 7-16.

5

CHAPTER

BIOLOGICAL CONTROL AGENTS AGAINST PATHOGENIC VIRAL STRAINS

Anna A. Slavokhotova, Ekaterina A. Istomina, Emma N. Andreeva, Tatyana V. Korostyleva, Vitalij A. Pukhalskij, Alexander N. Shijan, Tatyana I. Odintsova

Keywords: Cucumber Green Mottle Mosaic Virus, Tobamovirus, Attenuated Virus Strain, Cross-Protection, Biological Control

INTRODUCTION

During their lifetime plants have to counteract with various pathogens including viruses as one of the most common disease agents. Many viruses are able to systemically infect their hosts due to specific interactions between viral genome/proteins and the host proteins [1] . In turn, plants have developed induced mechanisms to inhibit virus replication and movement which finally prevents disease development. Induced resistance can also be activated by the so-called biocontrol or attenuated viral strains that are able to infect plants, but do not cause visible disease symptoms. In this case, the plant is already "prepared" for severe virus invasion due to specific defense mechanisms including RNA silencing that involves quick degradation of the RNA of the pathogenic strain [2] . Obviously, if resistant varieties are absent or unavailable, the development of novel biocontrol viral strains may be an effective measure to control viral diseases taking into account that chemical protection against viruses is usually inefficient.

Cucumber green mottle mosaic virus (CGMMV) is a member of the Tobamovirus genus. It causes one of the most common diseases of cucurbits, including cucumber,

Citation: Slavokhotova, A.A., Istomina, E.A., Andreeva, E.N., Korostyleva, T.V., Pukhalskij, V.A., Shijan, A.N. and Odintsova, T.I. (2016) Biological Control Agents against Pathogenic Viral Strains, 7, 724-732. http://dx.doi.org/10.4236/ajps.2016.75066.

pumpkin, watermelon, melon, squash, gourds, etc. CGMMV forms rod-shaped particles 300 × 18 nm in size and contains a single-strand, positive-sense, ~6-kb RNA [3] . The virus was originally described by Ainsworth in the United Kingdom in 1935 [4] . Later, CGMMV was identified in other European countries, such as Greece [5] , France [6] , and Russian Federation [7] . CGMMV is also widespread in Asia, including India [8] , Japan [3] , China [9] , Korea [10] and Pakistan [11] . Infected plants are characterized by stunted growth, mottling and mosaic on leaves, fruit mottling and distortion. Note that CGMMV symptoms may be difficult to recognize on young seedlings as well as on fruits that can, however, be internally discolored or necrotic. Like all members of the Tobamovirus genus, CGMMV might be preserved in greenhouses, equipment surfaces, and tools for a long time. It can be easily transmitted by seeds, sap or mechanical transfer. Outbreaks of the disease usually occur in greenhouses; however, the virus can also cause substantial yield losses in the field [12] . It should be noted that at present, only a limited number of cucumber varieties resistant to CGMMV are commercially available [12] . For other cucurbits, no resistant varieties have been produced so far. Hence, the development of biocontrol CGMMV strains with high level of cross- protection may represent an effective control measure to protect plants against pathogenic strains.

In the Tobamovirus genus, several attenuated viral strains used as biocontrol agents have been reported: L11A [13] , ToMV K [14] , and V-69 [15] strains of tomato mosaic virus (ToMV) for tomato plant protection; Tpa18ch, Pa18, C-1421, TPO-2-19 [16] strains against severe isolates of pepper mild mottle mosaic virus (PMMoV), and SH33b strain [17] of CGMMV for muskmelon protection. However, mild CGMMV strains used for cucumber plant protection have not been described so far. In this study, we for the first time describe the attenuated VIROG-43M CGMMV strain that can be used as a biocontrol agent for cucumber protection. We determined the full-length nucleotide sequences of this attenuated strain together with two pathogenic strains isolated from greenhouse-grown cucumber plants in Moscow region. Comparison of the attenuated and pathogenic viral genomes allowed us to detect mutations that led to attenuation in the CGMMV biocontrol strain. In addition, analyzing NCBI GenBank database, we retrieved all available CGMMV genome sequences and performed phylogenetic analysis.

MATERIALS AND METHODS

Isolation of CGMMV Strains

Pathogenic CGMMV strains MC-1 and MC-2 were isolated from leaves of cucumber plants during the outbreak of the infection in greenhouses in the Moscow region. The pathogenic CGMMV MC-1 strain produced green mosaic and mottling on cucumber leaves, while the pathogenic strain CGMMV MC-2 in addition to green mosaic and mottling, caused distortion and blistering.

The attenuated CGMMV strain VIROG-43M was obtained after a number of passages of the initial VIROG- 43 isolate; it represents a more stable analogue of the VIROG-43 strain. VIROG-43 was obtained from the pathogenic CGMMV strain by chemical mutagenesis using nitrous acid as a mutagen followed by selection of attenuated strains. The attenuated VIROG-43M strain was selected by a number of passages of the diluted virus VIROG-43 on plants with local necrotic (Chenopodium amaranticolor) and systemic (Cucumis sativus) reaction to infection. The presence of the attenuated virus in symptomless plants was controlled by DAS-ELISA. Finally, the virus was isolated from cucumber leaves and propagated on the Kurazh F1 variety (Gavrish , Russian Federation). Plants were inoculated with viral particles and grown for 3 weeks in a greenhouse under controlled temperature conditions (25°C day/18°C night). Viral particles were isolated and purified from infected cucumber leaves by stepwise precipitation with ammonium sulfate [7] .

Sequencing of CGMMV Strains

Viral RNA was isolated by RNA adsorption on silica gel in the presence of guanidine thiocyanate (GTC) [7] . An ammonium sulfate precipitate of viral particles was dissolved in 2.5 M GTC solution, containing 24% ethanol. High-molecular-weight RNA was adsorbed on glass fiber filters (GF/F, Whatman) and eluted with RNase-free water. The first cDNA strand was synthesized with RevertAid first strand cDNA synthesis kit (Fermentas , Lithuania) and random hexamer primers following the manufacturer's instructions. The second strand was synthesized, and the fragments were amplified using Taq DNA polymerase and specific primers. To design primers, we preliminarily analyzed nucleotide sequences of CGMMV isolates from the NCBI GenBank. Viral genomic RNA was sequenced at least two times in both directions from overlapping PCR fragments. If any variation in the primary structure of the studied strains occurred, the regions were sequenced more than three times. Amplified fragments were eluted from agarose gel with a Diatom DNA elution kit (Laboratoriya IzoGen , Russia) and sequenced using an ABI PRISM BigDye Terminator v. 3.1 kit. The reaction products were analyzed on an ABI PRISM 3100-Avant automated DNA sequencer. The sequence data for MC-1, MC-2 and VIROG-43M are available under NCBI GenBank accession numbers FJ848666, GQ495274, and GQ495275, respectively.

Evaluation of Phenotypic Stability, Interfering Capability and Cross-Protection Efficiency

The phenotypic stability of the VIROG-43M strain was assayed under laboratory greenhouse conditions. The experiment was repeated three times. In each replication, 33 cucumber plants were inoculated with the pathogenic MC-1 strain and 66 plants with the attenuated VIROG-43M. 0.5 μg/ml of each viral strain was used for inoculation. After two weeks, one half of plants initially inoculated with VIROG-43M were challenged with the pathogenic MC-1 strain. The number of symptomless plants was determined 1.0, 1.5, 2.0 and 2.5 months after first inoculation.

The degree of cross-protection was measured on the Kurazh F1 variety (Gavrish , Russian Federation) of cucumber plants grown in commercial greenhouses. For this experiment, stock solution of the attenuated VIROG- 43M strain was diluted with water, supplied with carborundum and sprayed on 10 - 12-day-old cucumber seedlings at 1 to 2 fully-expanded-leaf stage. One hectare of inoculated with the VIROG-43M strain and one hectare of non-inoculated control cucumber plants were analyzed throughout 3 months of vegetation during summer and winter crop rotations. The same trial was conducted in commercial greenhouses located in Leningrad, Krasnodar and Moscow regions with three other cucumber varieties: Tristan F1 (Enza Zaden, Holland), Rais F1 (Gavrish, Russia), and Mirabelle F1 (Seminis, Holland). The results are given in the corresponding section and Table 1 and Table 2.

DAS-ELISA

Cucumber plants of the Kurazh F1 variety (Gavrish, Russian Federation) were grown in laboratory greenhouse under controlled temperature and light conditions (25°C day/18°C night). When plants reached the stage of three fully expanded leaves, two bottom leaves were inoculated with 0.5 μg/ml of each viral strain. On the 3rd, 7th, 10th and 14th days post inoculation (d.p.i.), plant leaf discs were sampled and subjected to DAS-ELISA. For each sample (each CGMMV strain), ten plants were grown and inoculated with the corresponding strain. On the 3rd d.p.i., samples for DAS-ELISA were taken from six similarly developed plants. At one time point, two upper non-inoculated leaves were analyzed; from each leaf one leaf disc was taken. The experiment was repeated three times. ADGEN Identikit CGMMV (Neogen , Scotland , UK) was used for DAS-ELISA according to the

Table 1. Phenotypic stability and interfering capability of the attenuated VIROG-43M stain.

Treatment	Number of symptomless plants, %			
	1.5 months post inoculation	2.0 months post inoculation	2.5 months post inoculation	3.0 months post inoculation
VIROG-43M	100.0	100.0	98	98
VIROG-43M + MC-1	98.0	95.0	93.0	85.0
MC-1	0.0	0.0	0.0	0.0

Table 2. Suppression of CGMMV symptoms in cucumber plants inoculated with VIROG-43M.

n/a, not available.

Cucumber hybrid name, location of commercial greenhouse	Rotation	Sample	Number of plants with mosaic symptoms, %		
			1.0 months post inoculation	2.0 months post inoculation	3.0 months post inoculation
Kurazh F1 (Gavrish, Russia), Moscow region	Winter	VIROG-43M	0.0	1.0	1.5
		Control	2.0	7.5	15.0
	Summer	VIROG-43M	0.1	2.0	3.2
		Control	2.0	12.2	34.1
Tristan F1 (Enza Zaden, Holland), Leningrad region	Winter	VIROG-43M	n/a	n/a	0.5
		Control	n/a	n/a	15.0
Mirabelle F1 (Seminis, Holland), Moscow region	Winter	VIROG-43M	n/a	n/a	1.5
		Control	n/a	n/a	13.0
Rais F1 (Gavrish, Russia), Krasnodar region	Summer	VIROG-43M	n/a	n/a	3.0
		Control	n/a	n/a	19.0

manufacturer's manual. Reactions were read on an MR-96A Microplate Reader (Mindray , China) at 405 nm. For negative control, leaf discs from healthy plants were sampled at each designated time point. Systemically infected with a pathogenic strain and homogenized cucumber leaves supplied by ADGEN were used as positive control.

Phylogenetic Analysis of CGMMV Strains

The phylogenetic tree was built with MEGA6 software [18] on the base of CLUSTAL W alignment of CP nucleotide sequences. The evolutionary history was inferred by the Maximum Likelihood method based on the Tamura-Nei model [19] . Initial tree for the heuristic search was obtained by applying the Neighbor-Joining method to a matrix of pairwise distances estimated using the Maximum Composite Likelihood (MCL) approach. The tree was drawn to scale, with branch lengths measured in the number of substitutions per site. All positions with less than 95% site coverage were eliminated. That was, fewer than 5% alignment gaps, missing data, and ambiguous bases were allowed at any position. There were a total of 320 positions in the final dataset.

RESULTS

VIROG-43M, an Attenuated CGMMV Strain for Cucumber Protection

The procedure for VIROG-43M isolation was described in Material and Methods. The efficiency of the strain as a biocontrol agent was tested on the Kurazh F1 cucumber variety (Gavrish , Russian Federation) in laboratory and commercial greenhouses. Experiments in laboratory greenhouse allowed us to determine phenotypic stability of the VIROG-43M strain, its interfering capability against pathogenic strains, and cross-protection level. We discovered that VIROG-43M exhibited high phenotypic stability since 100% of cucumber plants inoculated with this attenuated strain showed no visible symptoms throughout two months of observation (Table 1). However, by the 3rdmonth of vegetation, the number of symptomless plants decreased to 98%, most likely due to mixed infection in 2% of plants. VIROG-43M also exhibited significant interfering capability, since throughout two months of vegetation, 95% of cucumber plants inoculated initially with VIROG-43M showed no disease symptoms. By the end of vegetation, the number of symptomless plants reduced, but still was rather high (85%) in comparison with 0% of symptomless plants inoculated initially with the pathogenic MC-1 strain (Table 1).

VIROG-43M was also tested in a number of commercial greenhouses on different cucumber varieties. The level of cross-protection was extensively studied on the Kurazh F1 variety (Gavrish ,Russia) in a commercial greenhouse located in the Moscow region during summer and winter rotations. The results are shown in Table 2. It is clearly seen from this table that at the end of vegetation in winter rotation only 1.5% of plants treated with VIROG-43M exhibited mosaic symptoms, while in control plants, this number increased to 15%. In summer rotation, the observed indicators increased to 3.2% for inoculated plants and to 34% for non-inoculated ones. We also tested other cucumber varieties grown in commercial greenhouses located in diverse regions of

Russia (Table 2). We found that by the end of winter rotation, Tristan F1 (Enza Zaden, Holland) and Marabelle F1 (Seminis, Holland) varieties treated with VIROG-43M exhibited mosaic symptoms only in 0.5% and 1.5% cases, in contrast to 15% and 13% in untreated plants. The Rais F1 variety (Gavrish , Russia) treated with VIROG-43M during summer rotation showed mosaic symptoms in 3.0% cases, in contrast to 19% in untreated plants. It should be also noted that using VIROG-43M during winter rotation, yield gain reached 19.8% on Kurazh F1 (Gavrish , Russia) and 16.5% on Mirabelle F1 (Seminis, Holland) varieties in comparison to untreated cucumber plants. The obtained results show that VIROG-43M can be applied in greenhouses for cucumber plant protection for two main reasons: this attenuated strain is phenotypically stable and shows high level of cross-protection against pathogenic strains resulting in significant yield gain.

Full-Length Genomic Nucleotide Sequences of Pathogenic MC-1 and MC-2 and Attenuated VIROG-43M Strains

The RNA genomes of MC-1, MC-2, and VIROG-43M CGMMV strains consist of 6422 nucleotides and have the overall structure similar to other tobamoviruses: the first and the second ORFs encode the short replicase (129-kDa) and the long replicase (186-kDa) proteins; the third and the fourth ORFs encode the 29-kDa movement protein (MP) and the 17.4-kDa CP, respectively (Figure 1).

Comparison of the genomes of MC-1 and MC-2 pathogenic strains showed that they differed in 8 nucleotides in the replicase gene and one mismatch in the MP gene. Only the C-2152 - T-2152 mismatch led to amino acid substitution P-698 → L in MC-2 strain in comparison to MC-1 strain (Figure 1). Comparison of the genomes of the pathogenic MC-1 and attenuated VIROG-43M strains revealed six mutations including two silent mutations located in the replicase and MP genes and four missense mutations in the replicase gene (Figure 1). Comparison of the nucleotide sequences of the pathogenic MC-2 and attenuated VIROG-43M strains revealed nine mutations in the replicase gene, six of which were silent and three were missense. It is interesting that these three missense mutations led to the same amino acid substitutions as those observed in the genome of the MC-1 strain (G-86 → S; S-534 → F; P-1362 → L). However, one MC-1-specific mutation leading to the P-698 → L substitution was absent from the MC-2 genome (Figure 1).

Accumulation Dynamics of VIROG-43M, MC-1 and MC-2 CGMMV Strains in Cucumber Plants

For VIROG-43M, MC-1 and MC-2, the accumulation dynamics in plants was determined by DAS-ELISA. Cucumber plants were grown, inoculated and sampled as described in Materials and Methods. The results of the experiment are presented in Figure 2. It follows from these results that on the 3rd d.p.i., the attenuated and pathogenic strains accumulated in plants in trace amounts, but even at this stage, the mean value for each strain already differed. From the 3rd till the 7th d.p.i., a significant increase in virion CP for all viral strains was detected; however, VIROG-43M accumulated in significantly less amounts than the pathogenic strains. The maximum rate of virus accumulation was observed from the 7th till the 10th d.p.i., when mean CP levels for all CGMMV strains reached a plateau. Moreover, on cucumber plants inoculated with the pathogenic strains, the first mosaic symptoms could be clearly observed. On the 14th d.p.i., a little decrease in CP level was recorded for MC-1 and VIROG-43M strains, although this value increased for MC-2 strain. Thus, from the 3rd till the 14th d.p.i., the attenuated VIROG-43M accumulated in plants in significantly less amounts than the pathogenic strains; the level of pathogenic strain accumulation differed from the 3rd till the 10th d.p.i. and by the 14th d.p.i., the difference between both pathogenic strains was insignificant.

Worldwide Distribution of CGMMV Strains

As mentioned above, CGMMV strains cause diseases in Cucurbitaceae plants in different countries. By analyzing NCBI GenBank data we retrieved 98 strains of CGMMV originated from 17 countries. The largest number of CGMMV strains (as great as 34) were described in China , next comes South Korea (13 isolates), followed by India (9 isolates) and Japan (7 isolates). In other countries less than 6 strains were characterized.

We performed phylogenetic analysis with all CGMMV strains retrieved from the NCBI GenBank. Note that nucleotide sequences of the coat protein (CP) genes were used for phylogenetic tree construction since this particular gene was sequenced in most strains deposited to GenBank. Eighty-three of 98 CGMMV strains were taken for analysis. The remaining 15 strains either did not have the nucleotide sequence of the CP gene or the CP gene was not completely sequenced. The phylogenetic tree built using MEGA software [18] is shown in Figure 3.

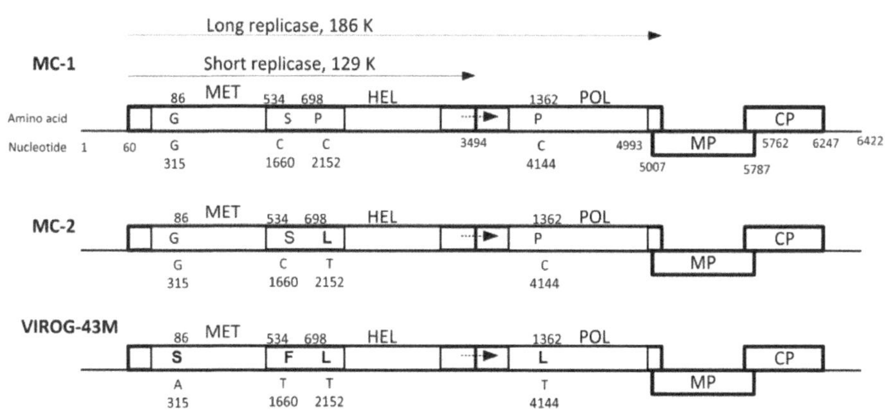

Figure 1. Schematic representation of the nucleotide mutations and amino acid substitutions found in the attenuated CGMMV strain VIROG-43M in comparison with pathogenic strains MC-1 and MC-2. The 5′ and 3′ UTRs are shown by single lines; the dotted arrow indicates the readthrough stop codon. The long and the short replicases, the movement protein (MP) and the coat protein (CP) are indicated with the bold bars. Three domains of replicases including MET—methyltransferase domain; HEL—helicase domain; POL—polymerase domain, are shown with regular bars.

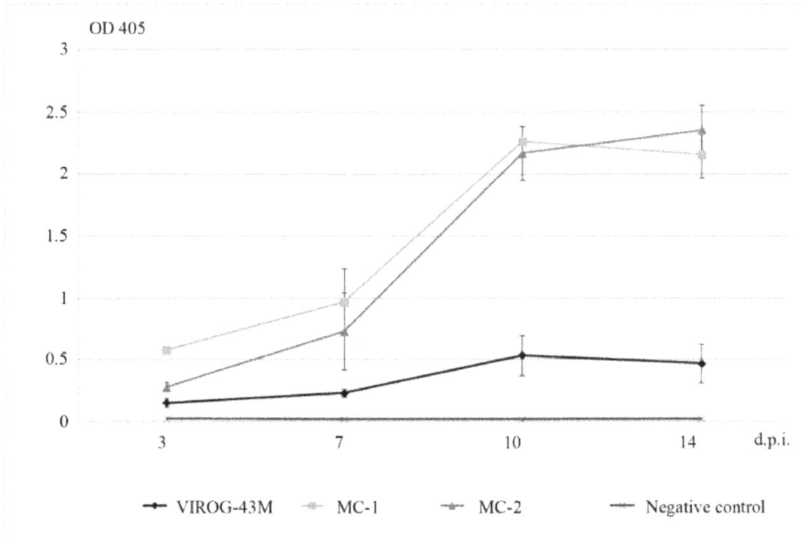

Figure 2. Accumulation dynamics of VIROG-43M, MC-1 and MC-2 strains in cucumber plants.

Figure 3. Molecular phylogenetic analysis of CGMMV strains by Maximum Likeli-

hood method. The phylogenetic tree was built with MEGA software [18] . The tree is drawn to scale, with branch lengths measured in the number of substitutions per site; all positions with less than 95% site coverage were eliminated. CGMMV strains are specified with NCBI accession number and the country in which they were isolated. Main clades of the tree are shaded bold, black and light gray.

It is evident that all CGMMV strains group into two main clades: the small clade with 8 representatives including Moscow region, Russian strains (the clade is indicated with a bold line in Figure 3) and the large clade with the remaining 75 representatives. Some subclades of the large clade are entirely Asian groups including CGMMV isolates from China, Taiwan, Japan, South Korea and Indonesia (the clade is indicated by bold and light grey line in Figure 3). Other clades represent combined groups with Asian, European, Canadian and Australian CGMMV isolates.

DISCUSSION

CGMMV causes a severe disease of cucurbits that can lead to significant yield losses. Analysis of the largest available NCBI database revealed 98 strains of CGMMV originated from 17 different countries. From these data it follows that CGMMV is extensively studied in China and other Asian countries, while in European countries it remains poorly studied. Interestingly, this virus successfully invades novel areas: USA, Canada, and Australia were free of CGMMV infection, but in 2013-2014 the virus was discovered in these countries as well [20] [21] . In the Russian Federation , CGMMV often occurs in commercial greenhouses; however, the virus is poorly studied. We isolated and sequenced full-length genomes of two pathogenic strains from infected leaves of cucumber plants grown in greenhouses of the Moscow region. Using phylogenetic analysis with 83 full-length CP genes of CGMMV, we found that Russian strains together with isolates from Spain , Greece , USA and Israel form a very compact clade distinct from most CGMMV isolates. According to the phylogenetic tree (Figure 3), there were at least two sources of CGMMV infection. The first source gave rise to a small group of CGMMV strains that infected, among others, cucurbits in the Moscow region. The second source of infection most probably originated in Asian countries and Europe, and then spread to Australia .Besides the analysis of pathogenic CGMMV strains, we for the first time characterized the attenuated strain VIROG-43M that can be used for cucumber protection. In a number of trials conducted in laboratory and commercial greenhouses in diverse regions of Russian Federation on different cucumber varieties we clearly showed that VIROG-43M can be applied for cucumber protection, since this attenuated strain possesses high phenotypic stability and level of cross-protection against the pathogenic strain; furthermore, it contributes to significant yield gain. However, by the end of vegetation, a few plants treated with VIROG-43M developed disease symptoms possibly because of mutations. As the number of such plants was very low, and they appeared at late plant growing stage, they had no impact on yields. However, it would be interesting to characterize the genetic structure of these mutants, and it will be done in our future work.Using the DAS-ELISA assay, we showed that VIROG-43M accumulated in cucumber plants in significantly less amounts than the pathogenic strains. Furthermore, we sequenced the genome of the attenuated

VIROG-43M strain and compared it with pathogenic strains MC-1 and MC-2. We found three missense mutations that could be crucial for attenuation. Interestingly, that Kubota et al. (2003) reported that the mutation crucial for the attenuated phenotype of ToMV (L11A strain) is located in the intervening region between MET and HEL domains (IR), and IR should be considered crucial for the symptoms' attenuation [22] . Additional amino acid substitutions found in this particular region changed the severity of symptoms, but did not lead to their elimination [22] . Furthermore, in several mild tobamovirus strains, the mutations crucial for attenuation are located in this particular region of the viral genome. For example, the attenuated M strain of tobacco mosaic virus (TMV) has numerous mutations in comparison to the pathogenic U1 strain; however, the nucleotide substitutions at two specific positions located in IR led to the induction of U1-TMV-like symptoms [23] . Moreover, in the attenuated V-69 strain of ToMV, the only nucleotide mutation responsible for attenuation was also in IR, as it was shown by a reverse genetics approach [15] . Finally, the attenuated C-1421 strain of PMMoV had two nucleotide mutations in comparison to the parental pathogenic J strain, although only one located in IR led to the amino acid substitution and was responsible for attenuation [24] . Considering these findings, we may suggest that the attenuated VIROG- 43M strain most likely has the only substitution crucial for attenuation at position 534 located in IR, and substitutions of residues 86 and 1362 in the replicase protein probably influence the severity of symptoms.According to current data, there are two mechanisms of cross-protection. The first one is based on inhibition of uncoating of a pathogenic strain in the presence of an attenuated strain [25] . The second mechanism involves posttranscriptional gene silencing (PTGS) induced in plants in response to attenuated virus inoculation. As a result, an incoming pathogenic strain is targeted by the PTGS-machinery, and the pathogenic RNA is degraded [26] . Kubota et al. showed that the ToMV short replicase protein, in addition to its main function, serves as a PTGS suppressor [22] . Moreover, the mutations crucial for attenuation and located in IR also led to defective PTGS suppressor activity, in other words, in symptomless plants the activity of the PTGS suppressor was lower or absent [22] . We suppose that cross-protection induced by VIROG-43M is also based on the PTGS mechanism, and the replicase protein of this strain acts as a suppressor with reduced activity.AcknowledgementsThe authors thank Elena V. Elisarova for assays of phenotypic stability and cross-protection and Dr. Andrew A. Shelenkov for useful suggestions in preparing the manuscript. This work was supported by RSF grant no. 16-16-00032. AAS is a recipient of the fellowship of the President of Russian Federation (MK-5568.2015.4).Cite this paperAnna A. Slavokhotova, Ekaterina A. Istomina, Emma N. Andreeva, Tatyana V. Korostyleva, Vitalij A. Pukhalskij, Alexander N. Shijan, Tatyana I. Odintsova (2016) An Attenuated Strain of Cucumber Green Mottle Mosaic Virus as a Biological Control Agent against Pathogenic Viral Strains. American Journal of Plant Sciences,07,724-732. doi: 10.4236/ajps.2016.75066

REFERENCES

1. Moffett, P. (2009) Mechanisms of Recognition in Dominant R Gene Mediated Resistance. Advances in Virus Research, 75, 1-3. http://dx.doi.org/10.1016/S0065-3527(09)07501-0

2. Nishiguchi, M. and Kobayashi, K. (2011) Attenuated Plant Viruses: Preventing Virus Diseases and Understanding the Molecular Mechanism. Journal of General Plant Pathology, 77, 221-229. http://dx.doi.org/10.1007/s10327-011-0318-x

3. Ugaki, M., Tomiyama, M., Kakutani, T., Hidaka, S., Kiguchi, T., Nagata, R., Sato, T., Motoyoshi, F. and Nishiguchi, M. (1991) The Complete Nucleotide-Sequence of Cucumber Green Mottle Mosaic-Virus (Sh-Strain) Genomic Rna. Journal of General Virology, 72, 1487-1495. http://dx.doi.org/10.1099/0022-1317-72-7-1487 [Paper reference:2]

4. Ainsworth, G.C. (1935) Mosaic Disease of Cucumber. Annals of Applied Biology, 22, 55-67. http://dx.doi.org/10.1111/j.1744-7348.1935.tb07708.x

5. Varveri, C., Vassilakos, N. and Bem, F. (2002) Characterization and Detection of Cucumber Green Mottle Mosaic Virus in Greece. Phytoparasitica, 30, 493-501. http://dx.doi.org/10.1007/BF02979754

6. Letschert, B., Adam, G., Lesemann, D., Willingmann, P. and Heinze, C. (2002) Detection and Differentiation of Serologically Cross-Reacting Tobamoviruses of Economical Importance by RT-PCR and RT-PCR-RFLP. Journal of Virological Methods, 106, 1-10. http://dx.doi.org/10.1016/S0166-0934(02)00135-0

7. Slavokhotova, A.A., Andreeva, E.N., Shijan, A.N., Odintsova, T.I. and Pukhalskij, V.A. (2007) Specifics of the Coat Protein Gene in Russian Strains of the Cucumber Green Mottle Mosaic Virus. Russian Journal of Genetics, 43, 1221-1226. http://dx.doi.org/10.1134/S1022795407110038

8. Vasudeva, R.S., Raychaudhuri, S.P. and Singh, J. (1949) A New Strain of Cucumis Virus 2. Indian Phytopathology, 2, 180-185.

9. Liu, Y., Wang, Y.A., Wang, X.F. and Zhou, G.H. (2009) Molecular Characterization and Distribution of Cucumber Green Mottle Mosaic Virus in China. Journal of Phytopathology, 157, 393-399. http://dx.doi.org/10.1111/j.1439-0434.2008.01509.x

10. Kim, S.M., Lee, J.M., Yim, K.O., Oh, M.H., Park, J.W. and Kim, K.H. (2003) Nucleotide Sequences of Two Korean Isolates of Cucumber Green Mottle Mosaic Virus. Molecules and Cells, 16, 407-412.

11. Ali, A., Natsuaki, T. and Okuda, S. (2004) Identification and Molecular Characterization of Viruses Infecting Cucurbits in Pakistan. Journal of Phytopathology, 152, 677-682. http://dx.doi.org/10.1111/j.1439-0434.2004.00915.x

12. Anonymous (2014) Cucumber Green Mottle Mosaic Virus. A Seed Production and Commercial Growers Guide. ASTA First-the seed. [Paper reference:2]

13. Nishiguchi, M., Kikuchi, S., Kiho, Y., Ohno, T., Meshi, T. and Okada, Y. (1985) Molecular-Basis of Plant Viral Virulence—The Complete Nucleotide-Sequence of an Attenuated Strain of Tobacco Mosaic-Virus. Nucleic Acids Research, 13, 5585-5590. http://dx.doi.org/10.1093/nar/13.15.5585

14. Yang, G., Qiu, B.S., Liu, X.G., Li, Y. and Wang, X.F. (2002) Nonsense Mutations of Replicase and Movement Protein Genes Contribute to the Attenuation of an

Avirulent Tomato Mosaic Virus. Virus Research, 87, 119-128. http://dx.doi.org/10.1016/S0168-1702(02)00025-4

15. Istomina, E.A., Snegireva, P.B. and Shiian, A.N. (2004) Construction of a Full-Length cDNA of Tobacco Mosaic Virus Strain V-69 Genome. Genetika, 40, 1637-1645. http://dx.doi.org/10.1007/s11177-005-0058-0 [Paper reference:2]
16. Ichiki, T.U., Nagaoka, E.N., Hagiwara, K., Uchikawa, K., Tsuda, S. and Omura, T. (2005) Integration of Mutations Responsible for the Attenuated Phenotype of Pepper Mild Mottle Virus Strains Results in a Symptomless Cross-Protecting Strain. Archives of Virology, 150, 2009-20. http://dx.doi.org/10.1007/s00705-005-0551-8
17. Tan, S.-H., Nishiguchi, M., Sakamoto, W., Ogura, Y., Murata, M., Ugaki, M., Tomiyama, M. and Motoyoshi, F. (1997) Molecular Analysis of the Genome of an Attenuated Strain of Cucumber Green Mottle Mosaic Virus. Annals of the Phytopathological Society of Japan, 63, 470-474. http://dx.doi.org/10.3186/jjphytopath.63.470
18. Tamura, K., Stecher, G., Peterson, D., Filipski, A. and Kumar, S. (2013) MEGA6: Molecular Evolutionary Genetics Analysis Version 6.0. Molecular Biology and Evolution, 30, 2725-2729. http://dx.doi.org/10.1093/molbev/mst197
19. Tamura, K. and Nei, M. (1993) Estimation of the Number of Nucleotide Substitutions in the Control Region of Mitochondrial-DNA in Humans and Chimpanzees. Molecular Biology and Evolution, 10, 512-526.
20. Ling, K.-S. and Li, R. (2014) First Report of Cucumber Green Mottle Mosaic Virus Infecting Greenhouse Cucumber in Canada. Plant Disease Journal, 98, 701.
21. Tian, T. and Posis, K. (2014) First Report of Cucumber Green Mottle Mosaic Virus on Melon in the United States. Plant Disease Journal, 98, 1163. http://dx.doi.org/10.1094/PDIS-02-14-0176-PDN
22. Kubota, K., Tsuda, S., Tamai, A. and Meshi, T. (2003) Tomato Mosaic Virus Replication Protein Suppresses Virus-Targeted Post-transcriptional Gene Silencing. Journal of Virology, 77, 11016-11026. http://dx.doi.org/10.1128/JVI.77.20.11016-11026.2003 [Paper reference:4]
23. Shintaku, M.H., Carter, S.A., Bao, Y. and Nelson, R.S. (1996) Mapping Nucleotides in the 126-kDa Protein Gene That Control the Differential Symptoms Induced by Two Strains of Tobacco Mosiac Virus. Virology, 221, 218-225.http://dx.doi.org/10.1006/viro.1996.0368
24. Hagiwara, K., Ichiki, T.U., Ogawa, Y., Omura, T. and Tsuda, S. (2002) A Single Amino Acid Substitution in 126-kDa Protein of Pepper Mild Mottle Virus Associates with Symptom Attenuation in Pepper; The Complete Nucleotide Sequence of an Attenuated Strain, C-1421. Archives of Virology, 147, 833-840. http://dx.doi.org/10.1007/s007050200030
25. Baulcombe, D.C. (1996) Mechanisms of Pathogen-Derived Resistance to Viruses in Transgenic Plants. Plant Cell, 8, 1833-1844. http://dx.doi.org/10.1105/tpc.8.10.1833

26. Qu, F. and Morris, T.J. (2005) Suppressors of RNA Silencing Encoded by Plant Viruses and Their Role in Viral Infections. FEBS Letters, 579, 5958-5964. http://dx.doi.org/10.1016/j.febslet.2005.08.041

6
CHAPTER

EFFECTIVENESS OF BACTERIAL AND FUNGAL ISOLATES TO CONTROL PHOMA LINGAM ON OILSEED RAPE BRASSICA NAPUS

Omar Hammoudi, Mazen Salman, Ruba Abuamsha, Ralf-Udo Ehlers

Keywords: Biological Control; Oilseed Rape; *Phoma lingam*; Bacteria

INTRODUCTION

Oilseed rape (OSR) Brassica napus is a major crop grown for its oil in many parts of the world including Canada, India, China and Europe [1]. In favourable conditions, B. napus is considered to have the highest yield potential of all Brassica oil crops [2]. Winter and spring varieties of the crop exist. Winter varieties are mainly grown in Europe and China and spring varieties in areas such as Australia and Canada. Winter OSR is an excellent starter crop in cereal rotations, due to its positive impact on disease control and soil improvement. Its cultivation favours the minimum tillage which is of great interest for sustainable agriculture [3].

In the last 20 years, an increase of 260% in OSR production has been recorded worldwide, driven by the enormous economic significance of oilseed crops as a renewable raw material used for a variety of applications in food and non-food areas [4]. In Germany, OSR is the most important oilseed crop, with acreage of more than 1.3 million hectares in 2005. Yields ranged between 2.9 and 3.5 tons per hectare. In 2005,

Citation: O. Hammoudi, M. Salman, R. Abuamsha and R. Ehlers, "Effectiveness of Bacterial and Fungal Isolates to Control Phoma lingam on Oilseed Rape Brassica napus," American Journal of Plant Sciences, Vol. 3 No. 6, 2012, pp. 773-779. doi: 10.4236/ajps.2012.36093.

5.1 million tons of rapeseed was harvested approximating a total seed-oil yield of 2 million tons [3].

With the increased production of oilseed rape (OSR) B. napus spp. napus, reports on pathogen causing diseases have increased simultaneously. Oilseed rape however, is susceptible to many fungal pathogens such as Pyrenopeuiza brassicae Raw. (anamorph C. concetricum Grev.), Leptosphaeria maculans (Desm.) Ges. Et De Not. (anamorph Phoma lingam (Tod.) Desm.), Sclerotiorum (Lib.) de Bary and Verticillium dahliae Kelb. Stem canker (also termed blackleg) caused by P. lingam is of major economic importance in the main OSR growing areas of Australia, Canada and Europe [5]. In some instances (e.g., when chemical fungicidal applications are not effective) heavy losses, as much as 95% might be caused by these fungi [1]. In Europe, the severity of blackleg disease was reported to be significant since 1966 [6]. The disease epidemic differs greatly between seasons, regions and crops. In some instances basal phoma stem canker can potentially cause total crop loss [5,7].

Fungicides are the most effective means for controlling stem canker, but their use must be optimised to achieve maximum economic response and to avoid unnecessary fungicide applications [8]. Beside the side effects to the environment and development of fungicide resistance, use of foliar fungicides in association with cultivars with little or no resistance has proved to be ineffective for controlling phoma stem canker, because the current fungicides have low eradicant activity and are effective as protectants for only a short period [5]. Biocontrol methods might be promising alternatives to reduce the use of fungicides. Various microbial antagonists, including strains of bacteria of the genera Baccillus, Pseudomonas and Serratia and strains of fungi such as Gliocladium and Trichoderma, have shown biocontrol activity against damping-off diseases and Botrytis on Photinia, in conventional Brassica seedling [9].

The aim of this work is to test the efficacy of different biocontrol antagonistic agents (BCA) against the Black leg disease in OSR.

MATERIALS AND METHODS

Cultivation and Maintenance of Antagonistic Biocontrol Agents

Bacterial isolates Serratia plymuthica HRO-C48, Bacillus subtilis B2g, Pseudomonas fluorescens E9, 1Re2-6, Paenibacillus polymyxa 1P1-2 and Pseudomonas chlororaphis MA342 were grown each in 250 ml Erlenmeyer flasks containing 50 ml of 20 g/l tryptic soy broth (TSB) (Difco Laboratories, USA) supplemented with 100 μg/ml refampicin. The flasks were placed on a rotary shaker (180 rpm) for 48 h at 25°C. Aliquots of 700 μl of each bacterial isolate were transferred into Eppendorf tubes containing 300 μl of 50% sterile glycerol. The cultures were stored after mixing at –80°C.

The fungus Gliocaldiom catenulatum isolate J1446 (commercially produced under the name Prestop® (Verdera, Finland) was grown on Malt extract agar (MEA) media containing 100 mg/l streptomycin and 50 mg/l red Bengal. Cultures were incubated at

22°C for one week. For spore production, MEA agar disks containing fungal mycelium were grown in 250 ml Erlenmeyer flasks containing 50 ml malt extract liquid media on a rotary shaker (140 rpm) for two weeks at 20°C. Spore concentration of the fungus was then adjusted to 6×10^8 cfu/ml in 0.85% normal saline.

In Vitro Antagonistic Tests

Antagonistic activity of the biocontrol agents against P. lingam was determined using dual culture technique. 100 µl of P. lingam spore suspension (2×10^7 Spore/ml) were spread on plates containing V8 agar media. Each antagonist was then streaked on the plate (3 streaks/plate) and incubated either at 10 or 22°C for 7 days. The control experiments were done by using sterile distilled water instead of antagonists. The effect of each BCA was determined by measuring the inhibition zones of fungal growth. The rating scale was: 4, inhibition zone > 10 mm; 3, inhibition zone 5-10 mm; 2, inhibition zone < 5 mm; 1, growth was stopped at the bacterial-streak line; 0, no inhibition zone and often P. lingam was growing over the bacterial streak [9]. The experiment was repeated using agar disks grown with the fungus instead of fungal spores.

Two V8 agar disks (7 mm diameter) grown with 7 days old P. lingam were placed about 20 mm from each of BCA streak. The plates were further incubated as mentioned above and zones of inhibition were determined according to the above scale.

To test the effect of G. catenulatum, WA agar disks grown with the fungus were used instead of bacteria.

Inoculation of Cotyledons with P. lingam

Cotyledons were punctured with a needle and 10 µl of pycnidiospore suspension (2×10^7 Spore/ml) was placed on the wound as described by Mithen et al. (1987) [10]. Control plants were inoculated with 10 µl sterilized distilled water. Plants were further incubated in growth chamber cabinets for 14 days. To prevent plant proliferation, growing leaves were removed every two days during the experimental time. Disease incidence was then measured according to Kutcher et al. (1993) [11] in a rating scale from 0 - 6 as follows: 0: no symptoms; 1: lesions on the infection site < 1.5 mm; 2: lesions on the infection site 1.5 - 3.5 mm; 3: lesions on the infection site > 3.0 mm; 4: gray to green tissue collapse 3.1 - 5.0 mm; 5: gray to green tissue collapse > 5.0 mm (≤10 pycnidia); 6: gray to green tissue collapse > 5.0 mm (>10 pycnidia) [12].

Inoculation of OSR Plantlets with P. lingam

The efficacy of the biocontrol agents on disease development at an earlier stage of plant growth was also tested. For this, OSR plantlets at growth stage of BBCH 14/15 [13] were inoculated at the stem base either with V8 agar disks (7 mm diameter) grown with P. lingam or with 40 µl pycnidiospors (2×10^7 spore/ml). Plants were incubated for 50 days in growth chamber under the above mentioned conditions. Infestation with P. lingam was then determined according to Kutcher et al. (1993) [14]. The volume of disease tissue (VDT) was assessed in a rating scale of 0 - 4 as illustrated in **Table 1**.

Colonization Studies

To study the population size of the antagonists in OSR rhizosphere, seeds were treated with the antagonists by soaking 1 g of seeds in 1-ml bacterial suspension for 5 h at 20°C. The seeds were grown in growth chambers at 22°C ± 1°C and 75% ± 5% relative humidity. After different periods of time (1, 10 and 30 days), 5 g of roots with adhering soils from five OSR seedlings of each treatment were placed in an Erlenmyer flask containing 40 ml sterile water and shaked for 1 h at 24°C. Serial dilutions (10^{-2}, 10^{-3} and 10^{-4}) were prepared and aliquots (100 μl) were spread on TSA media supplemented with 150 mg/ml refimpicin. Plates were further incubated at 25°C for 48 h and the number of bacteria was estimated as log cfu/g root fresh weight.

Table 1. Rating scale of P. lingam on OSR leaves.

Rating scale	Length of lesion (L)	Area of lesion (G)	Width of lesion (*P*)
0	No infection	No infection	No infection
1	1 - 9 mm	≤25%	≤25%
2	10 - 19 mm	25% - 50%	25% - 50%
3	20 - 29 mm	50% - 75%	50% - 75%
4	≥30 mm	75% - 100%	75% - 100%

VDT = $(1 - HR^2) \times G/4 \times L$ where HR= $1 - P/4$.
Based on the results of this equation, plant susceptibility was considered as resistant: 0 - 0.3 intermediate: 0.4 - 1.5 susceptible: 1.6 - 4.0.

Field Experiments

Experiments were conducted in naturally infested fields with P. lingam. The field which was located in Achterwehr, (54°19'N, 9°58'E) is owned by the institute of agronomy and crop science, Christian Albrecht University of Kiel. OSR was cultivated in a Randomized Block Design in which four blocks were used as replicates. The area of each block was 14 × 3 m and divided into smaller areas (4.5 × 3 m each).

Before sowing, OSR seeds were treated with the antagonists as follows: S. plymuthica, HRO-C48^{Rif}, S. plymuthica, HRO-C48^{Rif} combined with the fungicide Metconazol (applied two weeks after starting the experiment), G. catenulatum and G. catenulatum combined with the fungicide Metconazol (applied two weeks after starting the experiment). Untreated seeds were used as control. Seed coating with insecticides and fungicides was achieved as follows: seeds were first treated with S. Plymuthica or with G. Catenulatum as mentioned above. After that the insecticide Chinook was added (20.0 ml/kg seeds), followed by adding the fungicide DMM (Dimethomorph) (10 g/kg seeds). While rotating at 400 rpm, talcum blue was added until all liquid is bound to the seeds. The seeds were then dried under laminar flow for 1 h. The disease severity was assessed by counting the number of pycindiospores on OSR leaves.

Population size of S. plymuthica, HRO-C48 and G. catenulatum in OSR rhizosphere was studied under field conditions. 5 g of roots with adhering soils from five OSR seedlings of each treatment were placed in Erlenmyer flasks containing 40 ml sterile water and shaked for 1 h at 24°C. Serial dilutions (10^{-2}, 10^{-3} and 10^{-4}) were prepared and aliquots (100 μl) were spread on TSA media supplemented with 150 μg/ml refimpicin. Plates were further incubated at 25°C for 48 h and the number of bacteria was estimated as log cfu/g root fresh weight.

Statistical Analysis

Statistical analyses were done using XlStat program (Adinosoft). Analysis of variance (ANOVA) and significant differences for means were calculated after Tukeys HSD test at ($P \leq 0.05$).

RESULTS

In Vitro Inhibition of Mycelial Growth of P. lingam

All of the antagonistic isolates reduced the mycelial growth as well as the growth of pycnidospores of P. lingam on V8 media. The size of the inhibition zone varied according to antagonist and type of media used (**Table 2**). G. catenulatum was the most effective in inhibiting the mycelial growth of P. lingam on both types of media at 22°C and caused a very strong inhibition zones >10 mm. However, its effect on pycnidospores was lower (<5 mm). Except for B. subtilis, all bacteria isolates were able to inhibit P. lingam since the bacterium did not grow at 10°C. As shown in **Table 2** only the isolate P. fluorescens (E9) caused week inhibition zone (<5 mm) on both media and temperatures. The effect of P. chororaphis at 22°C on V8 media was more pronounced in inhibiting the growth of pycnidospores than that at 10°C (inhibition zone > 10).

Effect of Antagonists on Cotyledon Infestation with P. lingam

The antagonistic isolates S. plymuthica, P. chlororaphis were the most effective in reducing the infestation of cotyledons of OSR cultivar Talent grown under growth chamber conditions, and were able to reduce the disease infection by 43.3% (**Figure 1**). In addition to that, the number of healthy plants of cultivar Talnet under growth chamber at 22°C was significantly higher in the seeds treated with the isolate P. chlororaphis (**Figure 1**). Treatment with G. catenulatum resulted in a significant reduction of disease intensity (51.6%) compared to untreated control seeds.

Table 2. In vitro inhibition of P. lingam mycelium and pycndiospores grown on V8 and WA media at 10 and 22°C.

Isolate	Mycelia				Pycnidiospores			
	V-8		WA		V-8		WA	
	22°C	10°C	22°C	10°C	22°C	10°C	22°C	10°C
S. plymuthica	3	3	3	3	3	2	2	2
B. subtilis	4	0	4	0	4	0	4	0
Pa. polymyxa	3	2	2	3	3	2	1	1
P. fluorescens (1Re2-6)	2	3	3	4	4	2	4	4
P. chlororaphis	4	4	2	4	4	4	0	0
G. catenulatum	4	3	4	3	1	1	1	1
P. fluorescens (E9)	1	2	1	1	2	2	2	2

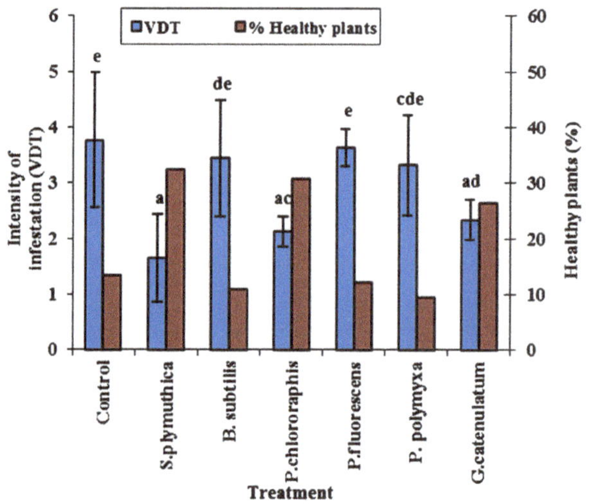

Figure 1. Effect of seed biopriming with different antagonists on P. lingam intensity on cotyledon of OSR cultivars Talent grown in growth chamber at 22°C. Disease intensity (scale 0-6) and the percent of healthy plants were recorded after inoculating the cotyledons 2×10^5 cfu/plant of a spore suspension of 2×10^7 cfu/ml of P. lingam. Results of different letters are significantly different after Tukey HSD test at $P < 0.05$, n = 45 plant per treatment).

The results showed a pronounced increase in the percent of healthy plants of OSR in the presence of P. chlororaphis. In seeds treated with S. plymuthica or G. catenulatum the percent of healthy plants was higher than that in seeds treated with the other antagonists (**Figure 1**).

Efficacy of Antagonists on P. lingam Infection on Stem Base

Data in **Figure 2** shows the effect of the BCAs on reduction of root collar infection. The volume of diseased tissue (VDT) in plants infested with spore suspension of the pathogen was significantly lower ($P < 0.05$) in OSR seeds treated with S. plymuthica, P. chlororaphis and G. catenulatum (1.14 ± 0.8, 1.08 ± 0.4 and 1.09 ± 0.7, respectively). In the presence of P. fluorescens the VDT (1.81 ± 0.7) did not differ significantly from the control untreated plants (VDT = 2.04 ± 1.1). In OSR plants infested with V8 agar pieces grown with P. lingam, the antagonists reduced significantly the VDT. The effect P. fluorescens was lower (but not significantly different) than the effects of the other antagonists (**Figure 2**).

Colonization of the Rhizosphere with S. plymuthica and G. catenulatum

Our results showed that the isolates S. plymuthica, P. chlororaphis and P. fluorescens are rhizosphere competent antagonist since they were able to colonize the rhizosphere of ORS in a short period of time. Root colonization was also recorded after seed treatment with B.

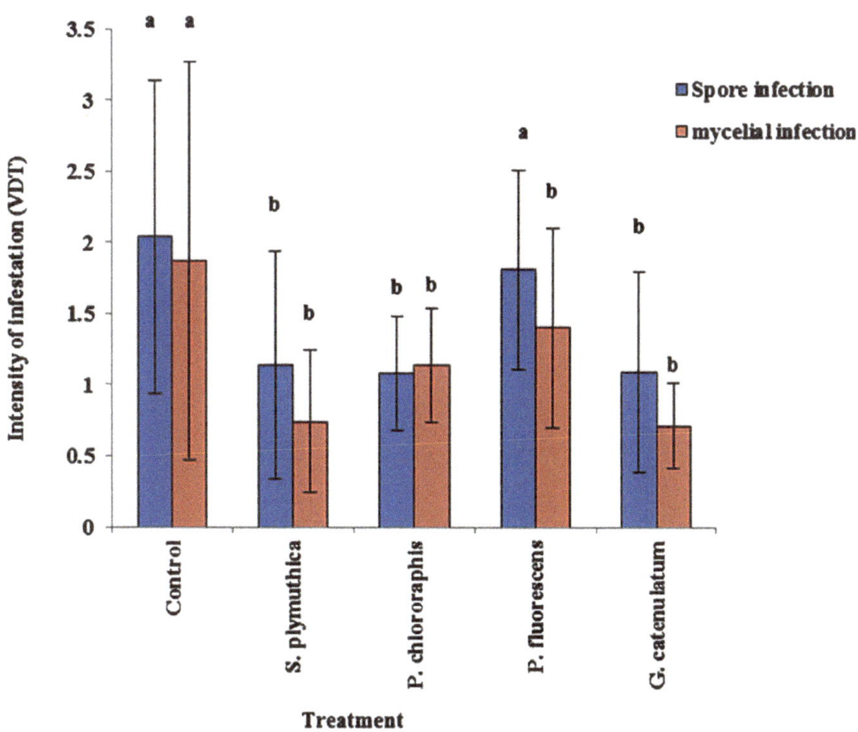

Figure 2. Effect of different BCAs on infection of stem base. OSR seeds cultivar Talent were treated with the antagonists at 10^6 cfu/seed and grown for 50 days in growth chamber at 12 h day light, 20°C - 24°C and 70% - 75% humidity. When plantlets reached the phonological stage of BBCH 14/15, P. lingam was applied as mycelium grown on V8

medium. Data of same letters are not significantly different after Tukey HSD test at $P < 0.05$.

subtilis und P. polymyxa (2 - 1.2 × 10^6 cfu/ml). However, the colonization was lower than that of S. plymuthica, P. chlororaphis and P. fluorescens. After 8 day of plant growth under greenhouse conditions, the average number of bacteria was 1.2 × 10^5 cfu/g root fresh weight. At the end of the experiment (30 days), colonization rate was 5.2 × 10^4 cfu/g root fresh weight (**Figure 3**).

Efficacay of Antagonists against P. lingam under Field Conditions

In plants treated with the antagonists, the disease intensity was lower than that of the control untreated plants. As shown in (**Figure 4**) the average number of pycnidiopore after two weeks of planting was 21.3 pycnidia/ plant in the control untreated plants and 2.5 and 1.5 pycnidia/plant in plants treated with S. plymuthica and G. catenulatum respectively. The combined treatments of S. plymuthica and G. catenulatum with the fungicides resulted in a pronounced reduction of disease infestation (0.0 and 0.9 pycnidiospore/plant respectevily).

The highest number of pycnidiospores was recorded after two weeks of fungicide application. After that the disease intensity decreased and no significant differences were recorded between the numbers of pycnidiospores in control plants and in plants treated with the antagonist or the antagonists combined with the fungicide.

Six months after sowing, the disease intensity was reduced and one month later no disease on leaves was recorded. The disease intensity was significantly lower in treated seeds compared to the control untreated seeds.

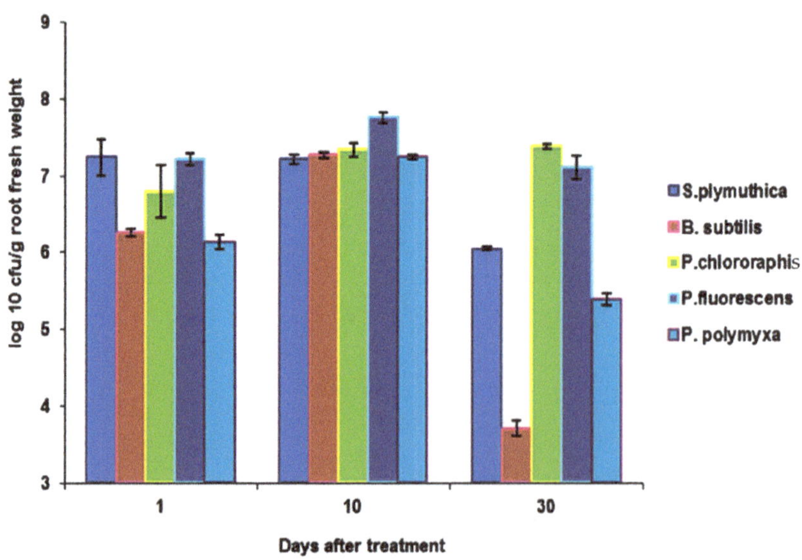

Figure 3. Comparison of root colonization of OSR roots by different antagonists. Data are the means (±SD) of 5 plants taken after 1, 10 and 30 days of incubation under

growth chamber cabins at 22°C and 12 h day/night regimes. Seeds were bioprimed with the antagonists by soaking the seeds in bacterial suspensions (2×10^{10} cfu/ml) for 5 h and dried at room temperature for two days.

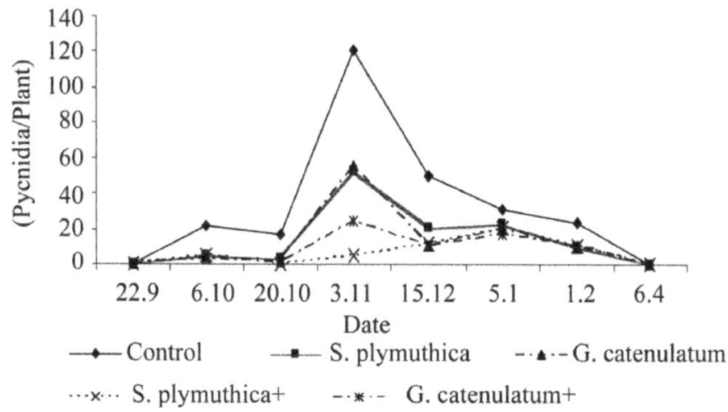

Figure 4. Effect of seed treatment with S. plymuthica and G. catenulatum on disease intensity of P. lingam on leaves of OSR grown under field conditions. The experiment was conducted in 2005/2006. OSR seeds were treated either with S. plymuthica (7.9×10^6 cfu/seed) or with G. catenulatum (8.5×10^5 cfu/seed). After two weeks of sowing the fungicide Caramba (60% Metconazol, 0.7 l/ha) was applied.

Interestingly, no significant difference in disease intensity was seen when the fungicide was combined with the biocontrol agents.

The disease incidence was also calculated at two time points (03.11 and 15.12). In control plants infested with P. lingam the disease incidence was 74% and 68% and in plants treated with the S. plymuthica combined with the fungicide the disease incidence was 12.5% and 25%.

DISCUSSION

In this study, several antagonistic microorganisms including bacteria and fungi were tested for their efficacy against the black leg disease caused by P. lingam on oilseed rape. The first step toward successful selection of effective biocontrol agents was done by testing the effect of the antagonists against the pathogen in dual culture assays.

As revealed from the dual culture assays, the biocontrol agents differ in their efficacy against P. lingam. The size of inhibition zone varied according to the incubation temperature. Although no general relationship exist between the ability of a microoganism to inhibit a pathogen in vitro and to suppress disease caused by that pathogen in ad planta [15,16], some authors [17,18] have shown that the antagonistic efficacy might be correlated with the in vivo effect of the antagonists. It was found that the results of dual culture assay depend highly on culture media [19,20]. In this work the size of the zone of inhibittion caused by the biocontrol agents was affected by the type of media used (V8 and WA media). Because the pathogen P. lingam infects mostly winter rape at low temperatures (10°C) the effect of temperature on antagonism was also tested. Except for B. subtilis and P. polymyxa most isolated with high efficacy at 22°C were also able to inhibit mycelial growth at 10°C.

P. lingam is able to infect all parts of OSR at all stages of plant development [21]. For example an infection in autumn might cause later an infection of root collar and stem base [22]. The fungus is able to spread from the infected leaves to other parts of the plant causing the disease symptoms [23]. For this reason inhibition of the fungus on cotyledons was studied.

To test the effect of the pathogen on emerging leaves, the leaves of the OSR plants in the physiological growth stage BBCH1-12 were injured mechanically by a sterile needle. A spore suspension was then spread on the wounds and the disease symptoms were monitored after 14 days [11,24]. In ad planta tests, more than 50% reduction of the disease caused by P. lingam on leaves of OSR seedling grown in growth chamber was recorded in seeds treated with S. plymuthica, P. chlororaphis, P. fluorescens and G. catenulatum. The percent of healthy plants increased significantly from less than 10% to 40% - 50% in the presence of the antagonists. However, the efficacy of the gram positive isolates was lower than that of the gram negative bacterial strains.

The isolate S. plymuthica was the most effective in controlling the disease caused by P. lingam on root collar and stem base. The disease intensity was reduced by 54% and 63% in plants grown in greenhouse and growth chamber conditions. In seeds treated with G. catenulatum, P. fluorescens und P. chlororaphis disease reduction was 52%.

For effective protection against plant pathogens, the antagonist must be able to colonize successfully the rhizosphere of the plant [25]. The antagonists must compete with other microorganisms in the root system of the plants in order inhibit the attack of pathogens [26]. It was found that colonization patterns differ between different antagonistic microorganisms [27]. In our experiments, OSR roots were highly colonized by isolates S. plymuthica, P. chlororaphis and P. fluorescens after seed treatments with $1.2 - 4 \times 10^7$ cfu/ml bacterial solutions. Sasse (1997) [20] found that root colonization of rape seed plants in seeds treated with fluorescent pseudomonads (8×10^5 cfu/seed) was 1.8×10^6 and 5.1×10^5 cfu/g root fresh weight after 10 and 30 day respectively. Other authors found that the concentration of P. fluorescens was reduced from 10^6 bis 10^8 cfu/g root fresh weight after 7 day of plant emergence to 10^4 bis 10^5 cfu/g root fresh weight after 60 day of plant growth [28]. The density of rhizospheric bacteria differ strongly accorcing to the root zone of plants as well as the age of the plant [29].

Berg et al. (1996) [9] found that Bacillus spp. were able to colonize the roots of Photinia after 28 day of growth (10^4 - 10^5 cfu/g root fresh weight). In in-vitro experiments Brückner (1998) [28] found that the number of bacterial cells of B. subtilis and P. fluorescens in the rhizosphere of winter oilseed rape, reduced from 10^6 and 10^8 cfu/g root fresh weight after 7 days of planting to 10^4 and 10^5 cfu/g root fresh weight after 60 of planting.

The total number of different bacterial cells was determined in the rhizosphere of OSR plants grown in growth chamber or in greenhouse. An average of 74×10^8 cfu/g root fresh weight was recorded. These results are in agreement with Kleeberger et al. (1983) [30], who recorded viable counts of bacteria of 10^7 - 10^8 cfu/g root fresh weight in roots of barley and wheat. Our results are also in agreement with Berg et al. (1996) [9] who determined a bacterial number of 1.5×10^8 cfu/g root fresh root weight of OSR plants.

ACKNOWLEDGEMENTS

The authors would like to thank Dr. Mohammed Abu-Eid, Director General, National Agriculture Research Center (NARC); Ministry of Agriculture, Palestine for his technical assistance to publish this work.

REFERENCES

1. R. Grison, B. Grezes-Besset, M. Schneider, N. Lucante, L. Olsen, J.-J. Leguay and A. Toppan, "Field Tolerance to fungal Pathogens of Brassica napus Constitutively Expressing a Chimeric Chitinase Gene," Nature Biotechnology, Vol. 14, No. 5, 1996, pp. 643-646.doi:10.1038/nbt0596-643
2. N. J. Mendham and P. A. Salisbury, "Physiology: Crop Development, Growth and Yield," In: D. S. Kimber and D. I. Mcgregor, Eds., Bmssica Oilseeds, Centre for Agricultural Bioscience International, Wallingford, 1995, pp. 11-64.
3. Eurostat, "Eurostat Website and National Data". http://epp.eurostat.ec.europa.eu
4. D. S. Kimber and D. I. McGregor, "The Species and Their Origin, Cultivation and World Production," In: D. S. Kimber and K. I. McGregor, Eds., Brassica Oilseeds, Production and Utilization, Centre for Agricultural Bioscience International, Wallingford, 1995, pp. 1-7.
5. J. S. West, P. D. Kharbanda, M. J. Barbetti and B. D. L. Fitt, "Epidemiology and Management of Leptosphaeria maculans (Phoma Stem Canker) on Oilseed Rape in Australia, Canada and Europe," Plant Pathology, Vol. 50, No. 1, 2001, pp. 10-27.doi:10.1046/j.1365-3059.2001.00546.x
6. M. L. Pilet, R. Delourme, N. Foisset and M. Renard, "Identification of Loci Contributing to Quantitative Field Resistance to Blackleg Disease, Causal Agent Leptosphaeria maculans (Desm.) Ces. et de Not., in Winter Rapeseed (Brassica napus L.)," Theorotical Applied Genetics, Vol. 96, No. 1, 1998, pp. 23-30. doi:10.1007/s001220050704
7. B. D. L. Fitt, H. Brun, M. J. Barbetti and S. R. Rimmer, "World-Wide Importance of Phoma Stem Canker (Leptosphaeria maculans and L. biglobosa) on Oilseed Rape (Brassica napus)," European Journal of Plant Pathology, Vol. 114, No. 1, 2006, pp. 3-15.doi:10.1007/s10658-005-2233-5
8. Y. Zhou, B. D. L. Fitt, S. J. Welham, P. Gladders, C. E. Sansford and J. S. West, "Effects of Severity and Timing of Stem Canker (Leptosphaeria maculans) Symptoms on Yield of Winter Oilseed Rape (Brassica napus) in the UK," European Journal of Plant Pathology, Vol. 105, No. 7, 1999, pp. 715-728. doi:10 .1023/A:1008761219493
9. G. Berg, "Rhizobacteria of Oilseed Rape Antagonistic to Verticillium dahliae var. longisporum STARK," Journal of Plant Disease and Protection, Vol. 103, No. 1,

1996, pp. 20-30.

10. R. F. Mithen, B. G. Lewis, R. K. Heaney, G. R. Fenwick, "Resistance of Leaves of Brassica Species to Leptosphaeria maculans," Transactions of the British Mycological Society, Vol. 88, No. 4, 1987, pp. 525-531. doi:10.1016/S0007-1536(87)80036-0
11. B. Volke, "Leptosphaeria maculans, der Erreger der Wurzelhalsund Stengelfäule an Raps: Verbreitung Verschiedener Pathogenitätsgruppen in Europa, Quantifizierung des Befalls und Schadwirkug im Freiland," Ph.D. Thesis, Göttingen, 1999.
12. P. A. Delwiche, "Genetic Aspects of Blackleg (Leptosphaeria maculans) Resistance in Rapeseed (Brassica napus)," Ph.D. Thesis, Wisconsin, 1980.
13. E. Weber and H. Blfiholder, "Erläuterungen zu den BBCHDezimal-Codes für die Entwicklungsstadien von Mais, Raps, Faba-Bohne, Sonnenblume und Erbse—Mit Abbildungen," Gesunde Pflanzen, Vol. 42, No. 9, 1990, pp. 308-321.
14. H. R. Kutcher, C. G. J. Van Denberg and S. R. Rimmer, "Variation in Pathogenicity of Leptosphaeria maculans on Brassica spp. Based on Cotyledon and Stem Reactions," Canadian Journal of Plant Pathology, Vol. 15, No. 4, 1993, pp. 253-258.doi:10.1080/07060669309501920
15. R. Campbell, "Biological Control of Soil-Borne Diseases: Some Present Problems and Different Approaches," Crop Protection, Vol. 13, No. 1, 1994, pp. 4-13.
16. J. W. Kloepper, "Development of in Vivo Assays for Prescreening Antagonists of Rhizoctonia solani on Cotton," Phytopathology, Vol. 81, No. 9, 1991, pp. 1006-1013.doi:10.1094/Phyto-81-1006
17. D. R. Fravel, "Role of Antibiosis in the Biocontrol of Plant Diseases," Annual Reviews of Phytopathology, Vol. 26, No. 1, 1988, pp. 75-91.doi:10.1146/annurev.py.26.090188.000451
18. R. Stanley, M. Brown, N. Poole, M, Rogerson D. C. Sigee, C. Knight, C. C. Ivuin, H. A. S. Epton and C. Leifert, "Biocontrol of Postharvest Fungal Diseases on Dutch White Cabbage by Pseudomonas and Serratia Antagonists in Storage Trials," Plant Pathology, Vol. 43, No. 4, 1994, pp. 605-611. doi:10.1111/j.1365-3059.1994.tb01597.x
19. M. Heupe, "Biologische Bekämpfung Bodenbürtiger Wurzelbranderreger der Zuckerrübe (Beta vulgaris L. ssp. Vulgaris var. Altissima Doell) Durch den Einsatz Mikrobieller Antagonisten," Ph.D. Thesis, Göttingen, 1992.
20. A. Sasse, "Untersuchungen zur Biologischen Bekämpfung des Rapswelkeerregers (Verticillum dahliae Kleb.) Durch den Einsatz Mikrobieller Antagonisten," Ph.D. Thesis, Göttingen, 1997, p. 133.
21. R. K. Gugel and G. A. Petrie, "History, Occurrence, Impact and Control of Blackleg of Oilseed Rape," Canadian Journal of Plant Pathology, Vol. 14, No. 1, 1992, pp. 36- 45.doi:10.1080/07060669209500904
22. K. E. Hammond, B. G. Lewis and T. Musa, "A Systemic Pathway in the Infection of Oilseed Rape Plants by Leptosphaeria maculans," Plant Pathology, Vol. 34,

No. 4, 1985, pp. 357-365. doi:10.1111/j.1365-3059.1985.tb01407.x

23. K. E. Hammond and B. G. Lewis, "The Establishment of Systemic Infection in Leaves of oilseed Rape by Leptosphaeria maculans," Plant Pathology, Vol. 36, No. 2, 1987, pp. 135-147. doi:10.1111/j.1365-3059.1987.tb02213.x

24. P. A. Delwiche and P. H. Williams, "Screening for Resistance to Blackleg of Crucifers in the Seedling Stage," Cruciferae News letter, Vol. 4, 1979, p. 24.

25. D. M. Weller, "Biological Control of Soilborne Plant Pathogens in the Rhizosphere with Bacteria," Annual Reviews of Phytopathology, Vol. 26, No. 1, 1988, pp. 379-407.doi:10.1146/annurev.py.26.090188.002115

26. Y. Elad and I. Chet, "Possible Role of Competition for Nutrients in Biocontrol of Pythium Damping-Off by Bacteria," The American Phytopathological Society, Vol. 77, No. 2, 1987, pp. 190-195. doi:10.1094/Phyto-77-190

27. E. A. Milus and C. S. Rothrock, "Rhizosphere Colonization of Wheat by Selected Soil Bacteria over Diverse Environments," Canadian Journal of Microbiology, Vol. 39, No. 3, 1993, pp. 335-341. doi:10.1139/m93-047

28. S. Brückner,"Möglichkeiten Einer Biologischen Kontrolle von Verticillium. Dahliae an Winterraps mit Hilfe Antagonistischer Rhitosphärenbakterien," Ph.D. Thesis, Universität Rostock, Rostock,1998.

29. E. Liljeroth, S. L. G. E Burgers and J. A. Van Veen, "Changes in Bacterial Populations along Roots of Wheat (Triticum aestivum L.) Seedlings," Biology and Fertility of Soils, Vol. 10, No. 4, 1991, pp. 276-280. doi:10.1007/BF00337378

30. A. Kleeberger, A. H. J. Carstorp and W. Klinmüller, "The Rhizosphere Microflora of Wheat and Barley with Special Reference to Gram-Negative Bacteria," Archives of Microbiology, Vol. 136, No. 4, 1983, pp. 306-311. doi:10.1007/BF00425222 O. Hammoudi, M. Salman, R. Abuamsha and R. Ehlers, "Effectiveness of Bacterial and Fungal Isolates to ControlPhoma lingam on Oilseed Rape Brassica napus," American Journal of Plant Sciences, Vol. 3 No. 6, 2012, pp. 773-779. doi: 10.4236/ajps.2012.36093.

7

CHAPTER

BIOTECHNOLOGY AND PLANT DISEASE CONTROL-ROLE OF RNA INTERFERENCE

Shabir H. Wani, Gulzar S. Sanghera, N. B. Singh

Keywords: RNAi, Viruses, Fungi, dsRNA, Gene Silencing

INTRODUCTION

Plant diseases are a threat to world agriculture. Significant yield losses due to the attack of pathogen occur in most of the agricultural and horticultural crop species. More than 70% of all major crop diseases are caused by fungi [1]. Plant diseases are usually handled with applications of chemicals. For some diseases, chemical control is very effective; but it is often non-specific in its effects, killing beneficial organisms as well as pathogens. Chemical control may have undesirable effects on health, safety and cause environmental risks [2]. Traditional plant breeding methods have been used to develop cultivars resistant to various diseases. However, this process is time-consuming and limited availability of genetic resources for most of the crops has left little room to continued improvement by these means. There are many reasons for the limited genetic resources available for breeding [3]. Two of the most important ones are: 1) loss of gene pools occurring during the domestication and breeding of crop plants [4] and 2) many of

Citation: S. Wani, G. Sanghera and N. Singh, "Biotechnology and Plant Disease Control-Role of RNA Interference," American Journal of Plant Sciences, Vol. 1 No. 2, 2010, pp. 55-68. doi: 10.4236/ajps.2010.12008.

the natural gene traits that may be beneficial in one plant tissue such as seeds and fruits, may be deleterious in other plant tissues such as vegetative tissues [5,6]. Over the past few decades, breeding possibilities have been broadened by genetic engineering and gene transfer technologies including gene mapping and identification of the genome sequences of model plants and crops. Modern technologies such as trancriptomics, proteomics, and metabolomics are now proved to be important in understanding plant metabolic pathways and the role of key genes associated with their regulation. This can facilitate new insights into the complex metabolite neighborhoods that give rise to a given phenotype and may allow discovery of new target genes to modify a given pathway. Such genes can then be subject to new metabolic engineering efforts and applications.

During the last decade, our knowledge repertoire of RNA-mediated functions has been greatly increased with the discovery of small non-coding RNAs which play a central part in a process called RNA silencing. Ironically, the very important phenomenon of co-suppression has recently been recognized as a manifestation of RNA interference (RNAi), an endogenous pathway for negative post-transcriptional regulation. RNAi has revolutionized the possibilities for creating custom "knock-downs" of gene activity. RNAi operates in both plants and animals, and uses double stranded RNA (dsRNA) as a trigger that targets homologous mRNAs for degradation or inhibiting its transcription or translation [7,8], whereby susceptible genes can be silenced. This RNA-mediated gene control technology has provided new platforms for developing eco-friendly molecular tools for crop improvement by suppressing the specific genes which are responsible for various stresses and improving novel traits in plants including disease resistance. It has emerged as a method of choice for gene targeting in fungi [9], viruses [10,11], bacteria [12] and plants [13] as it allows the study of the function of hundreds of thousands of genes to be tested [14]. It can silence a gene throughout an organism or in specific tissues [15], offer the versatility to partially silence or completely turn off genes, work in both cultured cells and whole organisms and can selectively silence genes at particular stages of the organism's life cycle [16]. Methods that introduce dsRNA into plant and animal cells have been enormously successful for decreasing cognate gene expression in vivo [17,18]. Due to all these elegant and unique features of RNAi, our review specifically focuses on 1) the current knowledge of RNAi concept and 2) its pathways and induction in plants and explore the possibilities for the applications of this technology in the development of disease resistance plants.

MECHANISM OF RNAI '

RNA interference' refers collectively to diverse RNA-based processes that all result in sequence-specific inhibition of gene expression at the transcription, mRNA stability or translational levels. It has most likely evolved as a mechanism for cells to eliminate foreign genes. The unifying features of this phenomena are the production of small RNAs (21-26 nucleotides (nt) that act as specific determinants for down-regulating gene expression [17, 19,20] and the requirement for one or more members of the Argonaute family of proteins [21]. RNAi operates by triggering the action of dsRNA intermediates, which are processed into RNA duplexes of 21-24 nucleotides by a ribonuclease III-like enzyme called Dicer [22-24]. Once produced, these small RNA molecules or short interfering RNAs (siRNAs) are incorporated in a multi-subunit complex called RNA induced silencing complex (RISC) [21,25]. RISC is formed by a siRNA and an endonuclease among other components. The siRNAs within RISC acts as a guide to target the degradation of complementary messenger RNAs (mRNAs) [21,25]. The host genome codifies for small RNAs called miRNAs that are responsible for endogenous gene silencing. The dsRNAs triggering gene silencing can originate from several sources such as expression of endogenous or transgenic antisense sequences, expression of inverted repeated sequences or RNA synthesis during viral replication [26]. When dsRNA molecules produced during viral replication trigger gene silencing, the process is called virus-induced gene silencing (VIGS) [27]. One interesting feature of RNA silencing in plants is that once it is triggered in a certain cell, a mobile signal is produced and spread through the whole plant causing the entire plant to be silenced [28]. After triggering RNA silencing, the mobile signaling molecules can be relay-amplified by synthesis of dsRNAs on the primary cleavage of product templates or by their cleavage into secondary siRNAs. This amplification leads to the transitory nature of silencing reaction that may spread along the mRNA, though initiated by a locally targeted single siRNA [29] and spreads in both the 5′and 3′ directions [25]. This bi-directional transition further have been witnessed by a process where both the 5′ and 3′ cleavage products of the initial target RNA act as aberrant mRNAs to trigger dsRNA synthesis [30], and induce secondary silencing reactions. This silencing process is also enhanced by the enzymatic activity of the RISC complex, mediating multiple turnover reactions [31,25]. Furthermore, production of the secondary siRNAs leads to enrichment of silencing via its spread from the first activated cell to neighboring cells, and systemically through the system [32]. The cell-to-cell spread can be mediated as passive spread of the small RNAs via plasmodesmata or by the silencing signal complex which is between 27 and 54 kDa [33]. The systemic spread in phloem is mediated by the 24 nt siRNAs [32], unloading of the systemic signal appears to be mediated via plasmodesmata, since it does not spread into meristematic cells [26]. The discovery of RNAbinding protein (PSRP1) in the phloem and its ability to bind 25 nt ssRNA species add further to the argument that siRNAs (24-26 nt) are the key components for systemic silencing signal. The extent of cell-to-cell movement is dependent on the levels of siRNAs produced at the site of silencing initiation, but is independent of the presence of siRNA target transcripts in either source or recipient cells [34].

METHODS TO INDUCE RNAI IN PLANTS

One of the biggest challenges in RNAi research is the delivery of the active molecules that will trigger the RNAi pathway in plants. In this system, a number of methods for delivery of dsRNA or siRNA into different cells and tissue include transformation with dsRNAforming vectors for selected gene(s) by an Agrobacterium mediated transformations [19,35], delivery cognate dsRNA of uidA GUS (β-glucuronidase) and TaGLP2a: GFP (green fluorescent protein) reporter genes into single epidermal cells of maize, barley and wheat by particle bombardment [36], introducing a Tobacco rattle virus (TRV)-based vector in tomato plants by infiltration [37], delivery of dsRNA into tobacco suspension cells by cationic oligopeptide polyarginine-siRNA complex;

infecting plants with viral vectors that produce dsRNA [38] and delivery of siRNA into cultured plant cells of rice, cotton and slash pine for gene silencing by nanosense pulsed laser-induced stress wave (LISW) [40]. Among these the most reliable and commonly used approaches for delivery of dsRNA to plants cells are agroinfiltration, micro-bombardment and VIGS. These are discussed in the following sections.

Agroinfiltration

Agroinfiltration is a powerful method to study processes connected with RNAi. The injection of Agrobacterium carrying similar DNA constructs into the intracellular spaces of leaves for triggering RNA silencing is known as agroinoculation or agroinfiltration [41]. In most cases agroinfiltration is used to initiate systemic silencing or to monitor the effect of suppressor genes. In plants, cytoplasmic RNAi can be induced efficiently by agroinfiltration, similar to a strategy for transient expression of T-DNA vectors after delivery by Agrobacterium tumefaciens. The transiently expressed DNA encodes either an ss- or dsRNA, which is typically a hairpin (hp) RNA. The infiltration of hairpin constructs are especially effective, because their dsRNA can be processed directly to siRNAs, while the constructs expressing ssRNA can also be useful to induce silencing [42-45] and for dissecting the mechanism of gene silencing, especially concerned with its suppressors, systemic silencing signal and also for simple protein purification [42-45]. Besides, they provide a rapid, versatile and convenient way for achieving a very high level of gene expression in a distinct and defined zone.

Micro-Bombardment

In this method, a linear or circular template is transferred into the nucleus by micro-bombardment. Synthetic siRNAs are delivered into plants by biolistic pressure to cause silencing of GFP expression. Bombarding cells with particles coated with dsRNA, siRNA or DNA that encode hairpin constructs as well as sense or antisense RNA, activate the RNAi pathway. The silencing effect of RNAi is occasionally detected as early as a day after bombardment, and it continues up to 3 to 4 days of post bombardment. Systemic spread of the silencing occurred 2 weeks later to manifest in the vascular tissues of the non-bombarded leaves of Nicotiana benthamiana that were closest to the bombarded ones. After one month or so, the loss of GFP expression was seen in non-vascular tissues as well. RNA blot hybridization with systemic leaves indicated that the biolistically delivered siRNAs induced due to de novo formation of siRNAs, which accumulated to cause systemic silencing [29].

Virus Induced Gene Silencing (VIGS)

Modified viruses as RNA silencing triggers are used as a mean for inducing RNA in plants. Different RNA and DNA viruses have been modified to serve as vectors for gene expression [46,47]. Some viruses, such as Tobacco mosaic virus (TMV), Potato virus X (PVX) and TRV, can be used for both protein expression and gene silencing [48-51]. All RNA virus-derived expression vectors will not be useful as silencing vectors because many have potent anti-silencing proteins such as TEV (Tobacco etch virus), that directly interfere with host silencing machinery [48,52]. Similarly, DNA viruses have not been used extensively as expression vectors due to their size constraints for movement [53]. However, a non-mobile Maize streak Virus (MSV)-derived vector has been successfully used for long-term production of protein in maize cell cultures [48]. Using viral vectors to silence endogenous plant genes requires cloning homologous gene fragments into the virus without compromising viral replication and movement. This was first demonstrated in RNA viruses by inserting sequences into TMV [54], and then for DNA viruses by replacing the coat protein gene with a homologous sequence [53]. These reports used visible markers for gene silencing phytoene desaturase(PDS) and chalcone synthase (CHS), providing a measure of the tissue specificity of silencing as these have been involved in carotenoid metabolic pathway. The PDS gene acts on the antenna complex of the thylakoid membranes, and protects the chlorophyll from photooxidation. By silencing this gene, a drastic decrease in leaf carotene content resulted into the appearance of photobleaching symptom [55,56]. Similarly, over expression of CHS gene causes an albino phenotype instead of producing the anticipated deep orange color [57]. As a result, their action as a phenotypic marker helps in easy understanding of the mechanism of gene silencing. Table 1 shows some general characteristics for currently available virus-derived gene silencing vectors. Most viruses are plus-strand RNA viruses or satellites, whereas Tomato golden mosaic virus (TGMV) and Cabbage leaf curl virus (CaLCuV) are DNA viruses. Though RNA viruses replicate in the cytoplasm DNA viruses replicate in plant nuclei using the host DNA replication machinery. Both types of viruses induce diffusible, homology-dependent systemic silencing of endogenous genes. However, the extent of silencing spread and the severity of viral symptoms can vary significantly in different host plants and host/ virus combinations. With the variety of viruses and the diversity of infection patterns, transmission vectors, and plant defenses it is not surprising that viruses differ with respect to silencing [58]. Because the continuing development of virus-based silencing vectors can extend VIGS to economically important plants, it is useful to consider some of the characteristics of successful VIGS vectors.

RNAI IN PLANT DISEASE MANAGEMENT

Despite substantial advances in plant disease management strategies, our global food supply is still threatened by a multitude of pathogens and pests. This changed scenario warrants us to respond more efficiently and effectively to this problem. The situation demands judicious blending of conventional, unconventional and frontier technologies. In this sense, RNAi technology has emerged as one of the most potential and promising strategies for enhancing the building of resistance in plants to combat various fungal, bacterial, viral and nematode diseases causing huge losses in important agricultural crops. The nature of this biological phenomenon has been evaluated in a number of host-pathogen systems and effectively used to silence the action of pathogen. Many of the examples listed below illustrate the possibilities for commercial exploitation of this inherent biological mechanism to generate disease-resistant plants in the future by taking advantage of this approach.

Management of Plant Pathogenic

Fungi RNA-mediated gene silencing (RNA silencing) is used as a reverse tool for gene targeting in fungi. Homologybased gene silencing induced by transgenes (co-suppression), antisense, or dsRNA has been demonstrated in many plant pathogenic fungi, including Cladosporium fulvum [59], Magnaporthae oryzae [60-62], Venturia inaequalis [63], Neurospora crassa [64], Aspergillus nidulans [65], and Fusarium graminearum [9] (Table 1), whether it is suitable for large scale mutagenesis in fungal pathogens remains to be tested. Hypermorphic mechanism of RNA interference implies that this technique can also be applicable to all those plant pathogenic fungi, which are polyploid and polykaryotic in nature, and also offers a solution to the problem where frequent lack of multiple marker genes in fungi is experienced. Simultaneous silencing of several unrelated genes by introducing a single chimeric construct has been demonstrated in case of Venturia inaequalis [63]. HCf-1, a gene that codes for a hydrophobin of the tomato pathogen C. fulvum [66], was co-suppressed by ectopic integration of homologous transgenes. Transformation of Cladosporium fulvum with DNA containing a truncated copy of the hydrophobin gene HCf-1 caused co-suppression of hydrophobin synthesis in 30% of the transformants. The co-suppressed isolates had a hydrophilic phenotype, lower levels of HCf-1 mRNA than wild type and contain multiple copies of the plasmid integrated as tandem repeats at ectopic sites in the genome.

Table 1. RNAi effects on targeted region in some fungal plant pathogens.

Pathogen	Targeted region	Result	References
Magnaporthae oryzae	*eGFP*	Sequence specific degradation of mRNA	[60]
Cladosporium falvum	*cgl* 1 and *cgl* 2	Blocking disease infection spread	[67]
Venturia inaequalis	*Multiple inverted repeats*	-	[63]
Fusarium graminearum	-	-	[9]
Blumeria graminis	*Mlo*	Immunity	[36]

The transcription rate of HCf-1 in the co-suppressed isolates was higher in the untransformed strains, which suggested that silencing acted at the post-transcriptional level. This was due to ectopic integration of the transgene next to promoters which initiate transcription to form antisense RNA and that this in turn determines down-regulation of HCf-1. But gene silencing was not associated with DNA cytosine methylation [59]. Similarly, the silencing of cgl1 and cgl2 genes using the cgl2 hairpin construct in Cladosporium fulvum has also been reported [67], though the effect was possibly restricted to highly homolougous genes (exons of cgl 1 and cgl 2 are 87% identical). However, the less homologus cgl 3 (53% overall identity to cgl 2) was not affected as the target specificity always depends upon the actual sequence alignment and more over, short regions of high density that led to unwanted off-targets effects. Such a strategy could be exploited for protecting the consumable products of vegetables and fruits crops from the post-harvest diseases caused by different plant pathogens in future. Hairpin vector technology resulted in simultaneous high frequency silencing of a green fluorescent protein (GFP) transgene and an endogenous trihydroxynaphthalene reductase gene (THN) in Venturia inaequalis [63] GFP transgene, acting as easily detectable visible marker while the trihydroxynaphthalene reductase gene (THN) playing role in melanin biosynthesis. High frequency gene silencing was achieved using hairpin constructs for the GFP or the THN genes transferred by Agrobacterium (71 and 61%, respectively). THN-silenced transformants exhibited a distinctive light brown phenotype and maintained the ability to infect apple. Silencing of both genes with this construct occurred at a frequency of 51% of all the transformants.

All 125 colonies silenced for the GFP gene were also silenced for THN [63]. Similarly, multiple gene silencing has been achieved in Cryptococcus neoformans using chimeric hairpin constructs [39] and in plants using partial sense constructs [68]. The first effort towards the systematic silencing of Magnaporthe grisea, a causal organism of rice blast was carried by using the enhanced green florescent protein gene as a model [60]. To assess the ability of RNA species to induce silencing in fungus, plasmid construct expressing sense, antisense and hairpin RNA were introduced into an eGFP-expressing transformants. The fluorescence of eGFP in the transformants was silenced much more efficiently by hairpin RNA of eGFP than by other RNA species. In the silenced transformants, the accumulation of eGFP mRNA was drastically reduced. But not methylation of coding or promoter region was involved. The small interfering RNA molecules of 19-23 nucleotides were observed in both sense and antisense strands of eGFP gene [60]. Later on a protocol for silencing the mpg1 and polyketide synthase-like genes was also developed [9]. Mpg1 gene is a hydrophobin gene which is essential for pathogenicity as it act as a cellular relay for adhesion and trigger for the development of appressorium [69]. Their work on this host-pathogen system revealed that they were successfully able to silence the above mentioned genes at varying degrees by pSilent-1-based vectors in 70–90% of the resulting transformants. Ten to fifteen percent of the silenced transformants exhibited almost "null phenotype". This vector was also efficiently applicable to silence a GFP reporter in another ascomycete fungus Colletotrichum lagenarium [9]. A novel highthroughput approach for gene function analysis using RNAi, which provides an alternative to the gene knock-out by homologous recombination was also described [70].

The authors developed an RNA silencing vector, pSilent-Dual1 (pSD1) that carries two convergent dual promoters, the Aspergillus nidulans tryptophan promoter (PtrpC) and the A. nidulans glyceraldehyde-3- phosphate dehydrogenase promoter (Pgpd). Both promoters have been used to drive constitutive gene expression in a large number of filamentous fungi. A multicloning site (MCS) has been inserted between two promoters. The greatest merit of the pSD1 system over others, such as hpRNA or intron spliced hairpin RNA (ihpRNA) silencing system is that it allows a single step cloning for generation of an RNAi construct. To facilitate efficient screening for silenced transformants, gfp gene was incorporated the into pSD1 system [70]. It allows expression of a chimeric RNA and assessment of gene silencing efficiency by utilizing a recipient strain that produces GFP and therefore, fluoresces green when using epifluorescence microscopy. A main bottleneck of this system is its lower silencing efficiency compared with hpRNA or ihpRNA-expressing RNA-silencing vectors. Formation of dsRNA in the pSD1 system requires physical annealing of two different RNA molecules in the target cells while that in the hpRNA systems is achieved by self-folding of inverted repeats within RNA molecule.

The difference in dsRNA formation between the systems can be a major cause of the different silencing efficiencies. The authors generated a series of knock-down mutants of almost all known calcium related genes in the genome of M. oryzae and examined for phenotypical defects. Gene knock-down requires relatively short stretches of sequence information. This is a major advantage for phytopathogens for which there is little sequence information available. As RNAi works at the mRNA level, its efficacy is not compromised by the presence of non-transformed nuclei or multicopy genes due to aneuploidy [71]. RNAi causes only a partial reduction, but not a complete loss of, in gene expression. Partial gene suppression is considered a main drawback of RNAi. However, it could be a merit where the effect of an essential gene on a phenotype is of interest. Gene knock-down offers a more convenient and effective tool, especially in combination with an inducible promoter that allows gene expression to be diminished at specific stages during development [72].

Another disadvantage of gene knock-down is, as it requires only a short sequence, that genes other than those targeted might be silenced. This causes unexpected changes in gene expression patterns (off-target effects). Testing for the possibility of off-target effects is simpler for phytopathogen species for which complete genome sequence data are available but remains elusive for those phytopathogens whose genomes have not been sequenced [71]. In another study, RNA interference (RNAi) strategy was used to specifically knockdown 59 individual rice genes encoding putative LRR-RLKs, and a novel rice blast resistance-related gene (designated as OsBRR1) was identified by screening T0 RNAi population using a weakly virulent isolate of Magnaporthe oryzae, Ken 54-04. Wild-type plants (Oryza sativa L. cv. 'Nipponbare') showed intermediate resistance to Ken 54-04, while OsBRR1 suppression plants were susceptible to Ken 54-04 [73]. Furthermore, OsBRR1-overexpressing plants exhibited enhanced resistance to some virulent isolates (97-27-2, 99-31-1 and zhong 10-8-14). OsBRR1 expression was low in leaves and undetectable in roots under normal growth conditions, while its transcript was significantly induced in leaves infected with the blast fungus (Ken 54-04) and was moderately affected by ABA, JA and SA treatment. Overexpression or RNAi suppression of OsBRR1did not cause visible developmental changes in rice plants.

Management of Plant Pathogenic Bacteria

One of the striking examples of bacterial disease management where RNAi showed a remarkable type of gene regulation has been documented [12]. They developed a crown gall disease management strategy that targets the process of tumourogensis (gall formation) by initiating RNAi of the iaaM and ipt oncogenes. Expression of these genes is a prerequisite for wild type tumor formation. Transgenic Arabidopsis thaliana and Lycopersicon esculentum transformed with RNAi constructs, targeting iaaM and ipt gene(s) showed resistance to crown gall disease. Transgenic plants generated through this technology contained a modified version of these two bacterial gene(s) required to cause the disease and was the first report to manage a major bacterial disease through RNAi. The extra genes recognize and effectively shut down the expression of the corresponding bacterial gene during infection, thus preventing the spread of infection. The incoming bacteria could not make the hormones needed to cause tumors and plants deficient in silencing were hyper-susceptible to A. tumefaciens [28]. Successful infection relied on a potent anti-silencing state established in tumors whereby siRNA synthesis is specifically inhibited. The procedure can be exploited to develop broadspectrum resistance in ornamental and horticultural plants which are susceptible to crown gall tumorigenesis. This approach can be advocated for the effective management of those pathogens which multiply very rapidly and results in tumor formation such as Albugo candida, Synchytrium endobioticum, Erwinia amylovora etc. The natsiRNA (nat-siRNAATGB2) was strongly induced in Arabidopsis upon infection by Pseudomonas syringae pv tomato and down-regulates a PPRL gene that encodes a negative regulator of the RPS2 disease resistance pathway. As a result, the induction of nat-siRNAATGB2 increases the RPS2-mediated race-specific resistance against P. syringae pv tomato in Arabidopsis [74]. Recently, the accumulation of a new class of sRNA, 30 to 40 nucleotides in length, termed long-siRNAs (lsiRNAs), was found associated with P. syringae infection. One of these lsiRNAs, AtlsiRNA-1, contributes to plant bacterial resistance by silencing AtRAP, a negative regulator of plant defense [75]. A Pseudomonas bacterial flagellin derived peptide is found to induce the accumulation of miR393 in Arabidopsis. miR393 negatively regulates mRNAs of F-box auxin receptors, resulting in increased resistance to the bacterium (P. syringae), and the overexpression of miR393 was shown to reduce the plant's bacterial titer by 5-fold [76].

Management of Plant Pathogenic Viruses

Antiviral RNAi technology has been used for viral disease management in human cell lines [77-80]. Such silencing mechanisms (RNAi) can also be exploited to protect and manage viral infections in plants [19,81]. The effectiveness of the technology in generating virus resistant plants was first reported to PVY in potato, harbouring vectors for simultaneous expression of both sense and antisense transcripts of the helper-component proteinase (HC-Pro) gene [82]. The P1/HC-Pro suppressors from the potyvirus inhabited silencing at a step down stream of dsRNA processing, possibly by preventing the unwinding of duplex siRNAs, or the incorporation into RISC or both [83]. The utilization of RANi technology has resulted in inducing immunity reaction against several other viruses in different plant-virus systems (Table 2). In phyto-pathogenic DNA viruses like geminiviruses non-coding intergenic region of Mungbean yellow mosaic India virus (MYMIV) was expressed as hairpin construct under the control of the 35S promoter and used as biolistically to inoculate MYMIV-infected black gram plants and showed a complete recovery from infection, which lasted until senescence [84]. RNAi mediated silencing of geminiviruses using transient protoplast assay where protoplasts were co-transferred with a siRNA designed to replicase (Rep)-coding sequence of African cassava mosaic virus (ACMV) and the genomic DNA of ACMV resulted in 99% reduction in Rep transcripts and 66% reduction in viral DNA [85].

Table 2. Effects of targeted region of RNAi in various plantvirus systems.

Host system	Virus	Targeted region	References
N. benthamiana, M. esculenta	*African cassava mosaic virus*	*pds, su, cyp79d2*	[98]
Barley, wheat	*Barley stripe mosaic virus*	*pds*	[99-101]
Soybean	*Bean pod mottle virus*	*Pds Actin*	[102-104]
Barley, rice, maize	*Brome mosaic virus*	*pds, actin 1, rubisco activase*	[105]
Arabidopsis	*Cabbage leaf curl virus*	*gfp, CH42, pds*	[56]
P. sativum	*Pea early browning virus*	*pspds, uni, kor*	[106]
N. benthamiana	*Poplar mosaic virus*	*gfp*	[107]
N. benthamiana, S. tuberosum	*Potato virus X*	*pds, gfp*	[108,109]
Nicotiana tabacum	*Satellite tobacco mosaic virus*	*Several genes*	[110]
N. benthamiana, N. tabacum	*Tobacco mosaic virus*	*pds, psy*	[48]
N. benthamiana, Arabidopsis, tomato, *Solanum* species, Chilli pepper, opium poppy, *Aquilegia*	*Tobacco rattle virus*	*Rar1, EDS1, NPR1/NIM1, pds, rbcS, gfp*	[111-115]
N. benthamiana	*Tomato bushy shunt virus*	*gfp*	[116]
N. benthamiana	*Tomato golden mosaic virus*	*su, luc*	[117]
N. benthamiana, Lycopersicon esculentum N. glutinosa, N. tabacum	*Tomato yellow leaf curl China virus-associated b DNA satellite*	*pcna, pds, su, gfp*	[118]

It was observed that siRNA was able to silence a closely related strain of ACMV but not a more distantly related virus. More than 40 viral suppressors have been identified in plant viruses [86]. Results from some of the well-studied virus suppressors indicated that suppressors interfere with systemic signaling for silencing [44]. During last few years, the p69 encoded by Turnip yellow mosaic virus has been identified as silencing suppressors that prevented host RDR-dependent secondary dsRNA synthesis [87]. P14 protein encoded by aureus viruses suppressed both virus and transgene-induced silencing by sequestering both long dsRNA and siRNA without size specificity [88]. Multiple suppressors have been reported in Citrus tristeza virus where p20 and coat protein (CP) play important role in suppression of silencing signal and p23 inhibited intracellular silencing [27].

Multiple viral components, viral RNAs and putative RNA replicase proteins were reported for a silencing or suppression of Red clover necrotic mosaic virus [89]. In this case, the RNA silencing machinery deprived of DICER-like enzymes by the viral replication complexes appears to be the cause of the suppression. Pns10 encoded by Rice dwarf virus suppressed local and systemic S-PTGS but not IR-PTGS suggesting that Pns10 also targets an upstream step of dsRNA formation in the silencing pathway [90]. A 273-bp (base pair) sequence of the Arabidopsis miR159 a pre-miRNA transcript expressing amiRNAs was used against the viral suppressor genes P69 and HC-Pro to provide resistance against Turnip yellow mosaic virus and Turnip mosaic virus infection, respectively [91]. In addition, a dimeric construct harboring two unique amiRNAs against both viral suppressors conferred resistance against these two viruses in inoculated Arabidopsis plants. Similarly, a different amiRNA vector was used to target the 2 b viral suppressor of the Cucumber mosaic virus (CMV), a suppressor that interacted with and blocked the slicer activity of AGO1 had also shown to confer resistance to CMV infection in transgenic tobacco [92]. A strong correlation between virus resistance and the expression level of the 2 b-specific amiRNA was shown for individual plant lines. It is evident from above-mentioned reports that the RNA components, such as single strand template RNA, dsRNA and/or siRNA of the silencing pathways are the preferred targets of most viral suppressors.

However, plant viruses are known to have evolved a counter-silencing mechanism by encoding proteins that can overcome such resistance [34,93]. These suppressors of gene silencing are often involved in viral pathogenicity, mediate synergism among plant viruses and result in the induction of more severe disease. Simultaneous silencing of such diverse plant viruses can be achieved by designing hairpin structures that can target a distinct virus in a single construct [93]. Contrarily, the RNAi system may cause an increase in the severity of viral pathogenesis and/or encode proteins, which can inactivate essential genes in the RNAi machinery [94] that helps them in their replication in the host genome [17].

Transgenic rice plants expressing DNA encoding ORF IV of Rice tungro bacilliform virus (RTBV), both in sense and in anti-sense orientation, resulting in the formation of dsRNA, were generated. Specific degradation of the transgene transcripts and the accumulation of small RNA were observed in transgenic plants. In RTBV-ODs2 line, RTBV DNA levels gradually rose from an initial low to almost 60% of that of the control at 40 days after inoculation [95]. For the effective control of PRSV and Papaya leaf-distortion mosaic virus (PLDMV), an untranslatable chimeric construct containing truncated PRSV YK CP and PLDMV P-TW-WF CP genes has been transferred into papaya (Carica papaya cv. 'Thailand') by Agrobacterium-mediated transformation via embryogenic tissues derived from immature zygotic embryos of papaya [96]. Based on sequence profile of silencing suppressor protein, HcPro, it was that PRSV-HcPro acts as a suppressor of RNA silencing through micro RNA binding in a dose dependent manner. In planta expression of PRSV-HcPro affects developmental biology of plants, suggesting the interference of suppressor protein in micro RNA-directed regulatory pathways of plants. Besides facilitating the establishment of PRSV, it showed strong positive synergism with other heterologous viruses as well [97].

Management of Plant Parasitic Nematodes

Several major plant parasitic nematodes such as the rootknot (Meloidogyne spp.) and cyst (Heterodera spp.) along with other minor nematodes cause significant damage to important agricultural crops such as legumes, vegetables and cereals in most parts of the world. Therefore, a natural, eco-friendly defense strategy that delivers a cost-effective control of plant parasitic nematodes is needed urgently which is difficult to achieve through conventional approaches. However, the origin of RNAi technology from classical C. elegans studies has shown the ways and means to explore the possibilities of this mechanism for protecting plants from nematode damage. In this context, two approaches have been advocated i.e., 1) relies on targeting plant genes that are involved with the infection process and 2) targets the essential genes within the nematode. RNAi can be induced in C. elegans by feeding it dsRNA, hence it was reasoned that expressing hpRNAs containing sequences of vital nematode genes in the host plant might deliver dsRNA to a feeding nematode to incapacitate or kill it. After the demonstration of gene silencing using siRNA duplexes in the nematode [22], the use of RNAi has rapidly emerged as the technique for plant nematologists to put their efforts, especially for nematode management in agriculture. RNAi-mediated suppression of a gene plays an indispensable role in hampering the nematode development and may adversely affect the progression of pathogenesis in direct or indirect ways. There are accumulating evidences for the efficacy of RNAi in plant parasitic nematode management and a wide range of genes have been targeted for silencing in cyst and rootknot nematode species (Table 3).

Table 3. RNAi effect on targeted region of plant parasitic nematodes.

Nematode	Targeted region	RNAi effect
M. incognita	Cysteine proteinase	Delayed development, Decrease in established nematodes population
	Dual oxidase	Decrease in established nematodes population and fecundity.
	Splicing factor, Integrase	Reduction in gall formation and Female nematode population
	Secreted peptide 16D 10	Reduction in gall formation and established nematode population
H. glycines	Cysteine proteinase	Increased male: female ratio.
	C-type lectin	Reduction in established nematodes population
	Major sperm protein	Reduction in mRNA level
	Aminopeptidase	Decrease in established nematodes population and increase in male: female ratio.
	β-1,4-endoglucanase	Decrease in established nematodes population
	Pectate lyase, Chorismate mutase	Increase in male: female ratio.
	Secreted peptide SYV46	Decrease in established nematode population
G. pallida	Cysteine proteinase	Increase in male: female ratio.
	FMR Famide-like peptides	Motility inhibited
G. rostochiensis	Chitin synthase	Delay in egg hatch
	β-1,4-endoglucanase	Decrease in established nematodes population
	Secreted amphid protein	Reduction in invasion ability to locate and invade plant roots
Heterodera schachtii	*Suc* transporter genes	Reduction of female nematode development
(Modified after [132])		

RNAi in the context of phyto-parasitic nematodes was used as early as the beginning of this century, when stimulation of oral ingestion by second-stage juveniles of cyst nematodes H. glycines, G. pallida [120] and rootknot nematode M. incognita was achieved by using octopamine [121]. Later on, resorcinol- and serotonininducing dsRNA uptake by second stage juvenile of M. incognita was found to be more effective than octopimine [122]. The genes targeted by RNAi to date are expressed in a range of different tissues and cell types. The ingested dsRNA can silence genes in the intestine [120, 123], female reproductive system [124], sperm [120, 125], and both subventral and dorsal oesophageal glands [121,122,126,127,]. Uptake of dsRNA from the gut is a proven route to systemic RNAi in C. elegans. The systemic nature of RNAi in plant parasitic nematodes following ingestion of dsRNA suggests that they share similar uptake and dispersal pathways. However, RNAi of a chitin synthase gene expressed in the eggs of Meloidogyne artiella was achieved by soaking intact eggs contained within their gelatinous matrix in a solution containing dsRNA [128]. The enzyme plays a key role in the synthesis of the chitinous layer in the eggshell. Depletion of its transcript by RNAi led to a reduction in stainable chitin in eggshells and a delay in hatching of juveniles from treated eggs. Similarly, RNAi targeting for cysteine proteinase transcripts did not reduce parasitic population of established nematodes on plants but result into the alteration of their sexual fate in favour of males at 14 days after invasion [120]. On the other hand H. glycines exposed to dsRNA corresponding to a protein with homology to C-type lectins did not affect sexual fate, but 41% fewer nematodes were recovered from the plants. But treatment with dsRNA corresponding to the major sperm protein (MSP) had no effect on nematode development or sexual fate 14 days after treatment. In addition to this, reduction in transcript abundance for targeted mRNAs in the infective juvenile and for MSP transcripts when males reached sexual maturity and sperm are produced was observed [120]. In further extension of such types of experiments showed efficient FITC uptake by soaking M. incognita, 90-95% of individuals swallowed the dye when the target was a dual oxidase (an enzyme comprised with a peroxidase domain EF-hands and NADPH oxidase domain and potentially involved in extracellular matrix development). The effect of RNAi was observed when root knot nematode (RKN) juveniles were fed on dual oxidase-derived dsRNA, the reduction in the number and size of established females at 14 and 35 days post- infection with an overall reduction of 70% in egg production was observed [121]. RNAi has also been induced for a chitin synthase gene that is expressed in the eggshells of M. artiella after soaking its developing eggs in a dsRNA [128]. Heterodera schachtii induces syncytial feeding structures in the roots of host plants, and this requires the up-regulation of Suc transporter genes to facilitate increased nutrient flow to the developing structure. Targeting these genes and downregulating them with RNA silencing resulted in a significant reduction of female nematode development [129]. Indeed, tobacco plants transformed with hpRNA constructs against two such root-knot nematode genes have shown such an effect: the target mRNAs in the plant parasitic nematodes were dramatically reduced, and the plants showed effective resistance against the parasite [130]. In another study mRNA abundances of targeted nematode genes were specifically reduced in nematodes feeding on plants expressing corresponding RNAi constructs. Furthermore, this host-induced RNAi of all four nematode parasitism genes led to a reduction in the number of mature nematode females. Although no complete resistance was observed, the reduction of developing females ranged from 23% to 64% in different RNAi lines [131]. These observations demonstrate the relevance of the targeted parasitism genes during the nematode life cycle and more importantly, suggest that a viable level of resistance in crop plants may be accomplished in the future by using RNAi technology against cyst nematodes.

CONCLUSIONS AND FUTURE PROSPECTS

RNAi and miRNA technologies of gene silencing are newly developed genomics tools that have great advantages over antisense and co-suppression due to their higher silencing efficiency and shorter time requirements for screening. These technologies are particularly useful in conjunction with the practice of gene or pathway discoveries through nutritional genomics, trancriptomics, proteomics and metabolomics in plants to improve human health. The RNA silencing has ability to reduce gene expression in a manner that is highly sequence specific as well as technologically facile and economical. Therefore, this technique has great potential in agriculture specifically for nutritional improvement of plants and the management of mascotous plant diseases. However, the major obstacles hindering its immediate applications include selection of targeting sequences and in the delivery of siRNA. The key issues are: 1) how to select silencing targets for a particular disease, and 2) how to efficiently deliver siRNAs into specific cell types in vivo. Tissue or organ-specific RNAi vectors have already been proven to be useful for targeted gene silencing in specific plant tissues and organs with minimal interference with the normal plant life cycle. New generation RNAi and miRNA vectors have been developed with high silencing accuracy and fewer side effects in plants. Genetic engineering of highly nutritional food crops requires both gene silencing and counter-silencing technologies. Besides, RNAi technology can be considered an eco-friendly, biosafe and ever green technology as it eliminates even certain risks associated with development of transgenic plants carrying first generation constructs (binary vectors and sense and antisense genes). As witnessed from earlier strategies for obtaining viral resistant plants, the expression of protein product from the transgene of interest risked hetero-encapsidation through protein-protein interactions between target and non-target viral gene product, resulted in the development of a non-aphid transmissible strain of Zucchini yellow mosaic virus to aphid-transmissible strain from a transgene expressing a plum pox capsid protein. Since RNAi triggers the formation of dsRNA molecules that target and facilitate the degradation of the gene of interest as well as the transgene itself to avoid problems arising from the synthesis of gene sequences as well as noncoding regions of gene, thus limiting undesirable recombination events. Keeping in view the potentialities of RNAi technology this technology has emerged to combat plant pathogens in the near future as it has already added new dimensions in the chapter of plant disease management. Further, development of vectors that can suppress the RNAi pathway but overexpress transgenes in a tissue-specific manner will revolutionize this field. Such vectors could be based on various viral RNA silencing suppressors and their derivatives. Future directions will focus on developing finely tuned RNAi-based gene silencing vectors that are able to operate in a temporally and spatially controlled manner. However, a better and comprehensive understanding of RNAi would allow the researchers to work effectively and efficiently in order to improve crop plants nutritionally and manage various mascotous intruders of crop plants.

REFERENCES

1. G. W. Chang, H.-W. Lin and S.-K. Chen, "Modeling Characteristics of Harmonic Currents Generated by HighSpeed Railway Traction Drive Converters," Transactions on Power Delivery, IEEE, Vol. 19, No. 2, April 2004, pp. 766-773.
2. L. Manczinger, Z. Antal and L. Kredics, "Ecophysiology and Breeding of Mycoparasitic Trichoderma Strains (a Review)," Acta Microbiologica et Immunologica Hungarica, Vol. 49, 2002, pp. 1-14.
3. D. Hoisington, M. Khairallah, T. Reeves, J. M. Ribaut, B. Skovmand, S. Taba and M. Warburton, "Plant Genetic Resources: What can they Contribute toward Increased Crop Productivity," Proceedings of the National Academy of Sciences of the USA, Vol. 96, 1999, pp. 5937-5943.
4. M. Lee, "Genome Projects and Gene Pools: New Germplasm for Plant Breeding," Proceedings of the National Academy of Sciences of the USA, Vol. 95, 1998, pp. 2001-2004.
5. V. Frankard, M. Ghislain and M. Jacobs, "Two Feedback-Insensitive Enzymes of the Aspartate Pathway in Nicotiana sylvestris," Plant Physiology, Vol. 99, 1992, pp. 1285-1293.
6. G. Galili, S. Galili, E. Lewinsohn and Y. Tadmor, "Genetic, Molecular and Genomic Approaches to Improve the Value of Plant Foods and Feeds," Critical Reviews in Plant Sciences, Vol. 21, 2002, pp. 167-204.
7. M. D. de Bakker, M. Raponi and G. M. Arndr, "RNA-Meditaed Gene Silencing in Non-Pathogenic and Pathogenic Fungi," Current Opinion in Microbiology, Vol. 5, 2002, pp. 323-329.
8. R, Almeida and R. C. Allshire, "RNA Silencing and Genome Regulation," Trends in Cell Biology, Vol. 15, 2005, pp. 251-258.
9. H. Nakayashiki, "RNA Silencing in Fungi: Mechanisms and Applications," Federation of European Biochemical Societies Letters, Vol. 579, 2005, pp. 5950-5970.
10. D. C. Baulcombe, "RNA silencing in Plants," Nature, Vol. 431, 2004, pp. 356-363.
11. S. H. Wani and G. S. Sanghera, "Genetic Engineering for Viral Disease Management in Plants," Notulae Scientia Biologicae, Vol. 2, 2010, pp. 20-28.
12. M. A. Escobar, E. L. Civerolo, K. R. Summerfelt and A.M Dandekar, "RNAi-Mediated Oncogene Silencing Confers Ressitance to Crown Gall Tumorigenesis," Proceedings of the National Academy of Sciences USA, Vol. 98, 2001, pp. 13437-13442.
13. P. Brodersen and O. Voinnet, "The Diversity of RNA Silencing Pathways in Plants," Trends in Genetics, Vol. 22, 2006, pp. 268-280.

14. M. R. Godge, A. Purkayastha, I. Dasgupta and P. P. Kumar, "Virus-Induced Gene Silencing for Functional Analysis of Selected Genes," Plant Cell Reporter, Vol. 27, 2008, pp. 209-219.

15. S. M. Shahinul Islam, T. Miyazaki, F. Tanno and K. Itoh, "Dissection of Gene Function by RNA Silencing," Plant Biotechnology, Vol. 22, 2005, pp. 443-446.

16. O. Milhavet, D. S. Gary and M. P. Mattson "RNA Interference in Biology and Medicine," Pharmacological Reviews, Vol. 55, 2003, pp. 629-648.

17. G. J. Hannon, "RNA Interference," Nature, Vol. 418, 2002, pp. 244-251.

18. H. Vaucheret, F. Vazquez, P. Crete, and D. P. Bartel, "The Action of ARGONAUTE1 in the miRNA Pathway and Its Regulation by the miRNA Pathway are Crucial for Plant Development," Genes and Development, Vol. 18, 2004, pp. 187-1197.

19. P. M. Waterhouse, M. B. Wang and T. Lough, "Gene Silencing as an Adaptive Defense against Viruses," Nature, Vol. 411, 2001, pp. 834-842.

20. A. S. Pickford and C. Cogoni, "RNA-Mediated Gene Silencing," Cellular and Molecular Life Science, Vol. 60, 2003, pp. 871-882.

21. S. M. Hammond, E. Bernstein, D. Beach and G. J. Hannon, "An RNA-Directed Nuclease Mediates Post-Transcriptional Gene Silencing in Drosophila Cells," Nature, Vol. 404, 2000, pp. 293-296.

22. A. Fire, S. Xu, M. K. Montgomery, S. A. Kostas, S. E .Driver and C. C. Mello, "Potent and Specific Genetic Interference by Double-Stranded RNA in Caenorhabditis elegans," Nature, Vol. 391, 1998, pp. 806-811.

23. E. Bernstein, A. A. Caudy, S. M. Hammond and G. J. Hannon, "Role for a Bidentate Ribonuclease in the Initiation Step of RNA Interference," Nature, Vol. 409, 2001, pp. 363-366.

24. S. V. Wesley, C. A. Helliwell, N. A. Smith, M. B. Wang, D. T. Rouse, Q. Liu, P. S. Gooding, S. P. Singh, D. Abbott, P. A. Stoutjesdijk, S. P. Robinson, A. P. Gleave, A. G. Green, and P. Waterhouse, "Construct Design for Efficient, Effective and High-Throughput Gene Silencing in Plants," Plant Journal, Vol. 27, 2001, pp. 581-590.

25. G. Tang, B. J. Reinhart, D. Bartel and P. D. Zamore, "A Biochemical Framework for RNA Silencing in Plants," Genes and Development, Vol.17, 2003, pp. 49-63.

26. O. Voinnet, "Non-Cell Autonomous RNA Silencing," Federation of European Biochemical Societies Letters, Vol. 579, 2005, pp. 5858-5871.

27. R. Lu, A. M. Martin-Hernandez, J. R. Peart, I. Malcuit and D. C. Baulcombe, "Virus Induced Gene Silencing in Plants," Methods, Vol. 30, 2003, pp. 296-303.

28. P. Dunoyer, C. Himber, V. Ruiz-Ferrer, A. Alioua and O. Voinnet, "Intra- and Intercellular RNA Interference in Arabidopsis thaliana Requires Components of the MicroRNA and Heterochromatic Silencing Pathways," Nature Genetics, Vol. 39, 2007, pp. 848-856.

29. U. Klahre, P. Crete, S. A. Leuenberger, V. A. Iglesias and F. Meins, "High Molecular Weight RNAs and Small Interfering RNAs Induce Systemic Post Transcriptional Gene Silencing in Plants," Proceedings of the National Academy

of Sciences of the USA, Vol. 99, 2002, pp. 11981-11986.

30. A. J. Herr, A. Molnar, A. Jones and D. C. Baulcombe, "Defective RNA Processing Enhances RNA Silencing and Influences Flowering of Arabidopsis," Proceedings of the National Academy of Sciences of the USA, Vol. 103, 2006, pp. 14994-15001.

31. G. Hutvagner and P. D. Zamore, "A microRNA in a Multiple-Turnover RNAi Enzyme Complex," Science, Vol. 297, 2002, pp. 2056-2060.

32. C. Himber, P. Dunoyer, G. Moissiard, C. Ritzenthaler and O. Voinnet, "Transitivity Dependent and Independent Cell-to-Cell Movement of RNA Silencing," The EMBO Journal, Vol. 22, 2003, pp. 4523-4533

33. K. Kobayashi and P. Zambryski, "RNA Silencing and Its Cell-to-Cell Spread during is Embryogenesis," The Plant Journal, Vol. 50, 2007, pp. 597-604.

34. F. Li, and S. W. Ding, "Virus Counter Defense: Diverse Strategies for Evading the RNA Silencing Immunity," Annual Review of Microbiology, Vol. 60, 2006, pp. 503- 531.

35. C. F. Chuang and E. M. Meyerowtiz, "Specific and Heritable Genetic Interference by Double-Stranded RNA in Arabidopsis thaliana," Proceedings of the National Academy of Sciences of the USA, Vol. 97, 2000, pp. 985- 4990.

36. P. Schweizer, J. Pokorny, P. Schulze-Lefert and R Dudler, "Double Stranded RNA Interference with Gene Functions at the Single Cell in Cereals," The Plant Journal, Vol. 24, 2000, pp. 895-903.

37. Y. L. Liu, M. Schiff and S. P. Dinesh-Kumar, "Virus Induced Gene Silencing in Tomato," The Plant Journal, Vol. 31, 2002a, pp. 777-786.

38. T. Dalmay, A. J. Hamilton, E. Mueller and D. C. Baulcombe, "Potato Virus X Amplicons in Arabidopsis Mediate Genetic and Epigenetic Gene Silencing," The Plant Cell, Vol. 12, 2000, pp. 369-380.

39. H. Liu, T. R. Cottrell, L. M. Pierini, W. E. Goldman and T. L. Doering, "RNA Interference in the Pathogenic Fungus Cryptococcus neoformans," Genetics, Vol. 160, 2002b, pp. 463-470.

40. W. Tang, D. A. Weidner, B. Y. Hu, R. J. Newton and X. Hu, "Efficient Delivery of Small Interfering RNA to Plant Cells by a Nanosecond Pulsed Laser-Induced Wave for Post Transcriptional Gene Silencing," Plant Science, Vol. 171, 2006, pp. 375-81.

41. J. M. Hilly, and Z. Liu, "An Overview of Small RNAs," In: C. L. Bassett, Ed., Regulation of Gene Expression in Plants, Springer-Verlag, Berlin, 2007, pp. 123-147.

42. L. K. Johansen, and J. C. Carrington, "Silencing on the Spot. Induction and Suppression of RNA Silencing in the Agrobacterium-Mediated Transient Expression System," Plant Physiology, Vol. 126, 2001, pp. 930-938.

43. O. Voinnet, "RNA Silencing as a Plant Immune System against Viruses," Trends in Genetics, Vol. 17, 2001, pp. 449-459.

44. S. Mlotshwas, O. Voinnet, M. F. Mette, M. Matzke, H. Vaucheret, S. W. Ding, G.

Pruss and G. B. Vance, "RNA Silencing and Its Mobile Silencing Signal," The Plant Cell, Vol. 14, 2002, pp. 289-301.

45. F. Tenllado, B. Martinez-Garcia, M. Vargas and J. R. Diaz-Ruiz, "Crude Extracts of Bacterially Expressed dsRNA can be Used to Protect Plants against Virus Infection," BMC Biotechnology, Vol. 3, 2003, pp. 3-14.

46. M. Timmermans, O. Das and J. Messing, "Geminivirus and Their Uses as Extrachromosomeal Replicons," Annual Review of Plant Physiology, Vol. 45, 1994, pp. 79- 112.

47. G. P. Pogue, J. A. Lindbo, S. J. Garger and W. P. Fitzmaurice, "Making an Ally from an Enemy: Plant Virology and the New Agriculture," Annual Review of Phytopathology, Vol. 40, 2002, pp. 45-74.

48. M. H. Kumagai, J. Donson, G. della-Cioppa, D. Harvey, K. Hanley and L. K. Grill, "Cytoplasmic Inhibition of Carotenoid Biosynthesis with Virus-Derived RNA," Proceedings of the National Academy of Sciences of the USA, Vol. 92, 1995, pp. 1679-1683.

49. S. M. Angell and D. C. Baulcombe, "Technical Advance: Potato Virus X Amplicon Mediated Silencing of Nuclear Genes," The Plant Journal, Vol. 20, 1999, pp. 357-362.

50. S. A. MacFarlane, and A. H. Popovich, "Efficient Expression of Foreign Proteins in Roots from Tobravirus Vectors," Virology, Vol. 267, 2000, pp. 29-35.

51. A. C. Mallory, G. Parks, M. W. Endres, D. Baulcombe and L. H. Bowman, "The Amplicon-Plus System for High- Level Expression of Transgenes in Plants," Nature Biotechnogy, Vol. 20, 2002, pp. 622-625.

52. K. E. Palmer and E. P. Rybicki, "Investigation of the potential of maize streak virus to act as an infectious gene vector in maize plants," Archives of Virology, Vol. 146, 2001, pp. 1089-1104.

53. S. Kjemtrup, K. S. Sampson, C. G. Peele, L. V. Nguyen and M. A. Conkling, "Gene Silencing from Plant DNA Carried by a Geminivirus," Plant Journal, Vol. 14, 1998, pp. 91-100.

54. M. J. Dallwitz and E. J. Zurcher, "Plant Viruses Online," In: A. A. Brunt, K. Crabtree, M. J. Dallwitz, A. J. Gibbs, L. Watson and E. J. Zurcher, Ed., Descriptions and Lists from the VIDE Database, CAB International, UK, 1996, pp. 1484.

55. Q. Liu, S. P. Singh, and A. G. Green, "High-Stearic and High-Oleic Cottonseed Oils Produced by Hairpin RNAMediated Post-Transcriptional Gene Silencing," Plant Physiology, Vol. 129, 2002c, pp. 1732-1743.

56. M. A. Turnage, N. Muangsan, C. G. Peele and D. Robertson, "Geminivirus-Based Vectors for Gene Silencing in Arabidopsis," The Plant Journal, Vol. 30, 2002, pp. 107-117.

57. C. Cogoni, N. Romano and G. Macino, "Suppression of Gene Expression by Homologous Transgenes," Antonie Leeuwenhoek International Journal of General Molecular Microbiology, Vol. 65, 1994, pp. 205-209.

58. P. Y. Teycheney and M Tepfer, "Virus Specific Spatial Differences in the

Interference with Silencing of the chs-A Gene in Non-Transgenic Petunia," Journal of Gen-eral Virology, Vol. 82, 2001, pp. 1239-1243.

59. W. Hamada and P. D. Spanu, "Co-Suppression of the Hydrophobin Gene Hcf-1 is Correlated with Antisense RNA Biosynthesis in Cladosporium fulvum," Molecular and General Genetics, Vol. 259, 1998, pp. 630-638.

60. N. Kadotani, H. Nakayashiki, Y. Tosa and S. Mayama, "RNA Silencing in the Pathogenic Fungus Magnaporthe oryzae," Molecular Plant-Microbe Interaction, Vol. 16, 2003, pp. 769-776.

61. J. A. Kim, K. Cho, R. Singh, Y. H. Jung, S. H. Jeong, S. H. Kim, J. E. Lee, Y. S. Cho, G. K. Agrawal, R. Rakwal, S. Tamogami, B. Kersten, J. S. Jeon, G. An and N. S. Jwa, "Rice OsACDR1 (Oryza sativa Accelerated Cell Death and Resistance 1) is a Potential Positive Regulator of Fungal Disease Resistance," Molecules and Cells, Vol. 30, 2009, pp. 431-439.

62. L. Chen, K. Shiotani, T. Togashi, D. Miki, M. Aoyama, H. L. Wong, T. Kawasaki and K. Shimamoto, "Analysis of the Rac/Rop Small GTPase Family in Rice: Expression, Subcellular Localization and Role in Disease Resistance," Plant and Cell Physiology, Vol. 51, 2010, pp. 585-595.

63. A. Fitzgerald, J. A. Van Kha and K. M. Plummer, "Simultaneous Silencing of Multiple Genes in the Apple Scab Fungus Venturia inaequalis, by Expression of RNA with Chimeric Inverted Repeats," Fungal Genetics and Biology, Vol. 41, 2004, pp. 963-971.

64. M. Goldoni, G. Azzalin, G. Macino and C. Cogoni, "Efficient Gene Silencing by Expression of Double Stranded RNA in Neurospora crassa," Fungal Genetics and Biology, Vol. 41, 2004, pp. 1016-1024.

65. T. M. Hammond and N. P. Keller, "RNA Silencing in Aspergillus nidulans is Independent of RNA-Dependent RNA polymerase," Genetics, Vol. 169, 2005, pp. 607-617.

66. P. Spanu, "HCf-1, a Hydrophobin from the Tomato PathoGen Cladosporium fulvum," Gene, Vol. 93, 1997, pp. 89-96.

67. G. C. Segers, W. Hamada, R. P. Oliver and P. D. Spanu, "Isolation and Characteristaion of Five Different Hydrophobin-Encoding cDNA from the Fungal Tomato Pathogen Cladosporium fulvum," Molecular and General Genetics, Vol. 261, 1999, pp. 644-652.

68. J. C. Abbott, A. Barakate, G. Pincon, M. Legrand, C. Lapierre, I. Mila, W. Schuch and C. Halpin, "Simultaneous Suppression of Multiple Genes by Single Transgenes. Down-Regulation of Three Unrelated Lignin Biosynthetic Genes in Tobacco," Plant Physiology, Vol. 128, 2002, pp. 844-853.

69. N. J. Talbot, M. J. Kershaw, G. E. Wakley, O. M. H. de Vries, J. G. H. Wessels and J. E. Hamer, "MPG1 Encodes a Fungal Hydrophobin Involved in Surface Interactions during Infection-Related Development of Magnaporthe grisea," Plant Cell, Vol. 8, 1996, pp. 985-999.

70. Q. B. Nguyen, N. Kadotani, S. Kasahara, Y. Tosa, S. Mayama and H. Nakayashiki,

"Systematic Functional Analysis of Calcium-Signalling Proteins in the Genome of the Rice-Blast Fungus, Magnaporthe oryzae, Using a High-Throughput RNA-Silencing System," Molecular Microbiology, Vol. 68, 2008, pp. 1348-1365.

71. R. J. Weld, K. M. Plummer, M. A. Carpenter, and H. J. Ridgway, "Approaches to Functional Genomics in Filamentous Fungi," Cell Research, Vol. 16, 2006, pp. 31-44.

72. H. Nakayashiki and Q. B. Nguyen, "RNA Interference: Roles in Fungal Biology," Current Opinion in Microbiology, Vol. 11, 2008, pp. 1-9.

73. H. Peng, Q. Zhang, Y. Li, C. Lei, Y. Zhai, X. Sun, D. Sun, Y. Sun and T. Lu, "A Putative Leucine-Rich Repeat Receptor Kinase, OsBRR1, is Involved in Rice Blast Resistance," Planta, Vol. 230, 2009, pp. 377-385.

74. S. Katiyar-Agarwal, R. Morgan, D. Dahlbeck, O. Borsani, J. A. Villegas, J. Zhu, B. J. Staskawicz and H. Jin, "A Pathogen-Inducible Endogenous siRNA in Plant Immunity," Proceedings of the National Academy of Sciences of the USA, Vol. 103, 2006, pp. 47-52.

75. S. Katiyar-Agarwal, S. Gao, A. Vivian-Smith and H. Jin, "A Novel Class of Bacteria-Induced Small RNAs in Arabidopsis," Genes and Development, Vol. 21, 2007, pp. 3123-3134.

76. L. Navarro, P. Dunoyer, F. Jay, B. Arnold, N. Dharmasini, M. Estelle, O. Vionnet and J. D. Jones, "A Plant miRNA Contributes to Antibacterial Resistance by Repressing Auxin Signaling," Science, Vol. 312, 2006, pp. 436-439.

77. V. Bitko and S. Barik, "Phenotypic Silencing of Cytoplasmic Genes with Sequence Specific Double Stranded Short Interfering RNA and Its Applications in the Reverse Genetics of Wild Type Negative Strand RNA Virus," BMC Microbiology, Vol. 1, 2001, pp. 34-44.

78. L. Gitlin, S. Karelsky, and R. Andino, "Short Interference Confers Intracellular Antiviral Immunity in Human Cells," Nature, Vol. 4, 2002, pp. 418-430.

79. J. M. Jacque, K. Triques and M. Stevenson, "Modulation of HIV-1 Replication by RNA Interference," Nature, Vol. 418, 2002, pp. 435-438.

80. C. D. Novina, M. F Murray, D. M. Dykxhoorn, P. J. Beresford, J. Riess, S. K. Lee, R. G. Collman, J. Lieberman, P. Shanker and P. A. Sharp, "siRNA-Directed Inhibition of HIV-1 Infection," Nature Mediterranean, Vol. 8, 2002, pp. 681-686.

81. E. Ullu, A. Djikeng, H. Shi, and C. Tschudi, "RNA Interference: Advances and Questions," Philosophical Transactions of the Royal Society of London British Biological Science, Vol. 29, 2002, pp. 65-70.

82. P. M. Waterhouse, M. W. Graham and M. B. Wang, "Virus Resistance and Gene Silencing in Plants can be Induced by Simultaneous Expression of Sense and Antisense RNA," Proceedings of the National Academy of Sciences of the USA, Vol. 95, 1998, pp. 13959-13964.

83. E. J. Chapman, A. I. Prokhnevsky, K. Gopinath, V. V. Dolja and J. C. Carrington, "Viral RNA Silencing Suppressors Inhibit the micro-RNA Pathway at an Interphase Step," Genes and Development, Vol. 18, 2004, pp. 1179- 1186.

84. M. Pooggin, P. V. Shivaprasad, K. Veluthambi and T. Hohn, "RNAi Targetting of DNA Viruses," Nature Bio-technology, Vol. 21, 2003, pp. 131-32.

85. R. Vanitharani, P. Chellappan and C. M. Fauquet, "Short Interfering RNA-Mediated Interference of Gene Expression and Viral DNA Accumulation in Cultured Plant Cells," Proceedings of the National Academy of Sciences of the USA, Vol. 100, 2003, pp. 9632-9636.

86. V. Ruiz-Ferrer and O. Voinnet, "Viral Suppression of RNA Silencing: 2b Wins the Golden Fleece by Defeating Argonaute," Bioassays, Vol. 29, 2007, pp. 319-323.

87. J. Chen, W. X. Li, D. Xie, J. R. Peng, and S. W. Ding, "Viral Virulence Protein Suppresses RNA Silencing-Mediated Defense but Upregulates the Role of microRNA in Host Gene Regulation," Plant Cell, Vol. 16, No. 5, 2004, pp. 1302-1313.

88. Z. Merai, Z. Kerenyi, A. Molnar, E. Barta, A. Valcozi, G. Bistray, Z. Havelda, J. Burgyan and D. Silhavy, "Aureusvirus P14 is an Efficient RNA Silencing Suppressor that Binds Double Stranded RNAs without Size specificity," Journal of Virology, Vol. 79, 2005, pp. 7217- 7226.

89. A. Takeda, M. Tsukuda, H. Mizumoto, K. Okamoto, M. Kaido, K. Mise and T. Okuno, "A Plant RNA Virus Suppressor RNA Silencing through RNA Replication," The EMBO Journal, Vol. 24, 2005 pp. 3147-3157.

90. X. Cao, P. Zhou, X. Zhang, S. Zhu, X. Zhong, Q. Xiao, B. Ding and Y. Li, "Identification of an RNA Silencing Suppressors from a Plant Double Stranded RNA Virus," Journal of Virology, Vol. 79, 2005, pp. 13018-13027.

91. Q. W. Niu, S. S. Lin, J. L. Reyes, K. C. Chen, H. W. Wu, S. D. Yeh and N. H. Chua, "Expression of Artificial microRNAs in Transgenic Arabidopsis thaliana Confers Virus Resistance," Nature Biotechnology, Vol. 24, 2006, pp. 1420-1428.

92. J. Qu, J. Ye and R. X. Fang, "Artificial microRNA-Mediated Virus Resistance in Plants," Journal of Virology, Vol. 81, 2007, pp. 6690-6699.

93. J. A. Díaz-Pendón and S. W. Ding, "Direct and Indirect Roles of Viral Suppressors of RNA Silencing in Pathogenesis," Annual Review of Phytopathology, Vol. 46, 2008, pp. 303-326.

94. S. M. Elbashir, W. Lendeckel, and T. Tuschl, "RNA Interference is Mediated by 21 and 22-Nucleotide RNAs," Genes and Development, Vol. 15, 2001, pp. 188-200.

95. H. Tyagi, S. Rajasubramaniam, M. V. Rajam and I. Dasgupta, "RNA Interference in Rice against Rice Tungro Bacilliform Virus Results in Its Decreased Accumulation in Inoculated Rice Plants," Transgenic Research, Vol. 17, 2008, pp. 897-904.

96. Y. J. Kung, T. A. Yu, C. H. Huang, H. C. Wang, S. L. Wang and S. D. Yeh, "Generation of Hermaphrodite Transgenic Papaya Lines with Virus Resistance via Transformation of Somatic Embryos Derived from Adventitious Roots of in vitro shoots," Transgenic Research, 2009, in press.

97. S. K. Mangrauthia, P. Singh and S. Praveen, "Genomics of Helper Component

Proteinase Reveals Effective Strategy for Papaya Ringspot Virus Resistance," Molecular Biotechnology, Vol. 44, 2010, pp. 22-29.

98. I. B. Fofana, A. Sangare, R. Collier, C. Taylor and C. M. Fauquet, "A Geminivirus-Induced Gene Silencing System for Gene Function Validation in Cassava," Plant Molecular Biology, Vol. 56, 2004, pp. 613-624.

99. S. Holzberg, P. Brosio, C. Gross and G. P. Pogue, "Barley Stripe Mosaic Virus-Induced Gene Silencing in a Monocot Plant," Plant Journal, Vol. 30, 2002, pp. 315- 327.

100. S. R. Scofield, L. Huang, A. S. Brandt and B. S. Gill, "Development of a Virus-Induced Gene Silencing System for Hexaploid Wheat and Its Use in Functional Analysis of the Lr21-Mediated Leaf Rust Resistance Pathway," Plant Physiology, Vol. 138, 2005, pp. 2165-2173.

101. C. Cakir and M. Tör, "Factors Influencing Barley Stripe Mosaic Virus-Mediated Gene Silencing in Wheat," Physiological and Molecular Plant Pathology, Vol. 74, 2010, pp. 246-253.

102. C. Zhang and S. A. Ghabrial, "Development of Bean Pod Mottle Virus-Based Vectors for Stable Protein Expression and Sequence-Specific Virus-Induced Gene Silencing in Soybean," Virology, Vol. 344, 2006, pp. 401-411.

103. C. Zhang, C. Yang, S. A. Whitham and J. H. Hill "Development and Use of an Efficient DNA-Based Viral Gene Silencing Vector for Soybean," Molecular Plant Microbe Interaction, Vol. 22, 2009, pp. 123-131.

104. C. Zhang, J. D. Bradshaw, S. A. Whitham and J. H. Hill "The Development of an Efficient Multi-Purpose BPMV Viral Vector Set for Foreign Gene Expression and RNA Silencing," Plant Physiology, 2010, in press.

105. X. S. Ding, W. L. Schneider, S. R. Chaluvadi, R. M. Rouf Mian and R. S. Nelson, "Characterization of a Brome Mosaic Virus Strain and Its Use as a Vector for Gene Silencing in Monocotyledonous Hosts," Molecular Plant Microbe Interaction, Vol. 19, 2006, pp. 1229-1239.

106. G. D. Constantin, B. N. Krath, S. A. MacFarlane, M. Nicolaisen, I. E. Johansen and O. S. Lund, "Virus-Induced Gene Silencing as a Tool for Functional Genomics in a Legume Species," Plant Journal, Vol. 40, 2004, pp. 622-631.

107. M. Naylor, J. Reeves, J. I. Cooper, M. L. Edwards and H. Wang, "Construction and Properties of a Gene Silencing Vector Based on Poplar Mosaic Virus (Genus Carlavirus)," Journal of Virology Methods, Vol. 124, 2005, pp. 27-36.

108. M. T. Ruiz, O. Voinnet and D. C. Baulcombe, "Initiation and Maintenance of Virus-Induced Gene Silencing," Plant Cell, Vol. 10, 1998, pp. 937-946.

109. O. Faivre-Rampant, E. M. Gilroy, K. Hrubikova, I. Hein, S. Millam, G. J. Loake, P. Birch, M. Taylor and C. Lacomme, "Potato Virus X-Induced Gene Silencing in Leaves and Tubers of Potato," Plant Physiology, Vol. 134, 2004, pp. 1308-1316.

110. V. V. Gossele, I. I. Fache, F. Meulewaeter, M. Cornelissen and M. Metzlaff, "SVISS-a Novel Transient Gene Silencing System for Gene Function Discovery and Validation in Tobacco," The Plant Journal, Vol. 32, 2002, pp. 859- 866.

111. F. Ratcliff, A. M. Martin-Hernandez and D. C. Baulcombe, "Tobacco Rattle Virus as a Vector for Analysis of gene Functions by Silencing," Plant Journal, Vol. 25, 2001, pp. 237-245.

112. G. Brigneti, A. M. Martin-Hernandez, H. Jin, J. Chen, D. C. Baulcombe, B. Baker, and J. D. Jones, "Virus-Induced Gene Silencing in Solanum Species," The Plant Journal, Vol. 39, 2004, pp. 264-272.

113. E. Chung, E. Seong, Y. C. Kim, E. J. Chung, S. K. Oh, S. Lee, J. M. Park, Y. H. Joung and D. Choi, "A Method of High Frequency Virus Induced Gene Silencing in Chili Pepper Capsicum annuum L. cv. Bukang)," Molecular Cell, Vol. 17, 2004, pp. 377-380.

114. L. C. Hileman, S. Drea, G. Martino, A. Litt and V. F. Irish, "Virus Induced Gene Silencing is an Effective Tool for Assaying Gene Function in the Basal Eudicot Species Papaver somniferum (Opium Poppy)," The Plant Journal, Vol. 44, 2005, pp.334-341.

115. B. Gould and E. M. Kramer, "Virus-Induced Gene Silencing as a Tool for Functional Analyses in the Emerging Model Plant Aquilegia (columbine, Ranunculaceae)," BMC Plant Methods, Vol. 12, 2007, pp. 6.

116. H. Hou and W. Qiu, "A Novel Co-Delivery System Consisting of a Tomato Bushy Stunt Virus and a Defective Interfering RNA for Studying Gene Silencing," Journal of Virology Methods, Vol. 111, 2003, pp. 37-42.

117. C. Peele, C.V. Jordan, N. Muangsan, M. Turnage, E. Egelkrout, P. Eagle, L. Hanley-Bowdoin and D. Robertson, "Silencing of a Meristematic Gene Using Geminivirus-Derived Vectors," Plant Journal, Vol. 27, 2001, pp. 357-366.

118. X. Tao and X. Zhou, "A Modified Viral Satellite DNA that Suppresses Gene Expression in Plants," The Plant Journal, Vol. 38, 2004, pp. 850-860.

119. M. R. Godge, A. Purkayastha, I. Dasgupta and P. P. Kumar, "Virus-Induced Gene Silencing for Functional Analysis of Selected Genes," Plant Cell Reporter, Vol. 27, 2008, pp. 209-219.

120. P. E. Urwin, C. J. Lilley and H. J. Atkinson, "Ingestion of Double-Stranded RNA by Pre-Parasitic Juvenile Cyst Nematodes Leads to RNA Interference," Molecular Plant-Microbe Interactions, Vol. 15, 2002, pp. 747-752.

121. M. Bakhetia, W. Charlton, H. J. Atkinson and M. J. McPherson, "RNA Interference of Dual Oxidase in the Plant Nematode Meloidogyne incognita," Molecular PlantMicrobe Interaction, Vol. 18, 2005, pp. 1099-1106.

122. M. N. Rosso, M. P. Dubrana, N. Cimbolini, S. Jaubert and P. Abad, "Application of RNA Interference to RootKnot Nematode Genes Encoding Esophageal Gland Proteins," Molecular Plant-Microbe Interactions, Vol. 18, 2005, pp. 615-620.

123. J. Shingles, C. J. Lilley, H. J. Atkinson and P. E. Urwin, "Meloidogyne incognita: Molecular and Biochemical Characterization of a Cathepsin L Cysteine Proteinase and the Effect on Parasitism Following RNAi," Experimental Parasitology, Vol. 115, 2007, pp. 114-120.

124. C. J. Lilley, S. A. Goodchild, H. J. Atkinson and P. E. Urwin, "Cloning and

Characterization of a Heterodera glycines Minopeptidase cDNA," International Journal of Parasitology, Vol. 35, 2005, pp. 1577-1585.

125. R. M. Steeves, T. C. Todd, J. S. Essig and H. N. Trick, "Transgenic Soybeans Expressing siRNAs Specific to a Major Sperm Protein Gene Suppress Heterodera glycines Reproduction," Functional Plant Biology, Vol. 33, 2006, pp. 991-999.

126. Q. Chen, S. Rehman, G. Smant and J. T. Jones, "Functional Analysis of Pathogenicity Proteins of the Potato Cyst Nematode Globodera rostochiensis Using RNAi," Molecular Plant-Microbe Interactions, Vol. 18, 2005, pp. 621-625.

127. G. Huang, R. Allen, E. L. Davis, T. J. Baum and R. S. Hussey, "Engineering Broad Root-Knot Resistance in Transgenic Plants by RNAi Silencing of a Conserved and Essential Root-Knot Nematode Parasitism Gene," Proceedings of the National Academy of Sciences of the USA, Vol. 103, 2006, pp. 4302-14306.

128. E. Fanelli, M. Di Vito, J. T. Jones and C. De Giorgi, "Analysis of Chitin Synthase Function in a Plant Parasitic Nematode, Meloidogyne artiellia, Using RNAi," Gene, Vol. 349, 2005, pp. 87-95.

129. Y. Hoffman, C. Aflalo, A. Zarka, J. Gutman, T.Y. James, and S. Boussiba, "Isolation and characterization of a novel chytrid species (phylum Blastocladiomycota), parasitic on the green alga Haematococcus," Mycological Research, Vol. 112, 2008, pp. 70-81.

130. D. J. Fairbairn, A.S. Cavalloro, M. Bernard, J. Mahalinga-Iyer, M. W. Graham and J. R. Botella, "Host-Delivered RNAi: An Effective Strategy to Silence Genes in Plant Parasite Nematodes," Planta, Vol. 226, 2007, pp. 1525-1533.

131. S. Sindhu, T R. Maier, M. G. Mitchum, R. S. Hussey, E. L. Davis and T. J. Baum "Effective and Specific in Planta RNAi in Cyst Nematodes: Expression Interference of Four Parasitism Genes Reduces Parasitic Success," Journal of Experimental Botany, Vol. 60, 2009, pp. 315-324.

132. M. Karakas, "RNA Interference in Plant Parasitic Nematodes," African Journal of Biotechnology, Vol. 7, 2008, pp. 2530-2534.

8
CHAPTER

INTEGRATED MANAGEMENT OF AFRICAN RICE GALL MIDGE , WEEDS AND DISEASES OF LOWLAND RICE

Akaamaa Clement Wada, Samuel Oladele Bakare, Mohammed Bashir, Mark Noye Ukwungwu

Keywords: African Rice Gall Midge; Diseases; Lowland Rice; Rice Nematode; IPM; Nigeria

INTRODUCTION

The National Cereals Research Institute, Badeggi has developed and released many high yielding, lowland rice varieties for increased sustainable crop production in the inland valley swamps. Many of the released varieties have been adopted for cultivation by many farmers in Nigeria. However, some of the varieties are becoming susceptible to the African Rice Gall Midge (AfRGM) [1]. FARO 37 is one of these released varieties. It is widely cultivated especially in and around Badeggi in the Southern Guinea Savannah ecological zone of Nigeria, and is being attacked by the pest. Chemical control of AfRGM has shown good results [2]. Also visual observations seem to indicate that alternating rice with a dry season crop may reduce the damage by AfRGM and incidence of diseases in lowland rice. Hoque [3] reported the practice of following rice by some short-duration non-rice crops in some developed swamps in some West

Citation: Wada, A. , Bakare, S. , Bashir, M. and Ukwungwu, M. (2012) Integrated management of African rice gall midge , weeds and diseases of lowland rice, 3, 486-490. doi: 10.4236/as.2012.34057.

African Countries including Nigeria. To ascertain the visual observations and exploit the report by Hogue [3], in the management of the AfRGM, weeds and diseases, a study was initiated in 2001 at Badeggi at two locations— Edozhigi and Sachi—to study the effect of cropping sequence, weeding regimes and minimal application of pesticides for the control of AfRGM and diseases in lowland rice with a view to identifying the most appropriate integrated method(s) for these pests.

MATERIALS AND METHODS

Two locations—Edozhigi and Sachi—were chosen to study the effect of following rice with another crop on the occurrence of African Rice Gall Midge weed interference and the incidence of diseases on FARO 37 in the 2001 and 2002 cropping seasons. Land clearing was done by slashing the two experimental sites with cutlasses. Bunds measuring 5 cm high were made to check movement of AfRGM from plot to plot. Eighteen plots at both locations left fallow after the 2001 rice harvest were cleared and heaps were made with hoes in early August 2002. The subsequent crop (Cowpea, IAR 48) planted at the on set of rains in May was harvested in the first week of August. Land leveling and puddling were done in late August for both the heaped plots and plots from where cowpea was harvested. Transplanting followed on 22nd September, 2002 at Sachi location.

The same operations were carried out at Edozhigi from mid September 2002 and transplanting done on 12 October 2002. Soil samples were collected from all the plots just before transplanting and after harvest for nematode extraction to wusnyigy nematode abundance. The nematode reproduction factor (Rf) was determined from the relation: Rf = Pf/Pi where Pi = initial nematode population.

Pf = final nematode population.

Transplanting was done at 21 days after seeding at both locations at the spacing of 20 cm × 20 cm in plot sizes of 4 m × 4.5 m. There were twelve treatments made up of:

T_1 = Sole rice + 1 hand weeding + isazofos @ 0.7 kg a.i. ha^{-1} + benomyl @ 1.0 kg a.i. ha^{-1}.

T_2 = Sole rice + 2 hand weeding + isazofos @ 0.75 kg a.i. ha^{-1} + benomyl @ 1.0 kg a.i. ha^{-1}.

T_3 = Sole rice + 2 hand weeding + isazofas @ 0.75 kg a.i. ha^{-1} + benomyl @ 1.0 kg a.i. ha^{-1}.

T_4 = Sole rice + no weeding + no isazofos + no benomyl applications.

T_5 = Sole rice + 1 hand weeding + no isazofos + no benomyl application.

T_6 = Sole rice + 2 hand weeding + no isazofos + no benomyl applications.

T_7 = Rice followed by cowpea + no weeding + isazofos @ 0.75 kg a.i. ha^{-1}.

T_8 = Rive followed by cowpea + 1 hand weeding + isazofos @ 0.75 kg a.i. ha^{-1} + benomyl @ 1.0 kg a.i. ha^{-1}.

T_9 = Rice followed by cowpea + 2 hand weeding + isazofos at 0.75 kg a.i. ha^{-1}.

T_{10} = Rice followed by cowpea + 1 hand weeding + no isazofos and benomyl applications.

T_{11} = Rice followed by cowpea + 1 hand weeding + no isazofos and benomyl applications.

T_{12} = Rice followed by cowpea + 2 hand weeding + no isazofos and benomyl applications.

These treatments were set up in randomized complete block design in three replicates. Fertilizer was applied at transplanting at the rate of 50:30:30:NPK and top dressing treatment was given at 7 weeks after transplanting (WAT). Miral insecticide and Benlate fungicide were applied at 20 and 40 days respectively after transplanting. Hand weeding was done at 14 and 30 days for plots receiving 2 hand weeding and 30 days after transplanting for plots receiving one hand weeding. A weed control rating was done on a scale of 0 - 10 where 0 = no weed control, 8 = acceptable weed control and 10 excellent weed control. Percentage weed cover in each plot was also determined (**Table 1**). African rice gall midge infestation was recorded by counting the number of silver shoots per hill for 20 hills per plot at 45 and 60 DAT. Galls were also collected and dissected for percentage parasitism at 45 and 60 DAT.

Disease incidence was assessed for the prevalent rice diseases at the two locations using the IRRI standard evaluation for rice, scale (0 - 9) at 50 DAT and at maturity.

The weed flora and weed control ratings were recorded at 30 and 60 DAT. At maturity, the number of effective tillers and plant height were taken for 20 hills per plot. At harvest, two outer rows were harvested as discard while the net plot size (3.95 m × 4 m) was harvested and weighed after winnowing. All the collected data except the nematodes extracted from each plot were subjected to analysis of variance and mean separation done by least significance differences. Economic analysis of the subsequent crop was also done.

RESULTS AND DISCUSSION

The same weed species widely occurring at each location in 2001 were recorded in 2002. These were Ludwigia decurens, Sphanoclea zeylanica Cyperus difformis and Leptochloa chinensis at Edozhigi and Cyperus difformis, Ipomea involucrate, I. aquatica and Fimbristylis litoralis at Sachi (**Table 1**).

Weedy plots recorded the least rice yields at all the locations (Tables 2 and 3). Yield was significantly greater from weeded and pesticide treated plots that were cropped with cowpea as subsequent crop to rice than the weedy checks. There was also low AfRGM infestation in all the locations both in treated and untreated plots. However, AfRGM infestation was lower from plots cropped with cowpea suggesting that the subsequent crop might have broken the pest's life cycle. Effect of leaf blast was not adverse on the yield of rice from all the locations, since this pathogen was inhibited by benomyl. The observed high incidence of brown leaf spot at the experimental sites for the two years suggested that benomyl was ineffective for brown leaf spot inhibition.

Table 4 shows the economic analysis of producing a subsequent crop on the same plot after rice before the next season of rice. There are clear benefits the farmer would derive from this rice-subsequent crop culture as shown from the net returns in Naira per kilogramme, of cowpea in addition to income from fallow plots after the previous year's rice harvest.

Table 5 shows the relative abundance of nematodes at the two locations in 2002. This aspect was however not investigated in 2001.

Table 1. Mean effect of treatments on percent weed cover and weed control on FARO 37 at Sachi and Edozhigi 2001.

Treatment	Sachi		Edozhigi	
	% Weed Cover	Weed Control	% Weed Cover	Weed Control
T_1	71.1	2.8	63.3	3.7
T_2	55.0	5.0	30.0	7.2
T_3	0.83	9.8	0.83	9.8
T_4	80.0	2.0	55.0	4.5
T_5	28.3	7.2	23.3	7.7
T_6	8.3	9.0	5.8	9.0
T_7	80.80	2.5	70.0	2.5
T_8	23.3	7.3	19.8	8.0
T_9	9.2	6.0	4.2	9.2
T_{10}	75.0	2.5	55.0	4.2
T_{11}	34.2	7.0	16.5	8.0
T_{12}	9.2	9.0	3.3	9.3
LSD (P = 05)	9.50	1.0	11.6	1.0
CV (%)	14.5	9.8	22.4	9.3

T_1 = Sole rice + 1 hand weeding + isazofos @ 0.7 kg a.i. ha^{-1} + benomyl @ 1.0 kg a.i. ha^{-1}; T_2 = Sole rice + 2 hand weeding + isazofos @ 0.75 kg a.i. ha^{-1} + benomyl @ 1.0 kg a.i. ha^{-1}; T_3 = Sole rice + 2 hand weeding + isazofas @ 0.75 kg a.i. ha^{-1} + benomyl @ 1.0 kg a.i. ha^{-1}. T_4 = Sole rice + no weeding + no isazofos + no benomyl applications; T_5 = Sole rice + 1 hand weeding + no isazofos + no benomyl application; T_6 = Sole rice + 2 hand weeding + no isazofos + no benomyl applications; T_7 = Rice followed by cowpea + no weeding + isazofos @ 0.75 kg a.i. ha^{-1}; T_8 = Rive followed by cowpea + 1 hand weeding + isazofos @ 0.75 kg a.i. ha^{-1} + benomyl @ 1.0 kg a.i. ha^{-1}; T_9 = Rice followed by cowpea + 2 hand weeding + isazofos at 0.75 kg a.i. ha^{-1}; T_{10} = Rice followed by cowpea + 1 hand weeding + no isazofos and benomyl applications; T_{11} = Rice followed by cowpea + 1 hand weeding + no isazofos and benomyl applications; T_{12} = Rice followed by cowpea + 2 hand weeding + no isazofos and benomyl applications.

Table 2. Occurrence of AfRGM and incidence of diseases and their effect on yield and yield related parameters of FARO 37 in an IPM trial of Edozhigi 2001 & 2002.

Treatments	AfRGM Occurrence (%) % Invested Tiller/Hill (DAT)		% Parasitism		Disease Score (0 - 9) SES Scale				
	45	60	45	60	Brown Leaf Spot	Leaf Blast	Plant Height (cm)	No of Effective Tillers/Hill	Yield (t.ha^{-1})
T_1	10.5	11.8	20.7	32.0	5	2	76.3	3.3	0.46
T_2	10.3	12.7	16.4	27.4	6	1	80.8	5.8	1.8
T_3	10.4	11.7	24.2	38.7	5	2	79.1	6.3	1.5
T_4	10.4	12.6	31.1	25.8	5	3	74.3	2.3	1.4
T_5	10.4	12.6	25.5	23.7	5	3	75.9	4.9	0.6
T_6	10.9	10.6	21.0	24.7	6	5	74.3	5.9	1.8
T_7	10.4	10.8	15.7	26.5	5	2	73.0	3.8	1.4
T_8	9.3	11.8	23.9	25.7	6	2	74.9	6.2	1.8
T_9	9.4	11.0	14.9	24.0	5	2	80.7	5.8	1.7
T_{10}	10.9	10.7	16.3	25.8	5	3	75.2	3.6	1.7
T_{11}	10.2	13.7	22.2	26.1	6	3	77.5	5.2	1.0
T_{12}	10.7	14.1	12.4	23.7	6	3	77.4	7.7	1.9
LSD (P = 05)						0.7	8.47	1.4	0.6
(P = 01)						0.95	7.26	1.9	0.7
CV (%)	29.2	27.9	46.52	23.47	9.6	19.5	28.6	18.2	22.7

T_1 = Sole rice + 1 hand weeding + isazofos @ 0.7 kg a.i. ha^{-1} + benomyl @ 1.0 kg a.i. ha^{-1}; T_2 = Sole rice + 2 hand weeding + isazofos @ 0.75 kg a.i. ha^{-1} + benomyl @ 1.0 kg a.i. ha^{-1}; T_3 = Sole rice + 2 hand weeding + isazofas @ 0.75 kg a.i. ha^{-1} + benomyl @ 1.0 kg a.i. ha^{-1}; T_4 = Sole rice + no weeding + no isazofos + no benomyl applications; T_5 = Sole rice + 1 hand weeding + no isazofos + no benomyl application; T_6 = Sole rice + 2 hand weeding + no isazofos + no benomyl applications; T_7 = Rice followed by cowpea + no weeding + isazofos @ 0.75 kg a.i. ha^{-1}; T_8 = Rive followed by cowpea + 1 hand weeding + isazofos @ 0.75 kg a.i. ha^{-1} + benomyl @ 1.0 kg a.i. ha^{-1}; T_9 = Rice followed by cowpea + 2 hand weeding + isazofos at 0.75 kg a.i. ha^{-1}; T_{10} = Rice followed by cowpea + 1 hand weeding + no isazofos and benomyl applications; T_{11} = Rice followed by cowpea + 1 hand weeding + no isazofos and benomyl applications; T_{12} = Rice followed by cowpea + 2 hand weeding + no isazofos and benomyl applications.

Table 3. Occurrence of AfRGM and Incidence of diseases and their effect on yield and yield related parameters of FARO 37 in an IPM trial at Sachi 2001 and 2002.

Treatments	AfRGM Occurrence (%) % Invested Tiller/Hill (DAT)		% Parasitism		Disease Score (0 - 9) SES Scale				
	45	60	45	60	Brown Leaf Spot	Leaf Blast	Plant Height (cm)	No of Effective Tillers/Hill	Yield (t.ha^{-1})
T_1	8.8	11.2	40.9	33.4	5.5	2	67.8	5.3	0.35
T_2	8.2	17.1	31.6	30.9	5.8	2	72.0	6.8	1.35
T_3	9.1	15.4	30.8	28.3	5.5	2	71.9	7.7	1.66
T_4	11.4	21.5	23.8	22.2	5.5	4	69.4	5.7	0.44
T_5	13.9	21.7	31.9	27.5	5.5	4	72.0	7.1	1.25
T_6	14.8	23.1	24.4	28.0	5.8	3.0	70.4	7.4	1.6
T_7	7.5	10.3	26.5	27.3	5.5	2.5	68.5	4.4	0.35
T_8	10.8	16.6	38.0	25.6	5.5	2.5	73.4	7.6	1.68
T_9	11.6	15.7	34.9	28.3	6.0	2.5	73.4	8.27	1.86
T_{10}	13.0	20.1	27.5	27.1	5.0	3.5	67.1	5.4	0.4
T_{11}	11.5	20.1	29.3	26.2	5.5	3.0	70.0	6.9	1.4
T_{12}	13.8	24.0	27.4	28.1	5.0	3.5	75.5	7.7	1.5
LSD (P = 05)		9.82	Ns	Ns	0.95	0.6	NS	1.43	0.42
(P = 01)		13.34	Ns	Ns		0.8	NS	1.95	0.60
CV (%)	26.9	25.7	33.2	24.6	8.0	15.9	5.8	23.8	23.7

T_1 = Sole rice + 1 hand weeding + isazofos @ 0.7 kg a.i. ha^{-1} + benomyl @ 1.0 kg a.i. ha^{-1}; T_2 = Sole rice + 2 hand weeding + isazofos @ 0.75 kg a.i. ha^{-1} + benomyl @ 1.0 kg a.i. ha^{-1}; T_3 = Sole rice + 2 hand weeding + isazofas @ 0.75 kg a.i. ha^{-1} + benomyl @ 1.0 kg a.i. ha^{-1}; T_4 = Sole rice + no weeding + no isazofos + no benomyl applications; T_5 = Sole rice + 1 hand weeding + no isazofos + no benomyl application; T_6 = Sole rice + 2 hand weeding + no isazofos + no benomyl applications; T_7 = Rice followed by cowpea + no weeding + isazofos @ 0.75 kg a.i. ha^{-1}; T_8 = Rive followed by cowpea + 1 hand weeding + isazofos @ 0.75 kg a.i. ha^{-1} + benomyl @ 1.0 kg a.i. ha^{-1}; T_9 = Rice followed by cowpea + 2 hand weeding + isazofos at 0.75 kg a.i. ha^{-1}; T_{10} = Rice followed by cowpea + 1 hand weeding + no isazofos and benomyl applications; T_{11} = Rice followed by cowpea + 1 hand weeding + no isazofos and benomyl applications; T_{12} = Rice followed by cowpea + 2 hand weeding + no isazofos and benomyl applications.

Table 4. Influence of cropping sequence on the occurrence of plant parasitic nematodes on FARO 37 at Edozhigi 2002.

Treatment	SACHI				EDOZHIGI			
	Grain yield ($Kg \cdot ha^{-1}$)	Production cost (ha^{-1})	Gross Benefit* (ha^{-1})	Net Benefit (ha^{-1})	Grain Yield ($Kg \cdot ha^{-1}$)	Production Cost (ha^{-1})	Gross Benefit* (ha^{-1})	Net Benefit (ha^{-1})
T_1	-	-	-	-	-	-	-	-
T_2	-	-	-	-	-	-	-	-
T_3	-	-	-	-	-	-	-	-
T_4	-	-	-	-	-	-	-	-
T_5	-	-	-	-	-	-	-	-
T_6	-	-	-	-	-	-	-	-
T_7	430	150	322.5	172.5	440	155	350	195.0
T_8	540	150	405.0	0	440	155	330	175.0
T_9	590	150	442.5	255292.5	560	155	420	265.0
T_{10}	740	150	555.0	405.0	440	155	330	175.0
T_{11}	670	150	502.5	352.5	470	155	352.5	195.5
T_{12}	770	150	577.5	427.5	560	155	420	265.0

*1 kg of cowpea sells at US$1.67; $T_1 - T_6$ = Left fallow after rice crop harvest of 2001; T_7 = Rice followed by cowpea + no weeding + isazofos @ 0.75 kg a.i. ha^{-1}; T_8 = Rive followed by cowpea + 1 hand weeding + isazofos @ 0.75 kg a.i. ha^{-1} + benomyl @ 1.0 kg a.i. ha^{-1}; T_9 = Rice followed by cowpea + 2 hand weeding + isazofos at 0.75 kg a.i. ha^{-1}; T_{10} = Rice followed by cowpea + 1 hand weeding + no isazofos and benomyl applications; T_{11} = Rice followed by cowpea + 1 hand weeding + no isazofos and benomyl applications; T_{12} = Rice followed by cowpea + 2 hand weeding + no isazofos and benomyl applications.

Table 5. Influence of cropping sequence on the occurrence of plant parasitic nematodes on FARO 37 at Edozhigi 2002.

Treatment	P_i*	Pf*	R1 (Pf/Pi)
T_1	(26.67) 96.33	(71.03) 337.67	(2.67) 3.51
T_2	(49.67) 151.33	(67.67) 259.0	(1.36) 1.71
T_3	(31.0) 102.67	(90.67)367.0	(2.92) 3.57
T_4	(22.0) 66.67	(70.67) 282.67	(3.21) 4.24
T_5	(22.33) 109.33	(72.33) 290.33	(3.24) 2.66
T_6	(34.0) 109.0	(87.0) 341.67	(2.56) 3.13
T_7	(47.0) 178.33	(90.67) 376.0	(1.93) 2.11
T_8	(57.0) 238.67	(110.0) 370.0	(1.93) 1.55
T_9	(821.33) 396.33	(196.33) 247.0	(2.38) 6.30
T_{10}	(87.0) 715.33	(150.33) 686.33	(1.73) 0.06
T_{11}	(136.0 770.0	(255.33) 1218.0	(1.88) 1.58
T_{12}	(195.67) 1322.67	(272.33) 1780.0	(1.39) 1.35

*Figures in parenthesis represent relative density of nematodes while those outside represent actual counts. T_1 = Sole rice + 1 hand weeding + isazofos @ 0.7 kg a.i. ha^{-1} + benomyl @ 1.0 kg a.i. ha^{-1}; T_2 = Sole rice + 2 hand weeding + isazofos @ 0.75 kg a.i. ha^{-1} + benomyl @ 1.0 kg a.i. ha^{-1}; T_3 = Sole rice + 2 hand weeding + isazofas @ 0.75 kg a.i. ha^{-1} + benomyl @ 1.0 kg a.i. ha^{-1}; T_4 = Sole rice + no weeding + no isazofos + no benomyl applications; T_5 = Sole rice + 1 hand weeding + no isazofos + no benomyl application; T_6 = Sole rice + 2 hand weeding + no isazofos + no benomyl applications; T_7 = Rice followed by cowpea + no weeding + isazofos @ 0.7 5kg a.i. ha^{-1}; T_8 = Rive followed by cowpea + 1 hand weeding + isazofos @ 0.75 kg a.i. ha^{-1} + benomyl @ 1.0 kg a.i. ha^{-1}; T_9 = Rice followed by cowpea + 2 hand weeding + isazofos at 0.75 kg a.i. ha^{-1}; T_{10} = Rice followed by cowpea + 1 hand weeding + no isazofos and benomyl applications; T_{11} = Rice followed by cowpea + 1 hand weeding + no isazofos and benomyl applications; T_{12} = Rice followed by cowpea + 2 hand weeding + no isazofos and benomyl applications.

At Edozhigi, the rice nematode Hirschmanniella sp. was not encountered but other nematode species notably Paratylenchus and Hoplolaimus were present. At Sachi location Hirschmanniella sp. was present amongst others. The trend of nematode population at the two locations could be due to the type of weeds available at each location. Weeds like Galinsoga parviflora, Crotalaris retusa, Desmodium sp., Enphorbia heterophylla, Portulace olearacea and Richardia bruasilensis are known to adversely influence nematode abundance in rice fields [4]. However, none of these weed species was among weeds recorded at the two experimental sites. Further studies are required to ascertain the role of weeds in AfRGM infestation intensity and nematode abundance as well as the possibility of weed species prevalent at each location which could influence nematode abundance as was earlier reported by Bendixen et al. [4].

ACKNOWLEDGEMENTS

The authors are grateful to Africa Rice Centre (WARDA) for approved funds for the study and to the Director, NCRI for smooth release of the said funds and logistic support and permission to publish the work.

REFERENCES

1. Ukwungwu, M.N. and Joshi, R.C. (1992) Evaluation of Asian gall midge resistant varieties for the African rice gall midge. International Rice Research Newsletter, 17, 12.
2. Ukwungwu, M.N. (1990) Efficacy of granular isazofos against rice gall midge in Nigeria. Crop protection, 9, 252-254. doi:10.1016/0261-2194(90)90100-L
3. Hoque, N.Z. (1987) Upland rice development in four west African Countries (Nigeria, Ghana, Liberia and the Gambia). International Rice Commission Newsletter, 36, 32.
4. Bendixen, L.E., Reynold, D.A. and Riedel, R.M. (1979) Annotated bibliography of weeds as reservoirs for organisms affecting different crops, 1. Nematodes. Ohio Agricultural Research Development Centre, Wooster.

9

CHAPTER

SYNTHETIC AND NATURAL INSECTICIDES FOR THE MANAGEMENT OF INSECT PEST CONTROL

Jayakrishnan Saimandir, Madhuban Gopal

Keywords: Thiacloprid; Indoxacarb; Methyl Parathion; Persistence; Integrated Pest Management; Dissipation

INTRODUCTION

Eggplant is one of the important vegetables in India which contributes 9% of total vegetables in India [1]. It is a good source of nutrients, minerals, antioxidants, vitamins, dietary fiber and body building factors and proteins [2,3]. The biggest threat to eggplant cultivation is the wide spread infestation of ESFB in India and it is difficult to control since the insect larvae harbor inside the plant tissue. The larval form causes maximum damage to the edible fruit [4]. The adult moth of ESFB lays eggs on the leaf and hatched larvae move into the fruits. It was hypothesized that contact insecticides can control and prevent hatched larvae getting entry into the fruits and shoots, whereas systemic insecticides can be effective in controlling early developmental stages of borer when it is inside the plant tissue. Therefore, there is a need to evaluate systemic as well as contact insecticide to control the pest. Thiacloprid (Bay YRC 2894) is used for foliar application with systemic properties and broad-spectrum efficacy against sucking and biting insects [5-11]. Thiacloprid selectively acts on the insect nervous system as an agonist of the nicotinic acetylcholine receptor (nAChR) [12]. Indoxacarb is a broad-

Citation: J. Saimandir and M. Gopal, " Synthetic and Natural Insecticides for the Management of Insect Pest Control, Vol. 3 No. 2, 2012, pp. 214-227. doi: 10.4236/ajps.2012.32026.

spectrum, contact, low risk pesticide with low mammalian toxicity and a benign profile for avian and aquatic toxicity [13-18]. It affects insects from direct exposure to spray droplets and through ingestion of treated foliage or fruit. Once absorbed, it kills by binding to a site on the sodium channel and blocking the flow of sodium ions and results in impaired nerve function, feeding cessation, paralysis, and death. Once indoxacarb is absorbed or ingested, feeding cessation occurs almost immediately even though it may take several days for the insect to die [19]. Along with these two, a commonly used OP insecticide, methyl parathion was also evaluated and its efficacy and residue behavior were compared with that of thiacloprid and indoxacarb. Methyl parathion is effective for controlling chewing and sucking insects in a wide range of crops, including cereals, fruit, vines, vegetables, ornamentals, cotton and field crops [20]. These chemicals have not been reported for managing ESFB of eggplant and there is no previous published work available on the residue behavior and persistence of these insecticides on eggplant fruits, we decided to study this aspect based on previous observations made about of efficacy of these insecticides.

The insecticides were applied systematically with a 14 day interval. To minimize the use of pesticides and in order to maintain a balanced natural enemy population, alternative approaches like cultural control, mechanical control and botanical insecticides were also used. Botanical pesticides like NSKE, Nimbo Bas™ (azadirachtin based formulation) and two Bt based bio-insecticides, namely PUSA Bt and Biolep™ were evaluated. The results of the study may be incorporated in IPM package for eggplant and may be able to reduce the cost of cultivation by reducing the quantity of pesticides which would be beneficial for both farmers as well as consumers.

MATERIALS AND METHODS

HPLC Analyses

A Lachrom High Pressure Liquid Chromatography (HPLC) instrument equipped with quaternary pump, UV detector connected with auto-injector and a computer (model Hitachi) was used for analysis. The stationary phase consisted of Lichrosphere RP-18 packed in a stainless steel column (250 mm × 5 mm i.d.). Chromatogram was recorded in a Windows' NT based program. Operational parameters of HPLC for indoxacarb were given below. Mobile phase—aetonitrile: water (9:1); flow rate—0.5 mL·min^{-1}; λ_{max}-213 nm; injection volume—10 µL; column and solvent temperature—ambient. Operational parameters for thiacloprid were, Mobile phase—acetonitrile: water (6:4 vv^{-1}); flow rate—0.5 mL min^{-1}; λ_{max}-213 nm; injection volume—10 µL; column and solvent temperature—ambient.

Gas Chromatographic Analyses

A Hewlett Packard 5890, Series II gas chromatograph was used for the analysis equipped with HP-1, megabore column (methyl silicon gum, 10 m × 0.53 mm; 2.65 µm film thickness) and ^{63}Ni ECD detector. Oven, injection and detection temperatures were 250°C, 250°C and 300°C, respectively. Carrier gas (Nitrogen) flow was 30 mL·min^{-1}.

Insecticides, Chemicals and Supplies

Thiacloprid, Calypso 240 SC, Bayer Crop Science India Limited, India; Indoxacarb, Avaunt 14.5 SC, E. I. DuPont India Limited, New Delhi, India; Methyl parathion, Metacid 50 EC, Bayer India Limited, India; Biolep™, commercially available B. thuriegensis var. Kurstaki, Strain Z-52, Serotype H-3a, 3b, water dispersible powder formulation, Biotech International Limited, New Delhi, India; PUSA Bt, a WDP formulation prepared by own methodology in our laboratory and Department of Entomology, IARI, New Delhi [21]; NSKE, water extract of Neem, Azadirachta indica, (Family Meliaceae) seed kernel powder at required concentration; Nimbo Bas 1500™, 0.15 EC, an Azadirachtin based formulation. All solvents and chemicals used were from Merck Limited, India. Adsorbents used in the experiments were obtained from Sigma Chemicals Inc., USA.

Field Experiment

The studies on the dissipation of synthetic insecticides namely, indoxacarb, thiacloprid and methyl parathion on eggplant were carried out through field experiments during two consecutive seasons in 2004 and 2005. Insecticides of biological origin namely Biolep, PUSA Bt, NSKE and Nimbo Bas™were considered only for efficacy evaluation during the same period. Eggplant seedlings, variety Hybrid-9, were transplanted on raised beds (plot size 4.2×3.75 m^2) during 2004 and 2005 at the experimental farm of Indian Agricultural Research Institute, New Delhi, India. The plant-to-plant and row-to-row spacing was 60 and 75 cm respectively, for both the years. Thiacloprid was applied at the rate of 30 and 60 g active ingredient per hectare (ai·ha^{-1}) for both seasons. Indoxacarb was applied at the rate of 75 and 150 g·ai·ha^{-1}. Methyl parathion was applied at the rate of 100 and 200 g·ai·ha^{-1}. For all synthetic pesticides a 4× dose were applied for phytotoxicity study. Bio pesticides like Biolep™ and PUSA Bt were applied at two concentrations i.e., 1 and 2 kg·ha^{-1} for both the seasons. NSKE was applied at 5% and 10% for both the seasons. Nimbo Bas™ was applied at two concentrations i.e., 1 and 2 L·ha^{-1} for the two seasons. The insecticides were sprayed on the eggplant crop just after some fruits had set in. Fourteen days later, a second spray was given. The fluid rate was 500 L·ha^{-1} for both the sprays. All the treatments were in Randomized Block Design (RBD) and taken in three replicates including untreated control. Each insecticide was applied in separate eggplant plots and in each plot, four random plants were tagged and periodic observations were made to estimate percent fruit damage throughout the cropping season. The parameter observed was the hole made by ESFB larvae on the fruit. The observation included the damage in terms of number of fruits and weight of fruits.

Field Sample Collection

For determining dissipation of thiacloprid, indoxacarb and methyl parathion from the eggplant fruits and to assess dietary intake exposure and consumer risk, the a representative fruit samples (500 g) were collected on 0 d (1 h after insecticide application), 1, 3, 5, 7, 10, 15, 20 and 25 d from both the trials. The samples were stored in freezer and processed on the same day. The fruits were cut into small pieces, mixed and a sub sample of 50 g was taken for further analysis. Based on the results of the recovery experiments, the optimum conditions suited were adopted for the analysis of field samples as per the method of Jayakrishnan [22]. The efficiency of extraction, cleanup and determinative steps for synthetic insecticides were checked by recovery experiments after fortifying the eggplant fruits.

Recovery Experiment for Indoxacarb

Untreated eggplant fruits (50 g) were spiked separately with 5, 25 and 50 μg of indoxacarb and kept for 24 h. Fortified samples were homogenized, extracted with acetone, filtered, rotary evaporated and the extracts were subjected for partitioning with acetonitrile (10 mL, 3 times), so that the pesticide moved into it from aqueous remain. Sodium chloride (3 g) was added to avoid emulsion formation. Acetonitrile layer was collected and rotary evaporated and 20 mL of distilled water was added. This aqueous layer was liquid-liquid partitioned with hexane (25 mL, 3 times). The hexane layer was removed completely and residue was dissolved in a small quantity of ethyl acetate and cleaned up by column chromatography using Florisil (10 g) and elution with 75 mL of hexane: ethyl acetate mixture (9:1 vv^{-1}). The cleaned extracts obtained after the above chromatographic separations were quantified by reverse phase HPLC after taking it in acetonitrile.

Recovery Experiment for Thiacloprid

Untreated eggplant fruits (50 g) were spiked separately with 5, 25 and 50 μg of thiacloprid and kept for 24 h before homogenization. Next day fortified samples were homogenized, extracted with 100 mL acetone, filtered through Whatman No. 1 filter paper and rotary evaporated at 40°C to remove acetone. The extracts were first liquid-liquid partitioned with hexane (10 mL, 3 times) in a separatory funnel to remove interfering co-extractives and pigments. The liquid-liquid partitioning was redone using dichloromethane (25 mL, 3 times), which was rotary evaporated and residues were dissolved in 5 mL of ethyl acetate: cyclohexane solution (1:1 vv^{-1}). This organic extract was cleaned up by column chromatography using Florisil (10 g), sandwiched between two layers of sodium sulfate (2 g) in a glass column with 2 cm diameter. It was first eluted with 10 mL of ethyl acetate: cyclohexane solution (1:1 vv^{-1}), followed by 50 mL of acetonitrile, which was collected and concentrated. The cleaned extract thus obtained after the above chromatographic separations were quantified by reverse phase HPLC.

Recovery Experiment for Methyl Parathion

Untreated eggplant fruits (chopped, 50 g) were spiked separately to get 0.1, 0.5 and 1.0 $mg{\cdot}kg^{-1}$of methyl parathion on the fruit. The fortified samples were kept for 24 h. Extraction was done by blending in a mixer-blender for 2 minutes with 100 mL acetone. The contents were transferred into 250 mL conical flask. The jar was rinsed two times with acetone (2 × 40 mL) and transferred into the flasks. The flasks were agitated on a mechanical shaker for 30 minutes. The supernatant liquid was then filtered under suction though Whatman No. 1 filter paper using Buchner funnel. The pooled extracts were concentrated using rotary evaporator to a minimum volume by distilling out the acetone. The remainder was transferred to 250 mL separatory funnel and 100 mL saturated aqueous sodium chloride solution was added to the contents. Then 75 mL of hexane was added and the contents were shaken vigorously for 2 minutes with intermittent release of pressure. After partitioning the contents were allowed to stand for 5 minutes. The upper layer i.e. hexane was collected in a conical flask after passing through activated anhydrous sodium sulphate. The left over aqueous layer was again partitioned twice with hexane (2 × 50 mL) following the same procedure. The hexane layers were pooled and concentrated to dryness using rotary vacuum evaporator. The residues were dissolved in 10 mL hexane-acetone mixture (9:1). Cleaned glass columns (60 cm long × 2 cm i.d.) were plugged at the bottom with cotton followed by addition of 5 g anhydrous sodium sulphate. Ten gram activated alumina (neutral) was poured over the layer of anhydrous sodium sulphate. Columns were tapped gently to have even packing. This packing was topped by small amount of cotton to avoid the disturbance of the alumina layer during addition of the solvent in the column. These columns were pre-washed with 25 mL solvent mixture (hexane-acetone, 9:1, vv^{-1}) to remove the impurities and the leachate was discarded. The concentrated extracts obtained from partitioning processes were poured quantitatively in the packed columns. The columns were eluted with 100 mL hexane: acetone mixture (9:1, vv^{-1}) which was collected in 250 mL conical flasks. The rate of elution was 3 mL $minute^{-1}$. The eluate thus collected was evaporated to dryness at 40°C in a round bottom flask with the help of a rotary evaporator. The remaining residue in the flask was dissolved in 5 mL hexane-acetone (9:1) mixture and transferred to 25 mL volumetric flasks and final volume was made up to the mark by using hexane-acetone (9:1) mixture for analysis by GLC.

Weather Data

During the period of the experiment in first year (2004) the minimum and maximum temperature varied between 13.3°C and 28.1°C respectively and mean relative humidity ranged from 58% - 70%. No rainfall was reported during the experiment. During second year (2005) the minimum and maximum temperature varied between 14.2°C and 29.3°C, respectively and mean relative humidity ranged from 58% - 68%. There was no rainfall reported during the experiment.

Statistical Analyses

All the treatments were in Randomized Block Design (RBD) and taken in three replicates including untreated control. The data was analyzed using Statistica (Stat Soft) and Minitab 14.

RESULTS AND DISCUSSION

Dissipation of Indoxacarb

The method involving Florisil gave highest average recovery (88%) of the pesticide when compared to other methods where average recoveries varied from 59% to 82% (**Table 1**). Highest recovery by Florisil (synthetic magnesium silicate) can be attributed to the fact that it is less reactive than alumina or silicagel in retaining the insecticide where as low recoveries on using charcoal were due to its high adsorptive power. The mean initial deposits of indoxacarb on eggplant fruits were 2.60 to 3.64 mg·kg^{-1} and 2.63 to 3.68 mg·kg^{-1} from 75 and 150 g·ai·ha^{-1} treatments from first and second year field trials, respectively (Tables 2 and 3). The reported concentration declined with time and reached non-detectable (<0.02 mg·kg^{-1}) after 15 - 20 days (**Figure 1**). Residues dissipated with half-life of 3.0 - 3.8 d from both first and second year sprays, respectively. There was no significant difference in dissipation pattern between the two rates of applications suggesting that the dissipation of indoxacarb was independent of the dose of insecticide. The residue concentration data fitted to a first order rate kinetics [23] and dissipation rate constants ranged from 0.090 and 0.133 for 75 and 150 g·ai·ha^{-1} treatments, respectively (**Table 8**).

The Theoretical Maximum Daily Intake (TMDI) for first year from maximum residues observed for first day for both the treatments was found to be 0.446 to 0.643 mg·person^{-1}·day^{-1} which is near the Maximum Permissible Intake (MPI) of 0.6 mg·person^{-1}·day^{-1} [24,25]. The TMDI was calculated considering recommended consumption of vegetables as 200 g for leafy or fruit vegetables other than roots and tubers [26,27] and MPI was calculated on the basis of ADI value of indoxacarb 0.01 mg·kg^{-1} body weight [28] and the average body weight of an average person was taken as 60 kg. Eggplant fruits are harvested frequently but first picking is intentionally done at least one d after the application of an insecticide. On the first day TMDI was more than MPI. The TMDI from 3 d residues was found to be less than the MPI (0.6 mg·person^{-1}·day^{-1}) calculated with the data of 3 d residues. This encompasses more margin of safety (MOS). In view of this a minimum of 3 d waiting period for harvest of fruits after insecticide application and processing is recommended to achieve maximum safety and minimum risk to consumers. Taking 0.5 mg·kg^{-1} MRL of indoxacarb on vegetables [28,29] for eggplant, a practical waiting period can be proposed, which can be practiced by the farmers.

Dissipation of Thiacloprid

The method involving Florisil gave highest average recovery (89%) of the pesticide when compared to other adsorbents, where average recoveries varied from 61% to 83% (**Table 1**). Neutral alumina and silicagel also gave satisfactory percent recoveries i.e., 83 and 80, respectively. The mean initial deposits of thiacloprid on eggplant fruits

Table 1. Recovery of indoxacarb, thiacloprid and methyl parathion from eggplant fruits.

Adsorbent Used	Fortification Level (mg·kg^{-1})	Indoxacarb		Thiacloprid		Methyl Parathion	
		Mean Amount Recovered (mg·kg^{-1})	Average Recovery (%)	Mean Amount Recovered (mg·kg^{-1})	Average Recovery (%)	Mean Amount recovered (mg·kg^{-1})	Average Recovery (%)
Florisil	0.1	0.09		0.09		0.09	
	0.5	0.43	88	0.44	89	0.44	89
	1.0	0.89		0.90		0.9	
Alumina (N)*	0.1	0.08		0.08		0.09	
	0.5	0.42	82	0.43	82	0.42	86
	1.0	0.81		0.81		0.85	
Silica Gel	0.1	0.08		0.08		0.09	
	0.5	0.39	79	0.40	81	0.43	87
	1.0	0.80		0.82		0.84	
Magnesium Oxide	0.1	0.07		0.08		0.08	
	0.5	0.38	73	0.38	77	0.37	76
	1.0	0.72		0.75		0.75	
Florisil + Charcoal (2%)	0.1	0.07		0.07		0.07	
	0.5	0.34	68	0.35	70	0.35	71
	1.0	0.66		0.69		0.72	
Alumina (N) + Charcoal (2%)	0.1	0.06		0.07		0.06	
	0.5	0.32	63	0.34	69	0.32	62
	1.0	0.64		0.68		0.62	
Silica gel + Charcoal (2%)	0.1	0.06		0.07		0.07	
	0.5	0.30	61	0.34	68	0.34	67
	1.0	0.63		0.66		0.64	
Charcoal	0.1	0.06		0.06		0.06	
	0.5	0.29	59	0.32	61	0.31	62
	1.0	0.58		0.60		0.63	

*Alumina (neutral).

Table 2. Persistence of indoxacarb residues from eggplant fruits: first year.

Time after application (days)	Treatment level									
	75 g·ai·ha^{-1} Residues (mg·kg^{-1})					150 g·ai·ha^{-1} Residues (mg·kg^{-1})				
	R_1	R_2	R_3	Mean	Per cent loss	R_1	R_2	R_3	Mean	Per cent loss
0	2.436	2.589	2.771	2.599 (±0.168)*	-	3.886	3.585	3.447	3.639 (±0.224)	-
1	2.134	2.168	2.279	2.194 (±0.076)	15.60	3.143	2.965	3.125	3.078 (±0.098)	15.43
3	0.986	1.214	1.129	1.110 (±0.115)	57.30	2.074	1.943	1.982	2.000 (±0.067)	45.05
5	0.783	0.791	0.804	0.793 (±0.011)	69.50	1.236	1.364	1.287	1.296 (±0.064)	64.39
7	0.421	0.563	0.586	0.523 (±0.089)	79.86	0.765	0.731	0.755	0.750 (±0.017)	79.38
10	0.324	0.312	0.275	0.304 (±0.026)	88.32	0.388	0.342	0.347	0.359 (±0.025)	90.13
15	0.143	0.095	0.084	0.107 (±0.031)	95.87	0.139	0.128	0.136	0.134 (±0.006)	96.31
20	ND	ND	ND	-	-	0.033	0.042	0.048	0.041 (±0.008)	98.87
25	ND	ND	ND	-	-	ND	ND	ND	-	-

ND = Non detectable; * = Standard deviation.

Table 3. Persistence of indoxacarb residues from eggplant fruits: second year.

Time after application (days)	Treatment level									
	75 g·ai·ha^{-1} Residues (mg·kg^{-1})					150 g·ai·ha^{-1} Residues (mg·kg^{-1})				
	R_1	R_2	R_3	Mean	Per cent loss	R_1	R_2	R_3	Mean	Per cent loss
0	2.662	2.612	2.629	2.634 (±0.025)*	-	3.913	3.606	3.512	3.677 (±0.210)	-
1	2.234	2.185	2.279	2.233 (±0.047)	15.24	3.164	2.992	3.215	3.124 (±0.117)	15.05
3	0.975	1.345	1.214	1.178 (±0.188)	55.28	2.115	2.041	2.012	2.056 (±0.053)	44.08
5	0.763	0.806	0.833	0.801 (±0.035)	69.60	1.265	1.392	1.263	1.307 (±0.074)	64.46
7	0.431	0.587	0.613	0.544 (±0.098)	79.36	0.809	0.768	0.794	0.790 (±0.021)	78.51
10	0.362	0.349	0.304	0.338 (±0.030)	87.16	0.412	0.373	0.361	0.382 (±0.027)	89.61
15	0.156	0.087	0.107	0.117 (±0.036)	95.57	0.127	0.137	0.119	0.128 (±0.009)	96.53
20	ND	ND	ND	-	-	0.052	0.064	0.048	0.055 (±0.008)	98.51
25	ND	ND	ND	-	-	ND	ND	ND	-	-

ND = Non detectable; * = Standard deviation.

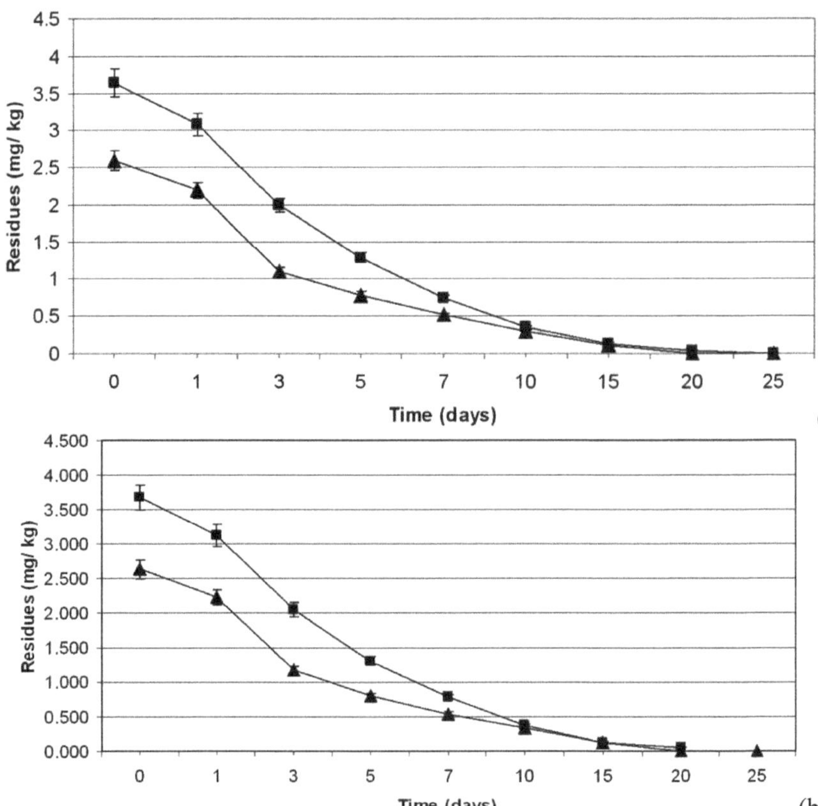

Figure 1. Persistence of indoxacarb residues from eggplant fruits: first year (a) and second year (b) at treatment levels 75 g·ai·ha^{-1} (■) and 150 g·ai·ha^{-1} (▲). Vertical bars represent standard deviation (n = 3).

were 3.39 to 5.40 mg·kg^{-1} and 3.40 to 5.39 mg·kg^{-1} from 30 and 60 g·ai·ha^{-1} treatments from first and second year field trails, respectively (Tables 4 and 5). The reported concentration declined with time but did not reach non-detectable (<0.01 mg·kg^{-1}) even after 20 d. (**Figure 2**). Residues dissipated with half-life of 11.1 and 11.6 d from both first and second year sprays, respectively (Table 8). There was no significant difference in dissipation pattern between the two rates of applications suggesting that the dissipation of thiacloprid was not dependent on the dose of insecticide applied. The residue concentration data fitted to a first order rate kinetics [30] and the dissipation rate constants ranged from 0.117 and 0.199 for 30 and 60 g·ai·ha^{-1} treatments for both years.

The TMDI from maximum residues observed for first day for both the treatments during first year trials was found to be 0.682 to 1.098 mg·person^{-1}·day^{-1} which is not lower than the MPI of 0.72 mg·person^{-1}·day^{-1}. The TMDI was calculated considering recommended consumption of vegetables as 200 g for leafy or fruit vegetables [26,27] and MPI was calculated on the basis of ADI value of thiacloprid 0.012 mg·kg^{-1} body weight [29] and the average body weight of an average person was taken as 60 kg. Eggplant fruits are normally harvested frequently but first picking is intentionally done one to three day after the spray of an insecticide. Considering this, 3 d residues were also examined. The TMDI even from 3 d residues for recommended dose i.e. 30 g·ai·ha^{-1} was not found less than the MPI value. There is a need for higher MOS when thiacloprid is applied on eggplant in the recommended dose. In view of this, 3 d waiting period for harvest of fruits after insecticide application cannot be suggested unless the fruits are decontaminated before consumption. Taking 0.02 mg·kg^{-1} MRL of thiacloprid on vegetables (onion) [31,32] a waiting period of more than 20 d is needed, even for smallest dose i.e., 30 g·ai·ha^{-1}, which is not practical for the farmers. This large waiting period may be attributed to the systemic nature of thiacloprid. Hence this insecticide cannot be included in good agricultural practice for growing eggplant. The MRL of 0.7 mg·kg^{-1} for thiacloprid under consideration by Codex eggplant has been questioned [33].

Table 4. Persistence of thiacloprid residues on eggplant fruits: first year.

Time after application (days)	Treatment level									
	30 g·ai·ha^{-1} Residues (mg·kg^{-1})					60 g·ai·ha^{-1} Residues (mg·kg^{-1})				
	R_1	R_2	R_3	Mean	Per cent loss	R_1	R_2	R_3	Mean	Per cent loss
0	3.356	3.398	3.412	3.389 (±0.029)*	-	5.366	5.488	5.352	5.402 (±0.075)	-
1	2.965	2.867	2.922	2.918 (±0.049)	13.90	4.845	4.861	4.731	4.812 (±0.071)	10.92
3	2.613	2.591	2.638	2.614 (±0.024)	22.87	4.383	4.422	4.337	4.381 (±0.043)	18.91
5	2.313	2.388	2.411	2.371 (±0.051)	30.05	3.862	3.853	3.813	3.843 (±0.026)	28.87
7	2.131	2.014	2.058	2.068 (±0.059)	38.99	3.323	3.543	3.286	3.384 (±0.139)	37.36
10	1.786	1.825	1.759	1.790 (±0.033)	47.18	2.785	2.763	2.745	2.764 (±0.020)	48.83
15	1.345	1.387	1.423	1.385 (±0.039)	59.13	2.135	1.956	1.984	2.025 (±0.096)	62.51
20	0.966	0.985	0.954	0.968 (±0.016)	71.43	1.256	1.362	1.315	1.311 (±0.053)	75.73

* = Standard deviation.

Table 5. Persistence of thiacloprid residues on eggplant fruits: second year.

Time after application (days)	Treatment level: 30 g·ai·ha^{-1} Residues (mg·kg^{-1})					60 g·ai·ha^{-1} Residues (mg·kg^{-1})				
	R_1	R_2	R_3	Mean	Per cent loss	R_1	R_2	R_3	Mean	Per cent loss
0	3.378	3.362	3.458	3.399 (±0.051)*	-	5.389	5.461	5.333	5.394 (±0.064)	-
1	2.959	2.838	2.949	2.915 (±0.067)	14.23	4.863	4.882	4.756	4.834 (±0.068)	10.39
3	2.638	2.568	2.659	2.622 (±0.048)	22.87	4.359	4.447	4.353	4.386 (±0.053)	18.68
5	2.333	2.364	2.429	2.375 (±0.049)	30.12	3.852	3.871	3.836	3.853 (±0.018)	28.57
7	2.147	2.089	2.106	2.114 (±0.030)	37.81	3.337	3.575	3.231	3.381 (±0.176)	37.32
10	1.752	1.838	1.777	1.789 (±0.044)	47.37	2.799	2.734	2.773	2.769 (±0.033)	48.67
15	1.361	1.353	1.482	1.399 (±0.072)	58.85	2.146	1.965	1.981	2.031 (±0.100)	62.35
20	0.929	0.937	0.968	0.945 (±0.021)	72.21	1.287	1.373	1.366	1.342 (±0.048)	75.12

* = Standard deviation.

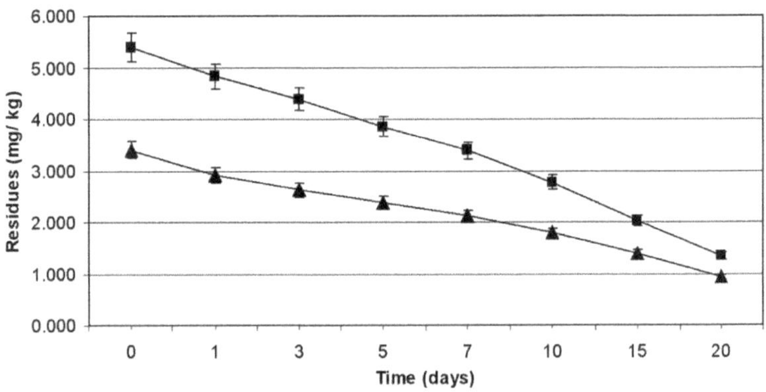

(a)

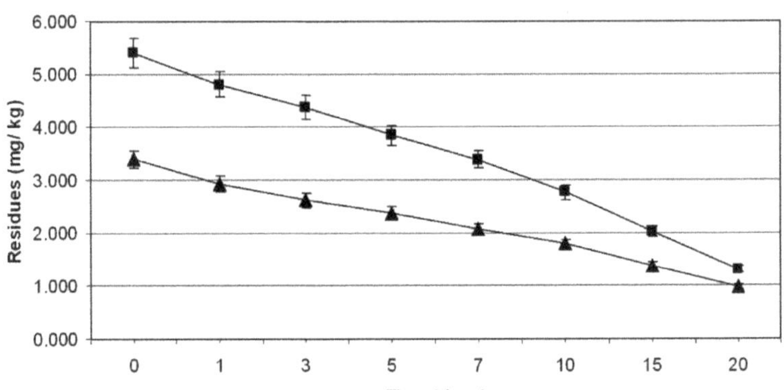

(b)

Figure 2. Persistence of thiacloprid residues from eggplant fruits: first year (a) and second year (b) at treatment levels 30 g·ai·ha^{-1} (■) and 60 g·ai·ha^{-1} (▲). Vertical bars represent standard deviation (n = 3).

Dissipation of Methyl Parathion

The method involving Florisil gave highest average recovery (89%) of the pesticide when compared to other methods where average recoveries varied from 62% to 87% (**Table 1**). Neutral alumina and silica gel also gave satisfactory per cent recoveries i.e., 86 and 87, respecttively. The initial residues on eggplant fruits from recommended dose declined from 3.60 to 3.12 mg·kg^{-1} in one day and 0.27 mg·kg^{-1} in twenty five days from lower rates of treatment i.e. 100 g·ai·ha^{-1} (**Table 6**). Similarly in the case of 200 g·ai·ha^{-1} treatment, the initial deposits declined from 4.02 to 0.34 mg·kg^{-1} in 25 days, respecttively (**Table** 7). The residue data exhibited similar pattern in second year also (**Figure 3**).

Table 6. Persistence of methyl parathion residues from eggplant fruits: first year.

Time after application (days)	Treatment level									
	30 g·ai·ha^{-1} Residues (mg·kg^{-1})					60 g·ai·ha^{-1} Residues (mg·kg^{-1})				
	R_1	R_2	R_3	Mean	Per cent loss	R_1	R_2	R_3	Mean	Per cent loss
0	3.436	3.567	3.786	3.596 (±0.177)*	-	3.938	3.966	4.169	4.024 (±0.126)	-
1	3.134	2.998	3.217	3.116 (±0.111)	13.34	3.434	3.492	3.517	3.481 (±0.043)	13.49
3	2.489	2.323	2.386	2.399 (±0.084)	33.28	3.175	3.048	2.994	3.072 (±0.093)	23.65
5	2.047	1.988	1.916	1.984 (±0.066)	44.84	2.448	2.568	2.516	2.511 (±0.060)	37.61
7	1.657	1.528	1.597	1.594 (±0.065)	55.67	1.957	1.928	2.097	1.994 (±0.090)	50.45
10	1.224	1.312	1.175	1.237 (±0.069)	65.60	1.554	1.612	1.575	1.580 (±0.029)	60.73
15	0.763	0.824	0.731	0.773 (±0.047)	78.51	0.963	1.024	0.931	0.973 (±0.047)	75.83
20	0.528	0.467	0.497	0.497 (±0.031)	86.17	0.588	0.667	0.597	0.617 (±0.043)	84.66
25	0.278	0.251	0.283	0.271 (±0.017)	92.47	0.398	0.322	0.303	0.341 (±0.050)	91.53

* = Standard deviation.

Table 7. Persistence of methyl parathion residues from eggplant fruits: second year.

Time after application (days)	Treatment level									
	30 g·ai·ha^{-1} Residues (mg·kg^{-1})					60 g·ai·ha^{-1} Residues (mg·kg^{-1})				
	R_1	R_2	R_3	Mean	Per cent loss	R_1	R_2	R_3	Mean	Per cent loss
0	3.478	3.592	3.805	3.625 (±0.166)*		4.097	3.895	4.117	4.036 (±0.123)	
1	3.172	3.122	3.269	3.188 (±0.075)	12.06	3.524	3.512	3.564	3.533 (±0.027)	12.45
3	2.524	2.421	2.437	2.461 (±0.055)	32.12	3.194	3.115	3.057	3.122 (±0.069)	22.65
5	2.103	2.058	1.964	2.042 (±0.071)	43.68	2.487	2.591	2.566	2.548 (±0.054)	36.87
7	1.682	1.581	1.613	1.625 (±0.052)	55.16	2.132	1.975	2.042	2.050 (±0.079)	49.22
10	1.274	1.354	1.225	1.284 (±0.065)	64.57	1.596	1.667	1.613	1.625 (±0.037)	59.73
15	0.778	0.843	0.794	0.805 (±0.034)	77.79	0.992	1.102	0.983	1.026 (±0.066)	74.59
20	0.552	0.493	0.517	0.521 (±0.030)	85.64	0.603	0.678	0.623	0.635 (±0.039)	84.27
25	0.265	0.288	0.313	0.289 (±0.024)	92.04	0.422	0.368	0.337	0.376 (±0.043)	90.69

* = Standard deviation.

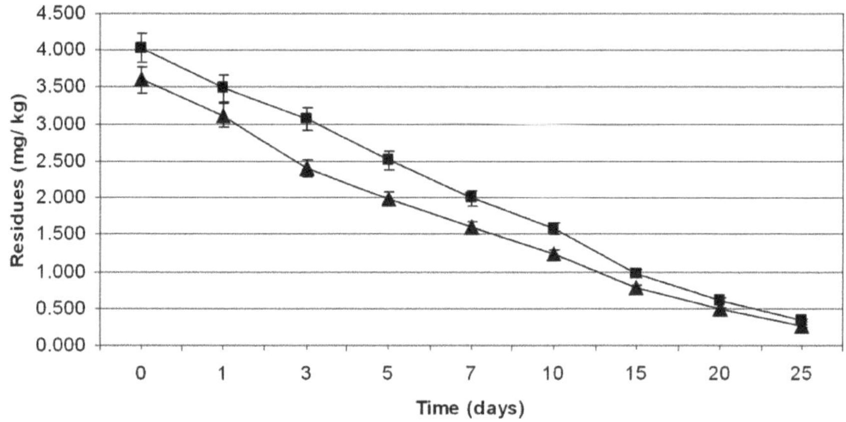

(a)

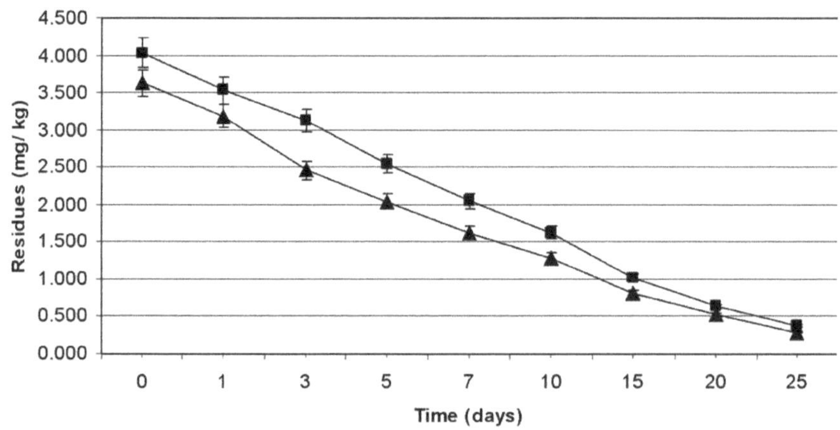

(b)

Figure 3. Persistence of methyl parathion residues from eggplant fruits: first year (a) and second year (b) at treatment levels 100 g·ai·ha^{-1} (■) and 200 g·ai·ha^{-1} (▲). Vertical bars represent standard deviation (n = 3).

Here also, there was no significant difference in dissipation pattern among the two treatments suggesting that the dissipation was independent of initial concentration. The residue data of methyl parathion when plotted for log residues versus time, fitted to a first order rate kinetics Y= C – mX, where "Y" is the logarithmic deposition of methyl parathion on eggplant fruits at time zero, "m" is the dissipation rate constant/slope of the regression line and "X" is the time in days. The dissipation rate constants were 0.123 and 0.140 for 100 and 200 g·ai·ha^{-1} treatments, respectively and the half life values ranged from 6.0 to 7.4 days (**Table 8**). The correlation coefficient values of –0.933 and –0.959 denotes that the data are significant. In the case of second year spray, the dissipation rate constants were 0.125 and 0.143, while the half life values were 6.02 and 7.39 days for 100 and 200 g·ai·ha^{-1} treatments, respectively and the values of correlation coefficient (–0.936 and –0.959) showed that the data is significant.

The MRL of methyl parathion on eggplant fruit is not available in India. However, FAO/WHO [34] has assigned 0.2 mg·kg^{-1} MRL on vegetable (Tomato). If this value is extended for eggplant, it would be seen (**Figure 3**; **Table 8**) that the residues were safe after 20 days from application of lower dose (100 g·ai·ha^{-1}). In the case of higher dose (200 g·ai·ha^{-1}), the residues reached at a safe level after 25 days. Therefore, there should be a minimum gap of 20 days after the last application of the insecticide to the harvest, which is impractical. This emphasizes the importance of decontamination of fruits and vegetables. This pesticide gives effective control of pest but has adverse residue profile. The prescribed acceptable daily intake (ADI) for methyl parathion is 0.02 mg·kg^{-1} body weight. Considering the average body weight of an Indian as 60 kg, the maximum permissible intake (MPI) was calculated to be 1.2 mg·person^{-1}·day^{-1}. The recommended vegetable in the balanced diet of an adult man is 200 g [26].

Residue data on eggplant showed that maximum residues were found on zero day samples and TMRC was calculated, based on 200 g eggplant fruits consumption and the residues encountered on zero day contaminated samples for all the treatments, by the following formula, TMRC = Residue level × Food factor × Daily food intake. The TMRC values ranged from 0.687 to 0.834 mg·person^{-1}·day^{-1} from eggplant. From the data, it was noticed that the calculated TMRC values were near the MPI value (1.2 mg·person^{-1}·day^{-1}) as this organophosphorus insecticide is applied in other crops as well. Methyl parathion (LD_{50} male rat 6 - 50 mg·kg^{-1}) treatment therefore is not safe for plant protection practices on vegetables, which are harvested frequently and for human consumption of eggplant fruits, unless suitable decontamination practices are adopted [35]. The MRL value of the methyl parathion on eggplant is not fixed, the data generated may help in examining the national MRL values. If the TMRC was calculated based on 3 day contaminated sample, the actual intake will be much less than that of 0 day contaminated samples. Even this was not enough to ensure safety, as the pesticide is far more toxic (lower LD_{50}) to mammals than indoxacarb (LD_{50} male rat 3619 mg·kg^{-1}) or thiacloprid (LD_{50} male rat 836 mg·kg^{-1}) and affects cholinesterase activity irreversibly [36].

Efficacy of Synthetic Insecticides

Indoxacarb. The per cent reduction on the weight and number basis of eggplant fruits showed almost similar effectiveness of indoxacarb treatments on ESFB population in both first year and second year (**Table 9**). Statistical analysis of the data showed that considering critical difference (CD) 5%, efficiency of double dose was not significantly different from that of the lower dose. The higher treatment rate i.e. 150 g·ai·ha^{-1} was marginally, but not statistically more effective and ESFB population could be controlled through out the experimental period even after 15 days from both the sprays. The data revealed better efficacy of indoxacarb against this insect even at lower dose and hence application of this insecticide at recommended dose could enhance the yield sig-

Table 8. Dissipation rates and half life values of indoxacarb, thiacloprid and methyl parathion on eggplant fruits.

Insecticide	Spray	Treatment level (g·ai·ha^{-1})	Correlation coefficient (r)	Half life (day)	Regression equation
Indoxacarb	First year	75	–0.822	3.9	Y = 1.71 – 0.090X
		150	–0.842	3.0	Y = 2.53 – 0.133X
	Second year	75	–0.827	3.8	Y = 1.75 – 0.093X
		150	–0.847	2.9	Y = 2.56 – 0.134X
Thiacloprid	First year	30	–0.986	11.1	Y = 3.07 – 0.117X
		60	–0.992	11.6	Y = 5.00 – 0.199X
	Second year	30	–0.985	11.1	Y = 3.08 – 0.118X
		60	–0.991	11.5	Y = 5.01 – 0.198X
Methyl parathion	First year	100	–0.933	6.0	Y =2.90 – 0.123X
		200	–0.959	7.4	Y = 3.41 – 0.140X
	Second year	100	–0.936	6.1	Y = 2.96 – 0.125X
		200	–0.959	7.3	Y = 3.47 – 0.143X

Table 9. Efficacy of synthetic and biopesticides against ESFB.

Insecticide	Treatment	Percent fruit damage			
		First year		Second year	
		Weight basis	Number basis	Weight basis	Number basis
Indoxacarb (g·ai·ha^{-1})	75	11.47	24.24	12.29	28.54
	150	8.53	16.27	7.20	19.47
	Control	28.68	41.35	31.54	47.28
	CD @ 5%	3.45	8.22	5.81	11.43
Thiacloprid (g·ai·ha^{-1})	30	14.23	34.48	12.74	32.76
	60	12.87	28.64	11.57	29.79
	120	11.56	25.73	9.88	26.45
	240	9.24	19.26	8.14	23.59
	Control	17.38	37.74	15.78	35.28
	CD @ 5%	1.24	2.88	1.39	2.67
Methyl parathion (g·ai·ha^{-1})	100	13.45	26.35	11.98	31.43
	200	9.89	17.46	10.34	16.26
	Control	31.78	36.46	27.46	34.35
	CD @ 5%	5.43	11.24	4.37	16.88
BiolepTM (kg·formulation·ha^{-1})	1.0	15.86	29.79	14.63	24.34
	2.0	12.74	21.44	12.88	20.62
	Control	22.82	34.68	24.89	36.73
	CD @ 5%	2.54	6.46	1.38	2.62
PUSA BtTM (kg·formulation·ha^{-1})	1.0	16.34	22.59	13.54	23.21
	2.0	12.94	18.82	11.68	19.56
	Control	19.62	29.78	25.44	28.43
	CD @ 5%	5.98	6.77	4.86	9.18
NSKE (Percent)	5	19.65	23.93	18.23	24.34
	10	16.43	18.42	12.88	20.61
	Control	22.87	26.73	24.90	36.72
	CD @ 5%	4.87	7.62	8.84	6.43
Nimbo BasTM (mL·formulation·ha^{-1})	100	17.65	21.23	18.24	22.34
	200	14.33	16.64	15.68	18.64
	Control	20.88	22.74	21.98	23.84
	CD @ 5%	1.56	2.34	2.271	2.12

nificantly i.e. 239.4 quintal·ha^{-1} as compared to 125.2 quintal·ha^{-1} from untreated plots, without any adverse effect on the quality of eggplant fruits. Indoxacarb at the experimented doses (75 and 150 g·ai·ha^{-1}) was effective in managing ESFB and did not pose residue problem in fruits. A waiting period of 3 d for harvest of the fruits after insecticide application and processing factor (PF) could ensure desirable margin of safety (MOS) and hence no health risk situations are encountered from the use of this insecticide on eggplant crop. Thiacloprid. The data on the per cent reduction on the weight and number basis of eggplant fruits showed almost similar action of thiacloprid treatments on ESFB population in both the years considering CD values, for both concentrations of application (**Table 9**). Statistical analysis showed that there was only marginal advantage in applying the higher dose, but fruit damage was statistically significant in control plots as compared to those collected from treated plots at all doses in both years. It is clear from observations that thiacloprid is not effective enough for controlling ESFB in eggplant crop, although it could reduce the Eggplant leaf hopper, Amrasca devastans infestation. Thiacloprid at the experimented doses (30 and 60 g·ai·ha^{-1}) was not effective in managing ESFB and was not safe for human consumption even after a waiting period of 3 days. A waiting period of 3 d for harvest of the fruits after insecticide application and PF could not ensure sufficient margin of safety (MOS). Health risk may be encountered from the use of this insecticide on eggplant crop. The study reiterates that new pesticides must be evaluated for safety, contrary to general belief that newer pesticides are safer to mammals and beneficial insects because they are effective at low dosages. No phytotoxicity was observed for 4× dose, i.e. 120 g·ai·ha^{-1} of thiacloprid in field. The yield data for treated plots were only marginally better i.e. 134.5 from treated plots to 125.2 quintal·ha^{-1} from untreated plots.

Methyl parathion. The data on the efficacy of methyl parathion at two treatment levels were evaluated based on the damaged fruit count by per cent weight and by per cent number. Although methyl parathion is not recommended in vegetables in India, farmers use this pesticide widely against ESFB. The data on the per cent reduction on the weight and number basis of eggplant fruits showed similar effect of methyl parathion treatments on ESFB population (**Table 9**). The statistical analysis showed that 200 g·ai·ha^{-1} was more effective and the ESFB population was controlled through out the experimental period. The higher dose was not statistically more effective than the lower dose (100 g·ai·ha^{-1}). The data revealed the efficacy of methyl parathion against ESFB even at lower dose, but since the residue profile study showed that methyl parathion residues are persistent on eggplant fruits above MRL (**Figure 3**; **Table 8**), there is health risk beyond doubt and the insecticide cannot be included in good agricultural practice for growing this crop.

Efficacy of Bio/Botanical Insecticides

Biolep. The data on the per cent reduction on the weight and number basis of eggplant fruits showed similar effect of Biolep treatments on ESFB population (**Table 9**). Analysis of data showed that both the doses were not statistically effective in managing ESFB.

PUSA Bt. The potency of the formulation was 50000 I.U. mg^{-1} minimum on a surrogate insect Helicoverpa armigera. The statistical analysis of data on the per cent reduction on the weight and number basis of eggplant fruits showed PUSA Bt was significantly effective on ESFB population (**Table 9**). Both the doses were effective in controlling the pest and it was found that PUSA Bt gave better result in controlling the key pest when compared to Biolep. The higher treatment rate (2 kg formulation·ha^{-1}) was slightly more effective when compared to 1 kg formulation·ha^{-1} dose.

NSKE. The data on the per cent reduction on the weight and number basis of eggplant fruits showed that both doses of NSKE were significantly effective on ESFB

(**Table 9**). The higher treatment rate gave better control than 5% dose. Nimbo Bas. The data on the per cent reduction on the weight and number basis of eggplant fruits showed no significant effect for both the doses of Nimbo Bas on ESFB (**Table 9**).

Correlation of Seven Treatments and Control

In field, all seven treatments were not statistically significant (CD at 5% 12.93) from each other when compared to the yield from control (28 kg). It showed slightly higher yields for synthetic insecticides (mean weight 36 - 38 kg) as compared to biopesticides (mean weight 25 - 31 kg). Among synthetic insecticides, methyl parathion and thiacloprid were found to be of persistent nature and left toxic residue above maximum residue limit (MRL). Therefore taking second criteria of persistence along with the first criteria of yield, indoxacarb is the only treatment which can be applied after fruiting in eggplant crop. Once fruiting has begun, only bio-pesticides like Bt or NSKE should be recommended to avoid residue hazard and to manage ESFB. NSKE was found superior to Neem oil based Nimbo Bas against this pest. However, Nimbo Bas was found to be effective for some of the sucking pests. Similar yield recorded in control plots is due to contribution of cultural practices adopted for plant protection. This involved monitoring of holes on the tip of the branches, removal of the infected twig and burying them. Application of bio-insecticides (Biolep or PUSA Bt) or botanical insecticide (NSKE and Nimbo Bas) could improve the yield due to management of this and other pests like leaf hopper. Application of synthetic insecticides afforded maximum yield of 239 quintal·ha^{-1} verses 125 quintal·ha^{-1} realized from control plots subjected to only good agricultural practice of the region. From the results of this three-year research, an IPM package is being proposed for validation and refinement (**Table 10**).

CONCLUSION

This study involved the efficacy studies of three chemical and four bio-pesticides for the management of ESFB.

Analytical methods were developed for indoxacarb, thia-cloprid and methyl parathion from eggplant and their dissipation pattern were evaluated. The half lives of these synthetic insecticides were around 3, 11, 6 days, respectively (**Table 8**). It was found that indoxacarb was effective against ESFB where as thiacloprid was not effective against this pest and left high residues on the fruits. Methyl parathion was found effective although it is not recommended against ESFB in eggplant in India, as it left toxic residue on the fruits. Methyl parathion and thiacloprid can be applied only before fruiting stage. Among the bio-pesticides PUSA Bt was found significantly effective than Biolep formulation. Also, NSKE was found significantly effective over Nimbo Bas formulation. There is a need for a new improved IPM package for the management of ESFB in India. The proposed IPM package needs to be refined and validated further.

Table 10. Proposed IPM schedule for eggplant cultivation.

Stage of crop	Operation	Targeted Pests/Advantages	Remarks
Before sowing	FYM @ 5 kg bed^{-1} (5 × 1 m^2) + 0.5 kg neem cake		Mandatory
Seeds			
	Treatment with Captan^{TM} @ 2g·L^{-1}	Fungal diseases/root rot	Mandatory
Nursery bed	Apply Chlorpyrifos to nursery beds	Soil insects	Mandatory
Field			
15 d before transplanting	FYM @10 tons·ha^{-1} in field		Mandatory
1 d before transplanting	NPK: Spacing given as 60 × 75 cm^2 (instead of 60 × 60 cm^2)	Bacterial and fungal diseases	Optional
	P and K @ 60 kg·ha^{-1} P_2O_5 and 40 kg·ha^{-1} K_2O		
Transplanted crop			
7 d after transplanting	Half N @ 50 kg·ha^{-1} urea	Reduce sucking pests like jassids, whitefly etc.	Mandatory
35 d after transplanting	Half N @ 50 kg·ha^{-1} urea	-do-	Mandatory
	Manual weeding	Weeds are hosts of pests and diseases.	Mandatory
	Removal of infected shoots, Neem cake 2.5 q·ha^{-1}	Shoot borer	Mandatory
	Spray with methyl parathion	Jassids	Optional
	Remove infected branches	Shoot borer	Mandatory
45 d after transplanting (Flowering starts)	Spray with thiacloprid @ 30·g·ai·ha^{-1}	Jassids and sap sucking pests	Optional
At fruit setting			
After first harvest	Spray indoxacarb@ 75 g·ai/ha	ESFB	Mandatory
60 d after transplanting	NSKE (5%) spray	ESFB	Mandatory
	Pheromone traps (100 number·ha^{-1})	ESFB	Mandatory
75 d after transplanting	Remove infected fruits/phytosanitation	Reduce pest population	Mandatory
	BiolepTM, PUSA Bt, Neem oil sprays @ 14 d interval	Plant protection/plant resistance	Optional
120 d after transplanting	NSKE (5%)	ESFB, Plant protection/plant resistance.	Mandatory

*FYM—Farm yard manure; NPK—Nitrogen, Phosphorus, Potassium; q—quintal.

REFERENCES

1. A. S. Sidhu and A. S. Dhatt, "International Conference on Indigenous Vegetables and Legumes-Prospectus for Fighting Poverty, Hunger and Malnutrition: Current Status of Brinjal Research in India," ISHS Acta Horticulturae, Vol. 752, 2007, pp. 243-248.
2. K. Matsubara, T. Kaneyuki, T. Miyake and M. Mori, "Antiangiogenic Activity of Nasunin, an Antioxidant Anthocyanin in Eggplant Peels," Journal of Agricultural and Food Chemistry, Vol. 53, No. 16, 2005, pp. 6272- 6275. doi:10.1021/jf050796r
3. G. Oboh, M. M. Ekperigin and M. I. Kazeem, "Nutritional and Hemolytic Properties of Eggplant (Solanum macrocarpon) Leaves," Journal of Food Composition and Analysis, Vol. 18, No. 2-3, 2005, pp. 153-160. doi:10.1016/j.jfca.2003.12.013
4. M. S. Rahman, M. Z. Alam, M. Haq, N. Sultana and K. S. Islam, "Effect of Some Integrated Pest Management (IPM) Packages against Brinjal Shoot and Fruit Borer," Journal of Biological Sciences, Vol. 2, No. 7, 2002, pp. 489-491.
5. M. Schuld and R. Schmuck, "Effects of Thiacloprid, a New Chloronicotinyl Insecticide on the Egg Parasitoid Trichogramma cacaoeciae," Ecotoxicology, Vol. 9, No. 3, 2000, pp. 197-205. doi:10.1023/A:1008994705074
6. P. J. Charmillot, D. Pasquier, C. Grela, M. Genini, R. Olivier, C. Ioriatti and

A. Butturini, "Resistance of Carprocapsid Cydia pomonella to Insecticides," Arboriculture et Horticulture, Vol. 35, 2003, pp. 363-368.

7. Z. Puciennik, R. W. Olszak and U. Tworkowska, "Assessment of the Efficacy of Selected Insecticides in Controlling the Plum Fruit Moth (Laspeyresia funebrana)," Zeszyty Naukowe Instytutu Sadownictwai Kwiaciarstwaw Skierniewicach, Vol. 11, 2003, pp. 125-128.

8. M. D. Doerr, J. F. Brunner and L. E. Schrader, "Integrated Pest Management Approach for a New Pest, Lacanobia subjuncta (Lepidoptera: Noctuidae) in Washington Apple Orchards," Pest Management Science, Vol. 60, No. 10, 2004, pp. 1025-1034. doi:10.1002/ps.912

9. B. H. Labanowska and Z. Partyka, "Monitoring and Control of currant Clearwing Moth (Synanthedon tipuliformis) on Black Currant," Journal of Plant Protection Research, Vol. 44, 2004, pp. 929-932.

10. B. H. Labanowska, C. Linder and J. V. Cross, "Pest Control in Blackcurrant in Poland Using the New Neonicotinoid, Thiacloprid as Calypso 480 SC," Bulletin of International Organization for Biological and Integrated Control of Noxious Animals and Plants, Vol. 27, 2004, pp. 101-106.

11. J. Stará and F. Kocourek, "Insecticidal Resistance and Cross-Resistance in Populations of Cydia pomonella (Lepidoptera: Tortricidae) in Central Europe," Journal of Economic Entomology, Vol. 100, No. 5, 2007, pp. 1587-1595. doi:10.1603/0022-0493(2007)100[1587:IRACIP]2.0.CO;2

12. M. Tomizawa and J. E. Casida, "Neonicotinoid Insecticide Toxicology: Mechanisms of Selective Action," Annual Review of Pharmacology and Toxicology, Vol. 45, 2004, pp. 247-268.

13. K. D. Wing, M. Sacher, Y. Kagaya, Y. Tsurubuchi, L. Mulderig, M. Connair and M. Schnee, "Bioactivation and Mode of Action of the Oxadiazine Indoxacarb in Insects," Crop Protection, Vol. 19, No. 8-10, 2000, pp. 537-545. doi:10.1016/S0261-2194(00)00070-3

14. X. P. Hu, "Evaluation of Efficacy and Non-Repellency of Indoxacarb and Fipronil-Treated Soil at Various Concentrations and Thicknesses against Two Subterranean Termites (Isoptera: Rhinotermitidae)," Journal of Economic Entomology, Vol. 98, No. 2, 2005, pp. 509-517. doi:10.1603/0022-0493-98.2.509

15. F. R. Musser and A. M. Shelton, "Bt Sweet Corn and Selective Insecticides: Impacts on Pests and Predators," Journal of Economic Entomology, Vol. 96, 2003, pp. 71- 80.doi:10.1603/0022-0493-96.1.71

16. J. Z. Zhao, Y. X. Li, H. L. Collins, L. Gusukuma-Minuto, R. F. L. Mau, G. D. Thompson and A. M. Shelton, "Monitoring and Characterization of Diamondback Moth (Lepidoptera: Plutellidae) Resistance to Spinosad," Journal of Economic Entomology, Vol. 95, No. 2, 2002, pp. 430- 436. doi:10.1603/0022-0493-95.2.430

17. C. L. Barr, "Fire Ant Mound and Foraging Suppression by Indoxacarb Bait," Journal of Agricultural and Urban Entomology, Vol. 20, No. 3, 2003, pp. 143-150.

18. A. G. Appel, "Laboratory and Field Performance of an Indoxacarb Bait against

German Cockroaches (Dictyoptera: Blattellidae)," Journal of Economic Entomology, Vol. 96, No. 3, 2003, pp. 863-870. doi:10.1603/0022-0493-96.3.863

19. S. F. McCann, G. D. Annis, R. Shapiro, D. W.Piotrowski, G. P. Lahm, J. K. Long, K. C. Lee, M. M. Hughes, B. J. Myers, S. M. Griswold, B. M. Reeves, R. W. March, P. L. Sharpe, P. Lowder, W. E. Barnette and K. D. Wing, "The Discovery of Indoxacarb: Oxadiazines as a New Class of Pyrazoline-Type Insecticides," Pest Management Science, Vol. 57, No. 2, 2001, pp. 153-164. doi:10.1002/1526-4998(200102)57:2<153::AID-PS288>3.0.CO;2-O
20. C. Kao, C. Hung and C. Sun, "Parathion and Methyl Parathion Resistance in Diamondback Moth (Lepidoptera: Plutellidae) Larvae," Journal of Economic Entomology, Vol. 82, No. 5, 1989, pp. 1298-1304.
21. T. V. Prasad, "Development of Water Dispersible Powder (WDP) Formulation of Bacillus thuringiensis var. kurstaki and Its Effect on Detoxification Enzymes of Helicoverpa armigera Hubner," Ph.D. Thesis, Indian Agricultural Research Institute, New Delhi, 2003.
22. J. Saimandir, "Evaluation of Pesticide Residues on Eggplant, Solanum melongena L. Subjected to Plant Protection Measures for Control of Shoot and Fruit Borer, Leucinodes orbonalis Guen.," Ph.D. Thesis, Indian Agricultural Research Institute, New Delhi, 2005.
23. W. M. Hoskins, "Mathematical Treatment of LOSS of Pesticide Residues," Food and Agriculture Organization Plant Protection Bulletin, Vol. 9, 1961, pp. 163-168.
24. M. Gopal and I. Mukherjee, "MRL of Fenvalerate, HCH and Endosulfan on Chickpea," Annals of Plant Protection Sciences, Vol. 3, 1995, pp. 105109.
25. J. Saimandir and M. Gopal, "Analysis of Indoxacarb Applied for Managing Shoot and Fruit Borer, Leucinodes orbonalis G. of Eggplant (Solanum melongena L.) and Its Decontamination by Chemicals," Journal of Environmental Science and Health, Part B. Pesticides, Food Contaminants, and Agricultural Wastes, Vol. 44, 2009, pp. 292- 301.
26. P. Subbian, K. Annadurai and S. P. Palaniappan, "Agriculture, Facts and Figures," Kalyani Publishers Ltd, New Delhi, 2000.
27. S. Sinha and M. Gopal, "Evaluating the Safety of BetaCyfluthrin Insecticide for Usage in Eggplant (Solanum melongena L.) Crop," Bulletin of Environmental Contamination and Toxicology, Vol. 68, No. 3, 2002, pp. 400-405. doi:10.1007/s001280268
28. C. D. S. Tomlin, "The Pesticide Manual, 7th Edition, A World Compendium," British Crop Protection Council, Hampshire, 2006.
29. Codex Alimentarius Commission, 2008. http://www.codexalimentarius.net/mrls/servlet/pesticides/jsp/pest_q-e.j&p
30. W. M. Hoskins, "Mathematical Treatment of Loss of Pesticide Residues," Food and Agriculture Organization Plant Protection Bulletin, Vol. 9, 1961, pp. 163-168.
31. Food and Agriculture Organization, 2008. http://www.fao.org/ag/AGP/AGPP/Pesticid/JMPR/Download/2006_rep/Thiacloprid.pdf

32. WHO (World Health Organization), "Guidelines for Predicting Dietary Intake of Pesticide Residues," Program of Food Safety and Food Aid, Geneva, Switzerland, 1997.

33. S. Jayakrishnan, M. Gopal and S. Walia, "Risk Assessment of Thiacloprid and Its Chemical Decontamination from Eggplant, Solanum melongena L.," Pest Management Science, Vol. 65, No. 2, 2009. pp. 210-215. doi:10.1002/ps.1670

34. Food and Agriculture Organization/World Health Organization, "Acceptable Daily Intakes, Acute Reference Doses, Short-Term and Long-Term Dietary Intakes, Recommended Maximum Residue Limits and Supervised Trials Median Residue Values," Geneva, Switzerland, 2005.

35. J. Saimandir, A. K. Dikshit, J. P. Singh and D. C. Pachauri, "Dissipation of Lambda-Cyhalothrin on Tomato (Lycopersicon esculentum Mill.) and Removal of Its Residues by Different Washing Processes and Steaming," Bulletin of Environmental Contamination and Toxicology, Vol. 75, No. 2, 2005. pp. 324-328. doi:10.1007/s00128-005-0756-2

36. N. K. Roy, "Chemistry of Pesticide," 1st Edition, Cbs Publishers & Distributors, New Delhi, India, 2002.

10
CHAPTER

EFFICACY OF PLANT EXTRACTS IN PLANT DISEASE MANAGEMENT

Malkhan Singh Gurjar, Shahid Ali, Masood Akhtar, Kangabam Suraj Singh

Keywords: Botanical Pesticides; Plant Protection; Chemotherapeutants; Organoleptic

INTRODUCTION

The ultimate aim of recent research in this area has been the development of alternative control strategies to reduce dependency on synthetic fungicides. Plants have ability to synthesize aromatic secondary metabolites, like phenols, phenolic acids, quinones, flavones, flavonoids, flavonols, tannins and coumarins [1]. The components with phenolic structures, like carvacrol, eugenol, and thymol, were highly active against the pathogen. These groups of compounds show antimicrobial effect and serves as plant defence mechanisms against pathogenic microorganisms [2]. The volatile antimicrobial substance allicin (diallyl thio sulphinate) is synthesized in garlic when the tissues are damaged and the substrate alliin (S-allyl-L-cysteine Sulphoxide) mixes with the enzyme alliin-lyase. Allicin is readily membrane-permeable and undergoes thiol-disulphide exchange reactions with free thiol groups in proteins. Allicin effectively controlled seed-borne Alternaria spp. in carrot, Phytophthora leaf blight of tomato and tuber blight of potato as well as Magnaporthe on rice and downy mildew of Arabidopsis

Citation: Gurjar, M. , Ali, S. , Akhtar, M. and Singh, K. (2012) Efficacy of plant extracts in plant disease management. Agricultural Sciences, 3, 425-433. doi: 10.4236/as.2012.33050.

thaliana [3]. Application of plant products especially essential oils is a very attractive method for controlling post harvest diseases. Essential oil extracted from lemon grass (Cymbopogon spp.) post harvest anthracnose of mango fruit [4]. The anti viral protein (AVP) extracts from Bougainvillea spectabilis and Prosopis chilinesis were found to be effective in reducing the sunflower necrosis virus (SFNV) infection both in cowpea and sunflower plants [5]. At present, scientists are investigating for plant products of antimicrobial properties. It would be advantageous to standardize methods of extraction and in vitro antimicrobial efficacy testing so that the search for new biologically active plant products could be more systematic. Thousands of phyto chemicals which have inhibitory effects on all types of microorganisms in vitro should be subjected in vivo testing to evaluate the efficacy in controlling the incidence of diseases in crops, plants, and humans [2].

BOTANICALS

Some plant contains components that are toxic to pathogens. When extracted from the plant and applied on infested crops, these components are called botanical pesticides or botanicals.

Commonly used botanicals:

Plant extracts: Neem (Azadirachta indica, A. Juss), Garlic (Allium sativum, Linn., Eucalyptus (Eucalyptus globulus, Labill., Turmeric (Curcuma Longa, Linn., Tobacco (Nicotiana tabacum, Linn., Ginger (Zingiber officinale, Rosc.

Essential oils: Nettle oil (Urtica spp.), Thyme oil (Thymus vulgaris, Linn.), Eucalyptus oil Eucalyptus globulus, Labill. Rue oil (Ruta graveolens, Linn.), Lemon grass oil (Cymbopogon flexuosus (Steud.) Wats. and Tea tree oil (Melaleuca alternifolia).

Gel and latex: Aloe vera (Tourn. Ex Linn.).

- **Why Consider Botanicals?**
- Sustainable solutions in agriculture
- Reduce crop losses
- Eco-friendly
- Easily bio-degradable
- Organic farming
- Cheaper
- Integrated Diseases Management

PLANT EXTRACTS

Methods of Plant Extract Preparation

Extraction methods involve separation of medicinally active fractions of plant tissue from inactive/inert components by using selective solvents and extraction technology (**Table 1**). Solvents diffuse into the solid plant tissues and solubilize compounds of similar polarity. Quality of plant extract depends on plant material, choice of solvents and the extraction methods.

Plant Material

Plants are collected either randomly or by following leads supplied by local healers in geographical areas where the plants are found. Fresh or dried plant materials can be used as a source for the extraction of secondary plant components. Many authors had reported about plant extract preparation from the fresh plant tissues [6]. The logic behind this came from the ethno medicinal use of fresh plant materials among the traditional and tribal people. But as many plants are used in the dry form (or as an aqueous extract) by traditional healers and due to differences in water content within different plant tissues, plants are usually air dried to a constant weight before extraction. Other researchers dry the plants in the oven at about 40°C for 72 h [7]. In most of the reported works, underground parts (root, tuber, rhizome, bulb etc.) of a plant were used extensively compared with other above ground parts in search for bioactive compounds possessing antimicrobial properties.

Choice of Solvents

Successful determination of biologically active compound from plant material is largely dependent on the type of solvent used in the extraction procedure. Properties of a good solvent in plant extractions include low toxicity, ease of evaporation at low heat, promotion of rapid physiologic absorption of the extract, preservative action and inability to cause the extract to complex or dissociate. As the end product in extraction will contain traces of residual solvent, the solvent should be non-toxic and should not interfere with the bioassay [8]. The choice will also depend on the targeted compounds to be extracted. Initial screening of plants for possible antimicrobial activities typically begins by using the crude or alcohol extractions and can be followed by various organic solvent extraction methods. Water is universal solvent, used to extract plant products with antimicrobial activity. Though traditional healers use primarily water but plant extracts from organic solvents have been found to give more consistent antimicrobial activity compared to water extract [9]. Also water soluble flavonoids (mostly anthocyanins) have no antimicrobial significance and water soluble phenolics only important as antioxidant compound a study reported that extraction of tannins and other phenolics were better in aqueous acetone than in aqueous methanol. In another study, among the twenty different solvents evaluated, chloroform was found to be the best solvent for the extraction of non-polar biological active compounds [10]. Since nearly all of the identified antimicrobial compounds from plants are aromatic or saturated organic compounds, they are most often obtained through initial ethanol or methanol extraction. Thus the most commonly used solvents for preliminary investigations of antimicrobial activity in

Table 1. Solvents used for active component extraction [1].

Water	Ethanol	Methanol	Chloroform	Dichloro-methanol	Ether	Acetone
Tannins	Alkaloids	Terpenoids	Terpenoids	Terpenoids	Alkaloids	Flavonols
Saponins	Tannins	Saponins	Flavonoids	-	Terpenoids	-
Terpinoides	Terpinoides	Tannins	-	-	Coumarins	-
-	Flavonol	Flavones	-	-	-	-

plants are methanol, ethanol and water. The other solvents used by researchers are dicholro-methane, acetone, and hexane. Some authors use a combination of these solvents to obtain the best solvent systems for extraction. [11] examined a variety of extractants for their ability to solubilize antimicrobials from plants, rate of extraction, ease of removal, toxicity in bioassay and acetone received the highest overall rating. Though there is a wide diversification in the usage of solvents, it is necessary to focus on a standardized extraction method and solvent system for a wide variety of researchers working in diverse settings to minimize the variability in the antimicrobial efficacy reports.

Extraction Methods

Variation in extraction methods are usually depend on the length of the extraction period, solvent used, pH of the solvent, temperature, particle size of the plant tissues and the solvent-to-sample ratio. The basic principle is to grind the plant material (dry or wet) finer, which increases the surface area for extraction thereby increasing the rate of extraction. In the study by [12], 5 min extractions of very fine particles of diameter 10 μm gave higher quantities than values obtained after 24 h in a shaking machine with less finely ground material. Earlier studies reported that solvent-to-sample ratio of 10:1 (v/w) solvent to dry weight ratio has been used as ideal [13]. The extraction method that has been widely used by researchers is plant tissue homogenization in solvent [9]. Dried or wet, fresh plant parts are ground in a blender to fine particles, put in a certain quantity of solvent and shaken vigorously for 5 - 10 min or left for 24 h after which the extract is filtered. The filtrate then may be dried under reduced pressure and re-dissolved in the solvent to determine the concentration. Some researchers however centrifuged (approximately 20,000 × g, for 30 min) the filtrate for clarification of the extract. Another common method is serial exhaustive extraction which involves successive extraction with solvents of increasing polarity from a non polar (hexane) to a more polar solvent (methanol) to ensure that a wide polarity range of compound could be extracted [13]. Other researchers employ soxhlet extraction of dried plant material using organic solvent. This method cannot be used for thermolabile compounds as prolonged heating may lead to degradation of compounds.

ANTIMICROBIAL SECONDARY METABOLITES

Plants have limitless ability to synthesize aromatic secondary metabolites, most of which are phenols or their oxygen-substituted derivatives. Important subclasses in this group of compounds include phenols, phenolic acids, quinones, flavones, flavonoids, flavonols, tannins and coumarins (**Figure 1**). These groups of compounds show antimicrobial effect and serves as plant defense mechanisms against pathogenic microorganisms (**Table 2**). Simple phenols and phenolic acid are bioactive phytochemicals consisting a single substituted phenolic ring. Phenolic toxicity to microorganisms is due to the site(s) and number of hydroxyl groups present in the phenolic compound. Quinones are characteristically highly reactive, colored compounds with two ketone substitutions in aromatic ring. Flavones, flavonoids and flavonols are phenolic structure with one carbonyl group. They are synthesized by plants in response to microbial infection and are often found effective in vitro as antimicrobial substance against a wide array of microorganisms (**Table 3**). Tannins are polymeric phenolic substances possessing the astringent property. These compounds are soluble in water, alcohol and acetone and give precipitates with proteins. Coumarins are phenolic substances made of fused benzene and α-pyrone rings. The crude sap, volatile and essential oil extracted from whole plant or specialised plant parts like roots, stem, leaves, flowers, fruits and seeds are widely used in preparing the antimicrobial compounds which are significantly used against the different plant pathogens/diseases (**Table 4**).

Six broad chemical groups are as follows.

- Flavonoides and isoflavonoides
- Saponins
- Steroides
- Tannins
- Phenolic and phenolic acids—Chlrogenic acid, protocatechuic acid, ferulic acid, caffeic acid
- Coumarins and Pyrones

Table 2. Mode of action of phyto chemicals [1].

Class	Sub-class	Mechanism
Phenolics	Simple phenols	Membrane disruption, substrate deprivation
Phenolic acids	Phenolic acids	Bind to adhesins, complex with cell wall, inactivate enzymes
Terpenoids, essential oils		Membrane disruption
Alkaloids		Intercalate into cell wall
Tannins		Bind to proteins, enzyme inhibition, substrate deprivation
Flavonoids		Bind to adhesins, complex with cell wall, Inactivate enzymes
Coumarins		Interaction with eucaryotic DNA
Lectins and polypeptides		Form disulfide bridges

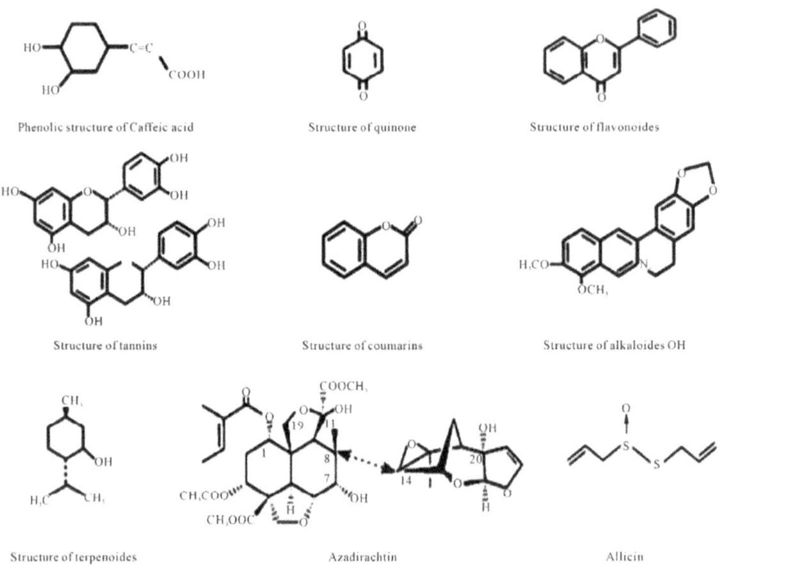

Figure 1. Structure of antimicrobial compounds [14].

Table 3. Botanicals produced by plants having antimicrobial activity.

Common name	Scientific name	Compound	Class	Activity
Apple	*Malus pumila* Mill.	Phloretin	Flavonoid derivative	General
Ashwagandha	*Withania somnifera* Dunal.	Withafarin A	Lactone	Bacteria, fungi
Bael tree	*Aegle marmelos* Linn.	Essential oil	Terpenoid	Fungi
Blue gum tree	*Eucalyptus globulus* Labill.	Tannin	Polyphenol	Fungi, Bacteria, Viruses
Onion	*Allium cepa* Linn.	Allicin	Sulfoxide	Fungi, Bacteria
Thyme	*Thymus vulgaris* Linn	Caffeic acid	Terpenoid	Fungi, Bacteria, viruses
Turmeric	*Curcuma longa* Linn.	Curcumin	Terpenoids	Fungi, Bacteria, protozoa
Thorn apple	*Datura stramonium* Linn.	Hyoscymine Scopolamine	Alkaloids	Fungi
Black pepper	*Piper nigrum* Linn.	Piperine	Alkaloid	Fungi
Castorbean	*Ricinus communis* Linn.	Ricinine Ricininoleic	Alkaloids	Fungi
Neem/Margosa tree	*Azadirachta indica* A.Juss.	Azadirachtin	Terpenoides	Fungi, Bacteria
Garlic	*Allium sativum* Linn.	Allicin	Solfoxide	Fungi, Bacteria

Table 4. Botanicals.

Plant	Part used	Preparations	Diseases/pathogen	References
Datura/thorn apple (*D.stamonium*) *Calotropis procera* (Ait.) R. Br. *Oscimum* spp.	Root, stem, Leaf, flowers	Crude extract	*Curvularia lunata*	[15]
Turmeric (*Curcuma longa* Linn.), Ginger (*Zingiber officinale* Rosc.)	Rhizome	Crude extract	*Phytopthora infestans, Fusarium solani, Pyricularia oryzae*	[16]
Perslane (*Portulaca olaracea* Linn.)	Leaf	Crude extract	*Helminthosporium maydis*	[17]
Hena (*Lowsonia inermis* Linn.)	Leaf	Crude extract	*Dreshslera oryzae*	[18]
Neem/Margosa (*Azadirachta indica* A.Juss.), Sugar apple (*Annona squamosa* Linn.), Holy basil (*Oscimum sanctum* Linn.)	Leaf, Stem Bark, root	Crude extract	Anthracnose of pepper	[19]
Neem/Margosa (*Azadirachta indica* A.Juss.),	Seed kernel	Oil	*A. alternata*	[20]
Ambrosioides Linn., *Oscimum* spp.	Leaf	Essential oils	*Aspergillus flavus*	[21]
Garlic (*Allium sativum* Linn.), Datura (*D. stramonium* Linn.)	Bulb, Leaf	Ethanol extracts	*Curvularia lunata*	[22]
Spearmint (*Mentha spicata* Linn.), Greek Sage (*Salvia fruticosa* Mill.), *Thymbra* spp.	Leaf	Essential oils	*Rhizoctonia solani, Sclerotium sclerotiorum*	[23]
Spanish flag (*Lantana camara* Linn.)	Leaf	Crude Extracts	Castor grey rot (*Botrytis ricini*)	[24]
Neem/Margosa (*Azadirachta indica* A.Juss.),	Seed, Leaf	Crude Extracts	Early blight of tomato	[25]
Madar (*Calotropis procera* (Ait.) R.Br.	Leaf	Crude Extracts	Tikka leaf spot disease of groundnut	[26]
Neem/Margosa (*Azadirachta indica* A.Juss.),	Seed	NSKE	Powdery mildew of pea	[27]
Spanish flag (*L.camara* Linn.), Pongam (*Pongamia pinnata* L.Pierre.)	Leaf	Crude extracts	Leaf blight of onion	[28]
Holy basil (*Oscimum sanctum*Linn.), peach (*Prunus persica* Linn.) Stokes.	Leaf	Essential oil	Grey mould (*Botrytis cinerea*) of grapes	[29]
Neem/Margosa (*Azadirachta indica* A.Juss.),	Leaf	Achook formulations (azadirachtina)	Sheath blight of rice	[30]
Neem/Margosa (*Azadirachta indica* A.Juss.),	Seed kernel	Neem oil	Rice tungro virus	[31]
Neem/Margosa (*Azadirachta indica* A.Juss.),	Leaf, Seed	Achook, Neemazal,	Bacterial blight of rice	[32]
Oregano (*Origanum hercleoticum* (weed species)	Leaf	Essential oils	*Fusarium oxysporum, Phoma tracheiphila*	[4]
Neem/Margosa (*Azadirachta indica* A.Juss.), Black cumin (*Nigelia sativa* Linn. Asfetida (*Ferula asafoetida* Linn.)	Seeds	Essential oils	*Fusarium oxysporum, A.niger, A.flavus*	[33]
Strawberry (*Fragaria* spp.)	Fruit	Volatile compounds	Anthracnose of strawberry	[34]
Raspberry (*Rubus* spp.) and Strawberry (*Fragaria* spp.)	Fruit	Volatile compounds	Post harvest decay fungi	[35]
Garden croton (*Codiaeum variegatum* Linn.)	Leaf	Phenolic compounds	*Alternaria alternate, Fusarium oxysporum*	[36]
Oleander (Nerium *oleander* Linn.)	Leaf	Crude extracts	Brown spot of rice (*Bipolaris oryzae*)	[37]
Indian aloe (*Aloe barbadensis* Mill.) Neem/Margosa (*Azadirachta indica* A.Juss.), Tobacco (*Nicotiana tabacum* Linn.)	Leaf	Crude extracts	Dry rot of yam *F. oxysporum, A.nizer*	[38]
Black pepper (*Piper nigrum* Linn., Clove (*Syzygium aromaticum* (Linn.) Merr. & Perry, Geranium (*Pelargonium graveolens* L'Herit), Nutmeg (*Myristica fragrans* Houtt.), (*Origanum vulgare* spp. *hirtum* (Link) Letsw. and thyme [*Thymus vulgaris* Linn.	Leaf	Volatile oil	Anti bacterial (gram positive and gram negative)	[39]

Metasequoia glyptostroboides	Leaf	Essential oil	*Fusarium oxysporum*, *Fusarium solani*, *Phytophthora capsici*, *Colletotrichum capsici*, *Sclerotinia sclerotiorum*, *Botrytis cinerea* and *Rhizoctonia solani*	[40]
Clove (*Syzigium aromaticum* Linn.), Turmeric (*Curcuma longa* Linn.), garlic (*Allium sativum* Linn.) and Holy Basil (*Ocimum Sanctum* Linn.)	Leaf, seed, fruit	Crude extract	*Aspergillus flavus*	[41]
Aspilia africana, *Chromolaena odorata*, *Musa paradisiaca* and *Tithonia diversifolia*	Leaf	Aqueous extract	*Cercospora* leaf spot of Seasame (*Sesamum indicum*) L.)	[42]
Lemon grass (*Cymbopogon* spp.), Thyme (*Thymus vulgaris* Linn.)	Leaf, root	Volatile compound	Black Mould Disease on Onion Bulbs (*Aspergillus niger*)	[43]
Brassica napus and Tomato (*Lycopersicon esculentum* Mill.)	Leaf, stem	Water extract	Bacterial disease on Onions	[44]
Ginger (*Zingiber officinale* Rosc.), aloe (*Aloe vera*), bitter kola (*Garcinia cola*) and neem (*Azadirachta indica* A.Juss.)	Leaf, Fruit and Seed	Crude	Root rot disease of cow pea (*Vigna unguiculata* L.)	[45]
Chloranthus japonicas, *Paulownia coreana*	Roots, stem	Crude extract	rice blast, rice sheath blight, and wheat leaf rust	[46]
Urtica dioica L., thyme *Thymus vulgaris* L., *Eucalyptus* spp., *Ruta graveolens* L. and *Achillea millefolium* L.	Leaf	Volatile oil	*Aletrnaria alternate*	[47]
Acacia nilotica, *Achras zapota*, *Datura stramonium*, *Emblica officinalis*, *Eucalyptus globules*, *Lawsonia inermis*, *Mimusops elengi*, *Peltophorum pterocarpum*, *Polyalthia longifolia*, *Prosopis juliflora*, *Punica granatum* and *Sygigium cumini*	Leaf, Seed, fruits	Aqueous extract	*Aspergillus* spp.	[48]
Eugenia aromatica , *Piper betle*), *Alpinia galanga* and *Sphaeranthus indicus*	Leaf	Crude	Stem rot disease of Vanilla	[49]
Ocimum gratissimum, *Aframomum melegueta*	Leaf	Crude	Post harvest yam (*Diascorea* spp.) rot	[50]
Neem (*Azadirachta indica* A.Juss.)	Leaf	Crude	Protozoan, Bacteria, Antifertility	[51]

Methods for Evaluation of Efficacy of Plant Extract

In vitro antimicrobial susceptibility testing (AST)

1) Diffusion test

a) agar well diffusion

b) agar disk diffusion

c) poison food technique

d) bio autography

2) Dilution methods

a) agar dilution

b) broth micro dilution assay

c) broth macro dilution assay

VOLATILE OILS

- Volatiles are small molecular weight organic compounds having appreciable vapour pressure at ambient temperature
- Play role in defense system
- Inhibit the growth of pathogen
- Generally Recognized As Safe (GRAS)
- Isolated by hydro distillation or steam
- Terpinoides—monoterpenes (C10), ses quiterpenes (C15), diterpines (C20)

Examples: Black pepper (Piper nigrum Linn.), Clove (Syzygium aromaticum Linn.) Merr. & Perry, Nutmeg (Myristica fragrans Houtt.), Oregano (Origanum vulgare Linn.) and Thyme (Thymus vulgaris Linn.)

ESSENTIAL OILS

- The potential essential oils used against post harvest pathogens.
- Fungistatic
- Terpenoides and aromatic compounds
- Component—carvacrol, thymol, cymene, terpine, phenylpropene derivatives, eucalyptol and anisol Examples: Nettle oil (Urtica spp.), Thyme oil (Thymus vulgais Linn.), Eucalyptus oil (Eucalyptus globules Labill., Rue oil (Ruta graveolens Linn.), Lemon grass oil (Cymbopogon flexuosus (Steud.) Wats.) and Tea tree oil (Melaleuca alternifolia Maiden & Betche).

LIMITATIONS OF BOTANICALS FOR PLANT DISEASE MANAGEMENT

- Extraction methods are not standardized
- Rapid degradation
- Most studies are in vitro efficacy
- Need the development of formulations
- Some chemical compounds are harmful to human and plants
- Less effective
- Less availability formulations

CONCLUSION

Plants contain thousands of constituents and are valuable sources of new and biologically active molecules possessing antimicrobial property (**Table 4**). The ethnobotanical study of plant is important for modern day medicine but its usefulness cannot be overemphasized if methods are not standardized to obtain comparable and reproducible results. At present, scientists are investigating for plant products of antimicrobial properties. It would be advantageous to standardize methods of extraction and in vitro antimicrobial efficacy testing so that the search for new biologically active plant products could be more systematic and interpretation of results would be facilitated. Thousands of phytochemicals which have inhibitory effects on all types of microorganisms in vitro should be subjected in vivo testing to evaluate the efficacy in controlling the incidence of disease in crops, plants, and humans. Efficient collaborations with pharmacologists and medical doctors, plant pathologists and microbiologists are crucial to see the complete development of an interesting lead compound into an exploitable product.

REFERENCES

1. Cowan, M.M. (1999) Plant products as antimicrobial agents. Clinical Microbiology Reviews, 12, 564-582.
2. Das, K., Tiwari, R.K.S. and. Shrivastava, D.K. (2010) Techniques for evaluation of medicinal plant products as antimicrobial agent: Current methods and future trends. Journal of Medicinal Plants Research, 4, 104-111.
3. Slusarenko, A.J., Patel, A. and Portz, D. (2008) Control of plant diseases by natural products: Allicin from garlic as a case study. European Journal of Plant Pathology, 121, 313-322. doi:10.1007/s10658-007-9232-7
4. Salomone, A., Scaritto, G., Sacco, A., Cabras, G. and Angioni, A. (2008) Inhibitory effects of the main compounds of oregano essential oil against some pathogenic fungi. Modern fungicide and Antifungal compounds V. Deutsche Phytomedizinische Gesellschaft, 44, 345-360.
5. Lavanya, N., Saravanakumar, D., Rajendran, L., Ramiah, M., Raguchander, T. and Samiyappan, R. (2009) Management of sunflower necrosis virus through anti-viral substances. Archives of Phytopathology and Plant Protection, 42, 265-276.doi:10.1080/03235400601037149
6. Tiwari, R.K.S., Chandravanshi, S.S. and Ojha, B.M. (2005) Efficacy of extracts of medicinal plant species on growth of Sclerotium rolfsii root rot in tomato. Journal of Mycology and Plant Pathology, 34, 461-464.
7. Salie, F., Eagles, P.F.K. and Lens, H.M.J. (1996) Preliminary antimicrobial screening of four South African Asteraceae species. Journal of Ethnopharmacology, 52, 27- 33.doi:10.1016/0378-8741(96)01381-5
8. Ncube, N., Afolayan, S.A.J. and Okoh, A.I. (2008) Assessment techniques of antimicrobial properties of natural compounds of plant origin: Current methods and future trends. African Journal of Biotechnology, 7, 1797-1806.
9. Parekh, J., Jadeja, D. and Chanda, S. (2005) Efficacy of aqueous and methanol extracts of some medicinal plants for potential antibacterial activity. Turkish Journal of Biology, 29, 203-210.
10. Harmala, P., Vuorela, H., Tornquist, K. and Hiltunen, R. (1992) Choice of solvent in the extraction of Angelica archangelica roots with reference to calcium blocking activity. Planta Medica, 58, 176-183. doi:10.1055/s-2006-961424
11. Eloff, J.N. (1998) A sensitive and quick microplate method to determine the minimum inhibitory concentration of plant extracts for bacteria. Planta Medica,

64, 711-713.doi:10.1055/s-2006-957563

12. Eloff, J.N. (1998) Which extractant should be used for the screening and isolation of antimicrobial components from plants? Journal of Ethnopharmacology, 60, 1-8.doi:10.1016/S0378-8741(97)00123-2

13. Green, R.J. (2004) Antioxidant activity of peanut plant tissues. Master Thesis, North Carolina State University.

14. Ciocan, I.D. and Băra, I. (2007) Plant products and antimicrobial agents. Universitatii ale Ştiinţifice Analele Alexandru Ioan Cuza, Tom VIII.

15. Manoharachary, C. and Gourinath, A. (1988) Effects of plant extracts on four pathogenic fungi. 5th International Congress of Plant Pathology, Kyoto.

16. Bandara, B.M.R., Kumar, N.S. and Samaranayake, K.M.S. (1989) An antifungal constituent from the stembark of Butea monosperma. Journal of Ethnopharmacology, 25, 73-75. doi:10.1016/0378-8741(89)90046-9

17. Noriel, L.M. and Robles, R.P. (1990) Fungicidal activity of Portulaca oleracea extract against Helminthosporium maydis Wisik and Miyake in corn (Zea mays L.). Philippine Journal of Weed Science, 17, 26-32.

18. Natarajan, M.R. and Lalithakumari, D. (1987) Antifungal activity of the leaf extract of Lawsonia inermis on Dreschslera oryzae. Indian Phytopathology, 40, 390-395.

19. Nduagu, C., Ekefan, E.J. and Nwankiti, A.O. (2008) Effect of some crude plant extracts on growth of Colletotrichum capsici (Synd) Butler & Bisby, causal agent of pepper anthracnose. Journal of Applied Biosciences, 6, 184-190.

20. Dharam, V. and Sharma R.K. (1985) Efficacy of fungicides XXIX studies on the fungicidal properties of neem oil. Indian Journal of Plant Pathology, 3, 241-242.

21. Mishra, A.K., Dwivedi, S.K. and Kishore, N. (1989) Antifungal activity of some essential oils. National Academy Science Letters, 12, 335-336.

22. Upadhyaya, M.R. and Gupta, R.C. (1990) Effect of extracts of some medicinal plants on the growth of Curvularia lunata. Indian Journal of Mycology and Plant Pathology, 20, 144-145.

23. Yegen, O., Begger, B. and Heitefuss, R. (1992) Studies on the fungitoxic effect for extracts of six selected plants from Turkey on phytopathogenic fungi. Zeitschrift fur Pflanzenkrankheiten und Pflanzenschütz, 99, 349-359.

24. Bhattiprolu, S.L. and Bhattiprolu, G.R. (2006) Management of castor grey rot disease using botanical and biological agents. Indian Journal of Plant Protection, 34, 101-104.

25. Patil, M.J., Ukey, S.P. and Raut, B.T. (2001) Evaluation of fungicides and botanicals for the management of early blight (Alternaria solani) of tomato. PKV Research Journal, 25, 49-51.

26. Srinivas, T., Rao, M.S., Reddy, P.S. and Reddy, P.N. (1997) Integrated management of leaf spot of groundnut (Arachis hypogaea L.) with botanicals and chemicals. Zeitschrift fur Pflanzenkrankheiten und Pflanzenschutz, 104, 528-530.

27. Surwase, A.G., Badgire, D.R. and Suryawanshi, A.P. (2009) Management of pea powdery mildew by fungicides, botanicals and bio-agents. Annals of Plant Protection Sciences, 17, 384-388.

28. Bhosale, M.V., Perne, R.R., Shete, M.H. and Pawar, N.B. (2008) Efficacy of botanicals, bioagents and fungicides on management of leaf blight (Alternaria porri) of onion. Journal of Plant Disease Sciences, 3, 88-89.

29. Tripathi, P., Dubey, N.K. and Shukla, A.K. (2008) Use of some essential oils as post harvest botanical fungicides in the management of grey mould of grapes caused by Botrytis cinerea. World Journal of Microbiology and Biotechnology, 24, 39-46.doi:10.1007/s11274-007-9435-2

30. Kandhari, J. (2007) Management of sheath blight of rice through fungicides and botanicals. Indian Phytopathology, 60, 214-217.

31. Muthamilan, M. and Revathy, N. (2007) Management of tungro virus disease of rice with antagonists and botanicals. Journal of Ecobiology, 19, 81-88.

32. Sunder, S., Singh, R. and Dodan, D.S. (2005) Management of bacterial blight of rice with botanical extracts and non-conventional chemicals. Plant Disease Research, 20, 12-17.

33. Sitara, U., Naseem, J. and Sultana, N. (2008) Antifungal effect of essential oils on in vitro growth of pathogenic fungi. Pakistan Journal of Botany, 40, 409-414.

34. Arroyo, F.T., Moreno, J., Daza P., Boianova, L. and Romero, F. (2007) Antifungal activity of strawberry fruit volatile compounds against Colletotrichum acutatum. Journal of Agricultural and Food Chemistry, 14, 5701- 5707. doi:10.1021/jf0703957

35. Vaughan, S.F., Spenser, G.F. and Shasha, B.F. (1993) Volatile compounds from raspberry and strawberry fruit inhibit postharvest decay fungi. Journal of Food Science, 58, 793-796. doi:10.1111/j.1365-2621.1993.tb09360.x

36. Naidu, G.P. (1988) Antifungal activity in Codieum variegatum leaf extract. Current Science, 57, 502-503.

37. Harish, S., Saravankumar, D. Ebenezar, E.G. and Seetharaman, K. (2008) Use plant extracts and biocontrol agents for the management of brown spot disease in rice. BioControl, 53, 555-567. doi:10.1007/s10526-007-9098-9

38. Taiga, A. (2009) Efficacy of selected plant extracts in the control of fungal dry rot of white yam (Dioscorea Rotundata) Tubers in Kogi State. American-Eurasian Journal of Sustainable Agriculture, 3, 310-313.

39. Dorman, H.J.D. and Deans, S.G. (2000) Antimicrobial agents from plants: Antibacterial activity of plant volatile oils. Journal of Applied Microbiology, 88, 308-316.doi:10.1046/j.1365-2672.2000.00969.x

40. Bajpai, V.K. and Kang, S.C. (2010) Antifungal activity of leaf essential oil and extract of Metaseqoia glystostroboides Miki ex Hu. Journal of the American Oil Chemists' Society, 87, 327-336. doi:10.1007/s11746-009-1500-6

41. Reddy, K.R.N., Reddy, C.S. and Muralidharan, K. (2009) Potential of botanicals and biocontrol agents on growth and aflotoxin production by Aspergillus

flavus infecting rice grains. Food Control, 20, 173-178. doi:10.1016/j.foodcont.2008.03.009

42. Enikuomehin, O.A. (2005) Cercospora leaf spot disease management in Seasame (Seasamum indicum L.) with plant extracts. Journal of Tropical Agriculture, 43, 19-23.
43. Abd-Alla, M.A., El Mohamedi, R.S.R. and Badeaa, R.I. (2006) Effect on some volatile compounds on black mould disease on onions bulbs during storage. Research Journal of Agriculture and Biological Sciences, 2, 384- 390.
44. Kowalska, B. and Smolinska, U. (2008) The effect of selected plant materials and extract on the development of bacterial disease on onion. Vegetable Crops Research Bulletin, 68, 33-45. doi:10.2478/v10032-008-0003-6
45. Suleiman, M.N. and Emua, S.A. (2009) Efficacy of four plant extract in the control of root rot disease of cow pea (Vigna unguiculata Linn. Walp). African Journal of Biotechnology, 8, 3806-3808.
46. Choi, G.J., Jang, K.S., Kim, J.S., Lee, S.W., Cho, J.Y., Cho, K.Y. and Kim, J.C. (2004) In vivo fungal activity of 57 plant extracts against six plant pathogenic fungi. Plant Pathology Journal, 20, 184-191.
47. Hadizadeh, I., Peivastegan, B. and Hamzehzarghani, H. (2009) Antifungal activity of essential oil from some medicinal plants of Iran against Aletrnarea alternata. American Journal of Applied Sciences, 6, 857-861. doi:10.3844/ajassp.2009.857.861
48. Satish, S., Raveesha, K.A. and Janardhana, G.R. (1999) Antibacterial activities of plant extract on phytopathogenic Xanthomonos compestris pathovars. Letters in Applied Microbiology, 28, 145-147. doi:10.1046/j.1365-2672.1999.00479.x
49. Suprapta, D.N. and Khalimi, K. (2009) Efficacy of plant extract formulation to suppress stem rot disease on vanilla seedlings. ISSA Journal, 15, 34-41.
50. Okigbo, R.N. and Ogbonnaya, U.O. (2006) Antifungal effect of two tropical plants leaf extracts (Ossimum gratissimum and Aframomum meleguata) on post harvest yam (Diascorea spp.) rot. African Journal of Biotechnology, 5, 727-731.
51. Atawodi, S.E. and Atawodi, J.C. (2009) Azadirachta indica (neem) a plant of microbial and pharmacological activity. Phytochemistry Reviews, 8, 601-620. doi:10.1007/s11101-009-9144-6

11
CHAPTER

PATHOGEN GENOMICS, DIVERSITY, EPIDEMIOLOGY, VECTOR DYNAMICS, AND DISEASE MANAGEMENT

Hong Lin and Neil C. Gudmestad

INTRODUCTION

Zebra chip (ZC) is a destructive disease of potato (Solanum tuberosum) that has emerged in North America as well as other potato-production regions in the world. The disease has caused significant economic losses to the U.S. potato industry. ZC was first discovered around Saltillo, Mexico in the 1990s (24,37,98). In the early 2000s, it was reported in commercial potato fields in the Pearsall and lower Rio Grande valley areas of Texas. By 2006, the disease had spread to most potato-production areas in Texas as well as other potato-production states, including Arizona, Kansas, Colorado, California, Nevada, and New Mexico (74). ZC was reported in New Zealand in 2008 (52) and, more recently, from Idaho, the Columbia Basin of Washington, and Oregon (18,22). ZC-affected potato plants exhibit a range of foliar symptoms that resemble those of the purple top wilt symptoms caused by the clover proliferation phytoplasma (37). Foliar symptoms include stunting, chlorosis, swollen nodes, proliferated axillary buds, aerial tubers, browning of the vascular system in stems, and leaf scorching (Fig. 1A and B). Belowground plant symptoms include enlarged lenticels of the underground stem, collapsed stolons, brown discoloration of the vascular ring, necrotic flecking on internal

Citation: Lin H, Gudmestad NC., Pathogen Genomics, Diversity, Epidemiology, Vector Dynamics, And Disease Management, doi: 10.1094/PHYTO-09-12-0238-RVW.

tuber tissues, and, occasionally, streaking of the medullary ray tissues (Fig. 1C) (37). At the time the disease was identified, the causal agent was unknown. To investigate the possible causal agents, insect transmission and graft experiments were conducted to determine whether infectious agents were transmissible to healthy plants or due to nonbiological factors such as heat stress (heat necrosis) or toxicity (psyllid toxin). In addition, research also was conducted to determine whether the symptoms were caused by nutritional deficiency, a toxin or other abiotic factor, or some combination of these factors. Although researchers were not able to consistently associate the disease with any specific causes, results from graft- and insect-transmission experiments suggested that there was likely a transmissible factor, probably a vascular plant pathogen, virus, or prokaryote or bacterial-like organism (BLO), that was associated with ZC (24,71).

These results were further supported by transmission electron microscopy studies, where BLOs were observed in the vascular tissue of infected plant parts that were similar to the phloemrestricted proteobacteria which are associated with the "basses richesses" syndrome of sugar beet (24). Because the symptoms of ZC resembled those of phytoplasma-associated diseases, studies were subsequently carried out to determine whether leafhoppers or potato-tomato psyllid Bactericera cockerelli (Šulc) was a primary vector responsible for transmission of a ZC-causing agent. Results from cage trial studies demonstrated that plants infested by potato psyllids developed ZC foliar symptoms and characteristic ZC discoloration in tubers (71,74). However, at that time, the causal agent of ZC was still unknown. To identify the causal pathogen or pathogens, polymerase chain reaction (PCR) primers detecting pathogens associated with vascular-restricted diseases, including cucurbit yellow vine disease, marginal chlorosis of strawberry, Pierce's disease of grapevine, and "stolbur-like" phytoplasma, were used to screen DNAs extracted from ZCaffected and healthy potato. However, attempts to amplify specific known disease-prokaryote products from a suspect ZC-associated prokaryote failed to obtain consistent results from ZC-diseased potato or psyllids or other insect vectors (37).

A breakthrough came in April 2008, when researchers at the Plant Health and Environment Laboratory, MAF Biosecurity, New Zealand reported that a new disease of glasshouse-grown tomato (S. lycopersicum) and pepper (Capsicum annuum) plants was identified using universal prokaryote 16S ribosomal DNA primer pairs that amplified a PCR product specifically from symptomatic plants (52,54). This result, along with transmission electron microscopy evidence (24,99), confirmed the presence of a phloem-inhabiting BLO in diseased plants, but not in healthy plants. The same New Zealand group subsequently confirmed that the same pathogen was associated with ZC of potato (52,54). Sequence analysis of the amplicon indicated that it shared high similarity with the bacterial species associated with citrus huanglongbing (HLB) (also known as citrus greening). Subsequently, it was given the name 'Candidatus Liberibacter solanacearum' (55). In August 2008, a research group at the University of California– Riverside also reported a new citrus "HLB species" bacterium named 'Ca. L. psyllaurous' responsible for tomato yellow disease caused by a bacterium transmitted by the tomato/potato psyllid (39). Because the sequences of 'Ca. L. solanacearum' and 'Ca. L. psyllaurous' share >99% similarity in the 16S rRNA gene, it is likely these two bacterial names are synonymous (99,110). 'Ca. L. solanacearum' is recommended as the reference to this bacterium because this name and a description of the bacterium were published in the International Journal of Systematic and Evolutionary Microbiology (55), a preferred journal for descriptions of uncharacterized bacterial species. Since the discovery that 'Ca. L. solanacearum' was responsible for ZC disease, various studies have been carried out (1,53,56, 76,99,110).

Figure 1 Disease symptomatology associated with A to C, zebra chip (ZC) and D to F, psyllid yellows (PY) of potato. A, Early symptom development of 'Candidatus Liberibacter solanacearum' infections in potato foliage; B, advanced foliar symptoms of ZC; and C, medullary streaking and flecking in tuber flesh. D, Early symptom development of PY in potato foliage; E, advanced symptom development of PY; and F, mild vascular discoloration of potato tubers due to PY.

All results were consistent with the conclusion that 'Ca. L. solanacearum' is the putative causal agent of ZC of potato. Due to its fastidious nature, the 'Ca. L. solanacearum' bacterium is, as of yet, unculturable and, thus, Koch's postulates have not been fulfilled. Consequently, conclusive detailed information regarding the general biology, physiology, and pathogenicity of 'Ca. L. solanacearum' is lacking. In spite of these limitations, using DNA enrichment techniques and next-generation sequencing technology, the whole genome sequence of 'Ca. L. solanacearum' has been obtained (58). Annotation of genome information provides insights into gene inventory account for biological functions, mode of pathogenicity, metabolic pathways, and nutritional requirements for this obligated intracellular pathogenic bacterium.

Evolution, speciation, and genetic diversity

The genus of 'Candidatus Liberibacter' belongs to the Rhizobiaceae family of α-Proteobacteria. There are four plant-pathogenic species of bacteria in this genus (8,46,55). Among the four pathogenic Liberibacter spp., three are associated with citrus HLB worldwide and each is named based on its presumptive origin: 'Ca. L. asiaticus' from Asian countries, 'Ca. L. africanus' from South Africa, and 'Ca. L. americanus' from South America (8). HLB is a destructive disease of citrus that represents a major threat to the citrus industry worldwide. 'Ca. L. solanacearum' is the species associated with potato ZC disease. Phylogenetic analyses of the 16S rRNA gene indicated that 'Ca. Liberibacter' spp. belong to the α-Proteobacteria group, and their closest cultivated relatives are members of the group of bacteria known as the α-2 subgroup (46). Duan et al (27) constructed a phylogenetic analysis using 94 nonhorizontal transfer, orthologous gene sequences from ≈50 species of α-Proteobacteria. Results indicated that 'Ca. L. asiaticus' is an early-branching member of the Rhizobiaceae family. This multi-gene-based sequence analysis is in agreement with other gene-based analyses (8,26). These results also suggested that the genome of 'Ca. L. asiaticus' had evolved rapidly (27). Such rapid genome evolution is typical of host-restricted symbionts or pathogens (67). Discovery of 'Ca. L. solanacearum' as the putative causal agent of ZC sparked a research interest in understanding the origin and evolutionary relationships of this bacterium with other Liberibacter spp. A phylogenetic tree constructed using the neighbor-joining method clustered all three HLB-associated 'Ca. Liberibacter' spp. along with 'Ca. L. solanacearum' into a monophyletic group based on 16S rRNA gene sequences (56). Interestingly, both 'Ca. L. americanus' and 'Ca. L. solanacearum' were closely positioned proximal to the basal node followed by other two species, 'Ca. L. asiaticus' and 'Ca. L. africanus', indicating early divergence of 'Ca. L. americanus' followed by 'Ca. L. solanacearum' (56) (Fig. 2).

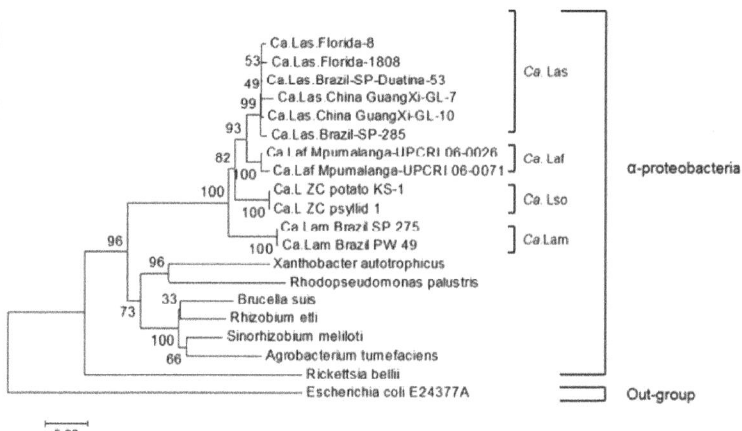

Figure 2 Phylogenetic relationships of plantpathogenic 'Candidatus Liberibacter sp.' in relation to other bacterial members of the Rhizobiaceae family based on 16S rRNA gene sequence data.

Shortly after the report of 'Ca. L. solanacearum' in New Zealand, a new strain of 'Ca. L. solanacearum' associated with carrot (Daucus carota), a nonsolanaceous crop, vectored by carrot psyllid (Trioza apicalis), was identified in Finland (72). This was followed by another report of an apparently nonpathogenic Liberibacter sp. named 'Ca. L. europaeus' naturally vectored by pear psyllid pest Cacopsylla pyri (L.) in Italy (88). Based on these research findings, it is expected that 'Ca. Liberibacter' spp. have undergone rapid evolutionary divergence, speciation, and adaptation to a wider range of hosts since the divergence from their ancestors. Genetic diversity of 'Ca. L. solanacearum' is probably much larger than previously expected (56). To determine the genetic diversity of 'Ca. L. solanacearum' in agricultural ecosystems, haplotypes of 'Ca. L. solanacearum' were characterized based on single nucleotide polymorphism (SNP) typing in the sequences of 16S rRNA, 16S/23S ISR, and 50S rplJ-rplL ribosomal gene loci (82). Two haplotypes were identified in North and Central America, and in New Zealand, which agreed with the reports based on clade relationships of 'Ca. L. solanacearum' isolates based on 16s rRNA gene sequence analyses (Table 1) (99,110). A third haplotype was only found in a single isolate of 'Ca. L. solanacearum' vectored by carrot psyllid in northern Europe (82). It is not clear whether this haplotype can also cause potato ZC disease. With the availability of the complete genome sequence of 'Ca. L. solanacearum', Glynn et al. (34) developed multilocus sequence typing (MLST) markers for this bacterium. MLST marker systems have been widely used for typing and strain differentiation (34).

This marker system is designed from a series of genomic housekeeping gene loci (usually 7 to 10 loci) located throughout the genome of the organism of interest (44). Housekeeping genes are selected for MLST because they undergo relatively little change in sequence over time. However, sequence variations due to insertion/deletion (INDEL) or SNPs within these loci may reflect pathogen adaptation. For example, mutation by synonymous nucleotide substitution will not alter amino acid sequences due to degeneracy of the genetic code and, therefore, such mutations will not be subject to natural selection whereas nonsynonymous mutation causing changes in amino acids will result in increased or decreased fitness. Using a panel of 10 MLST markers, two sequence types (ST-1 and ST-2) were identified, representing two major lineages in U.S. populations, whereas only one type was detected in Mexico (ST-1) and New Zealand (ST-2) isolates. Interestingly, two lineages of 'Ca. L. solanacearum' isolates identified by this MLST marker system agreed with the analyses based on conserved genes such as rRNA and other gene loci (Table 1) (82,99,110). In addition to the markers mentioned above, another marker system, named simple sequence repeat (SSR) (also known as variable number tandem repeat or microsatellite), was developed (57). A genome-wide sequence search led to the identity of SSR loci in the 'Ca. L. solanacearum' genome representing mono-, di-, tri-, and tetra-repeat units in perfect or imperfect repeat motifs. Due to the compact nature of bacterial genomes, some SSR loci reside in or near coding regions or in upstream regulatory regions and, therefore, have functional and structural properties. In contrast to conserved gene-based markers, this marker system possesses high discrimination power for differentiating strains within populations that have evolved over a relatively short time scale.

The sequence variations in SSR loci among individual isolates can be used to assess genetic diversity within a clonal complex (96). Not surprisingly, two major lineages were also identified in U.S. 'Ca. L. solanacearum' populations (Table 1). Although genetic analyses detected two lineages in agricultural population systems, the biological implication is yet to be determined. Interestingly, genetic analyses indicated that the occurrence of identical haplotypes was not restricted to any particular geographical area or host cultivar. These overlapping patterns were also observed by SNP-based typing analyses (82). In contrast to natural populations, it is expected that frequent exchange of plant materials in agricultural productions could facilitate the movement of ZC-affected materials, which could largely account for such non-pattern geographical distribution among identical haplotypes. Additionally, SSR DNA markers associated with hypervariable sequence regions can provide sufficient resolution for differentiating closely related isolates and for tracking genotypes of interest. Thus, these markers could help identify the source of invasive strains. Pathogens with higher diversity and evolutionary potential will likely pose a higher risk of overcoming host resistance or evolving to reduce the efficacy of disease control measures (65). Understanding genetic diversity and evolutionary potential of 'Ca. L. solanacearum' populations is critical for development of disease management strategies and tactics. Currently, information regarding the population structure and genetic diversity, as well as evolutionary and epidemiological relationships, among 'Ca. L. solanacearum' isolates in agricultural populations is limited. Pathogen genetic analyses, in combination with advances in the understanding of the spatial and temporal dynamics of the epidemiology of the disease, will facilitate development of effective integrated pest and disease management for the ZC problem.

Table 1 Genetic variations of 'Candidatus Liberibacter solanacearum' associated with zebra chip[a]

Sequencing data on rRNA	Simple sequence repeat	Multilocus sequence typing
Clade 1 or haplotype A	Lineage 2 (type 1)	ST 2
Clade 2 or haplotype B	Lineage 1 (type 2)	ST 1

[a] From Glynn et al. (34), Lin et al. (57), Nelson et al. (82), and Wen et al. (110).

General features of 'Ca. L. solanacearum' genome.

The genome of 'Ca. L. solanacearum' is composed of 1.26 Mbp (National Center for Biotechnology Information [NCBI] GenBank accession number CP002371) in a circular chromosome which encodes a total of 1,192 putative proteins coding sequence. In all, ≈68% of them are predicted to be homologous with known putative functions, and the rest of them are assigned to hypothetical proteins with unknown functions (Table 2). These numbers are similar to the 'Ca. L. asiaticus' genome. 'Ca. L. solanacearum' and 'Ca. L. asiaticus' have a G+C content of 35.24 and 36.48%, respectively, which is considerably lower than the ≈60% G+C content observed for most bacterial species in the Rhizobiaceae family (16,35,111). Sequence analysis identified three copies of complete rRNA operons (16S, 23S, and 5S), 45 genes encoding for tRNAs, and ≈35 genes identified as pseudogenes in the 'Ca. L. solanacearum' genome (Table 2). No plasmids were found in either Liberibacter genome. Comparative analyses of the 'Ca. L. solanacearum' and 'Ca. L. asiaticus' genomes revealed that ≈884 protein-coding sequences are common to both organisms (58). Notably, 236 coding sequences from 'Ca. L. solanacearum' have no corresponding orthologs in 'Ca. L. asiaticus'. On the other hand, 186 coding sequences from 'Ca. L. asiaticus' have no corresponding orthologs in 'Ca. L. solanacearum'. Comparative sequence analyses of 'Ca. L. solanacearum' and 'Ca. L. asiaticus' genomes show a high level of synteny, suggesting that both circular chromosomes are descendants of a common ancestor. Intriguingly, analyses of two Liberibacter genomes revealed that nearly 30% of open reading frames in 'Ca. L. solanacearum' and 'Ca. L. asiaticus' have no known function. This implies that a number of biochemical pathways remain to be identified. An incredible biological diversity among novel protein families has been observed in bacterial genomes (87).

Table 2 General genome features of 'Candidatus Liberibacter solanacearum' and 'Ca. L. asiaticus'

Genome features	'*Ca.* L. solanacearum'	'*Ca.* L. asiaticus'
Size (bp)	1,258,278	1,227,204
GC (%)	35.2	36.5
Protein coding genes	1,192	1,136
Hypothetical proteins	405	362
rRNA operons	3	3
tRNA genes	45	44
Pseudogenes	35	32

Functional determination of these novel genes will improve our knowledge for better understanding of these disease-associated, putative pathogenic bacteria. Recent advances in sequencing technology led to the discovery of a high correlation between the number of genes and the genome size of bacteria. Most bacterial genomes range in size from 0.5 Mbp to upward of 10 Mbp, with the average of 1 kb per gene. Molecular characterization of various microbial genomes has revealed that many pathogenic and mutualistic intracellular bacterial species have smaller genomes than their free-living relatives (5,33,68). The loss of a great number of genes results in the reduction of genome size that is often associated with the adaptation to an obligate lifestyle, in which many molecules can be directly acquired from the host (33,83). The phloem-restricted Liberibacter bacteria have small genomes, ≈1.2 Mbp. Likewise, the reduction of genome size through gene loss is associated with the evolutionary adaption to a lifestyle as an obligate intracellular pathogen that largely depends on intermediate metabolites acquired from its hosts. Bacteriophages, or phages, are bacterial viruses. Two homologous segments of prophage sequences were identified in the 'Ca. L. solanacearum' genome (P-I and P-II) (Fig. 3) (58). Sequence analysis showed that one of the prophage segments in 'Ca. L. asiaticus' strain psy62 has a high degree of similarity with the prophage sequences of 'Ca. L. solanacearum' genome but no circular phage sequence was identified in either Liberibacter genome (27,58).

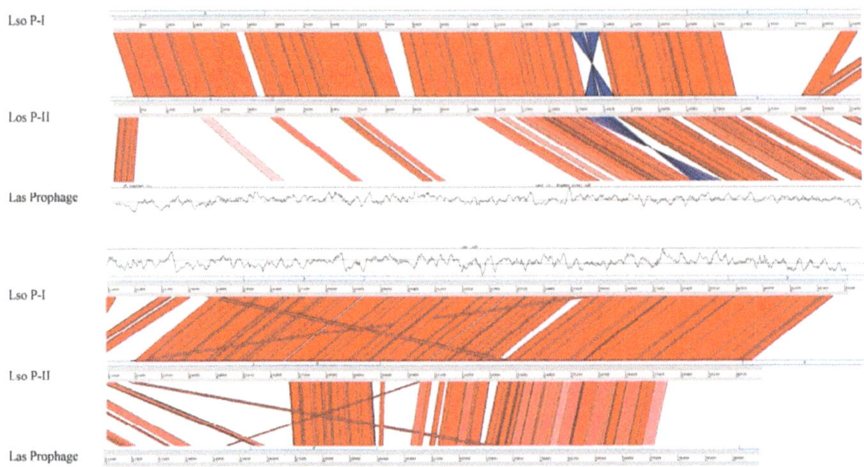

Figure 3 Syntenic conservation between prophage sequences of 'Candidatus Liberibacter solanacearum' (P-I and P-II) and 'Candidatus Liberibacter asiaticus' (Las Prophage). Colored bands represent tBLASTx similarity: collinear (red) and inverted (blue).

More recently, two circular phage genomes (SC1 and SC2) were identified in 'Ca. L. asiaticus'-infected periwinkle (Catharanthus roseus) and dodder plants (Cuscuta campestris) (115). In the Asian citrus psyllid, however, both were found integrated in tandem in the 'Ca. L. asiaticus' UF506 genome as prophage. Sequence analysis showed that SC2 appears to be derived from SC1 but lacks lysis genes and functions as lysogenic conversion. SC1, on the other hand, seemed to be a fully functional, temperate phage with a lytic cycle that appears to be activated when the 'Ca. L. asiaticus' host infects plants but not when inhabiting psyllids (115). It is not clear whether a lack of a circular phage genome in 'Ca. L. solanacearum' is due to the fact that sequenced genomes were derived from 'Ca. L. Solana cearum'-infective psyllids or to some other factor. It remains to be determined experimentally whether such lysogenic and lytic conversions are also the case in an infected host, such as when 'Ca. L. solanacearum' infects potato plants. Alignment of the two Liberibacter genomes suggests that several recombination events have occurred since the divergence of these two species from a common ancestor (58). Phage integration events appear to play a key role in the rearrangements in the Liberibacter genomes. In addition to the two prophage genome sequences, there are a number of prophage-like elements and phage remnants distributed throughout the 'Ca. L. solanacearum' genome, presumably derived from multiple ancestral bacteriophage integration events.

Virulence determinants

One of the most powerful approaches for the analysis of pathogen genomes involves computational bioinformatics identifying genes homologous to known virulence factor genes. Sequence-similarity-based searches such as BLAST and position-specific iterated BLAST against a reference database facilitate identification of orthologous genes with the predicted putative functions. Annotation of the 'Ca. L. solanacearum' genome sequence provides a basis for understanding the host– pathogen interaction that leads to disease development. Many gram-negative plant-pathogenic bacteria possess a type III secretion system (T3SS) to inject effector proteins directly into host cells. T3SSs are encoded by hypersensitive response and pathogenicity genes, which cause disease in susceptible plants but elicit the hypersensitive response in resistant plants (14,59). Interestingly, no T3SS was identified in the 'Ca. L. solanacearum' genome, nor were offensive virulence factors such as cell-walldegradation enzymes, cellulose, pectinase, xylanase found (58). Lack of a T3SS apparatus has been found in some insectmediated transmission and vascular-restricted bacteria, including 'Ca. L. asiaticus' and Xylella fastidiosa (27,102,107). Presumably, this secretion apparatus might not be necessary because the initial infection of this intracellular pathogen is mediated via direct injection into the host cell by its insect vector. In spite of missing a classical gene-for-gene model in the 'Ca. L. solanacearum' genome, other genes associated with putative roles in virulence or pathogenicity were identified, including genes involved in ion transport. For example, the zinc transport system is composed of a cluster of three genes (ZnuABC) that encode a prokaryotic-type ABC-family protein complex (30).

It is predicted that the putative zinc transport system is a high-affinity cation carrier that shuttles zinc ions across the inner membrane of 'Ca. L. solanacearum' (4). Results of a recent study of 'Ca. L. asiaticus' suggested that the disruption of this system may present a viable target not only to mitigate HLB symptom severity by making more free zinc available to the plant but also by providing a means to decrease the overall virulence of the bacterium itself (106). ZC-affected potato plants often have yellowing symptoms resembling zinc deficiency, likely associated with the presence of this putative high-affinity zinc carrier. Therefore, the silencing of this gene or the identification of specific inhibitors of this bacterial transporter may be a way to mitigate ZC symptoms. Sequence analysis predicted that the 'Ca. L. solanacearum' genome encoded iron-transport and accumulation (ITA) systems. The ITA system consists of a five-gene cluster found in the 'Ca. L. solanacearum' genome and some other pathogenic microbes; however, this system is missing from the citrus HLB-associated 'Ca. L. asiaticus' (58). It is hypothesized that this gene cluster encodes proteins involved in iron scavenging, transport, and sequestration based on their similarity to known iron transporters and gene arrangement (28,89). Because ZC-affected plants sometimes mimic plants with iron deficiency, perhaps this phenotype is related to a high-affinity iron uptake system possessed by 'Ca. L. solanacearum'. Thus, functional confirmation of one or more of the genes in the ITA cluster will be necessary to identify the target to develop a gene-based therapy for an effective ZC disease treatment. Another pathogenicity gene that was identified in the 'Ca. L. solanacearum' genome is a locus encoding an NahG-like salicylate hydroxylase.

Overwhelming evidence demonstrates that salicylic acid (SA) accumulation is essential for expression of multiple modes of plant disease resistance (9). Transgenic tobacco and Arabidopsis thaliana expressing the bacterial enzyme salicylate hydroxylase cannot accumulate SA, because it is rapidly degraded to catechol by the introduced enzymatic activity (7,113). This defect not only makes the plants unable to induce systemic acquired resistance (SAR) but also leads to increased susceptibility to viral, fungal, and bacterial pathogens. Therefore, a functional defect of the salicylate hydroxylase gene in 'Ca. L. solanacearum' may enhance host SAR in response to 'Ca. L. solanacearum' infection. Liberibacter spp. have not been successfully cultured in vitro. Consequently, conventional gene knock-out techniques cannot be applied to determine gene function. Elucidation of the function of these virulence genes will likely provide further fundamental insights into how this pathogenic bacterium survives with its hosts and elicits disease development, as well as facilitate development of new gene-based therapeutic treatment strategies for this potato disease.

Metabolic properties

'Ca. L. solanacearum' is environmentally restricted to the sieve cells of phloem vessels of potato plants or in the salivary glands of potato psyllids. Thus, its lifestyle reflects adaptive features to nutritional niches in both hosts. The complete genome sequence of 'Ca. L. solanacearum' allows reconstruction of detailed predictive metabolism models based on bioinformatics. Annotation of the 'Ca. L. solanacearum' genome revealed that it encodes a large number of membrane-transporterrelated proteins. Membrane transports are essential for the movement of substrates across a permeable barrier of the cell membrane. Transporters enable the acquisition of organic nutrients and inorganic elements, efflux of toxin compounds, ion homeostasis, signal sensing, energy production, and various important cellular functions (84). 'Ca. L. solanacearum' contains a nearly complete set of genes encoding enzymes in glycolytic pathways and enzymes required to convert pyruvate to acetyl-coA, an important biochemical molecule for fatty acid metabolism and entry into the tricarboxylic acid cycle (58). Apparently, these enable it to effectively utilize carbon sources available in phloem sieve cells of the host for catabolic and anabolic reactions (58). Although there are a few genes for production of enzymes found in the 'Ca. L. solanacearum' genome involving de novo biosynthesis of amino acids, at least three complete transport systems were identified. These transport systems function as a broad-range amino acid transport capability, including the L-amino acid ABC transporter, proline glycine-betaine ABC transporter system, and DctA-family dicarboxylate transporter. All of these transporter systems are also present in the members of Rhizobiaceae. Because both 'Ca. L. solanacearum' and 'Ca. L. asiaticus' lack the key enzymes involved in the terminal stages of oxidative phosphorylation pathway, it is concluded that both 'Ca. L. asiaticus' and 'Ca. L. solanacearum' have limited capacity for aerobic respiration, which seems to be consistent with oxygen-limited phloem cells where they have become adapted. Additionally, genes encoding for sulfur metabolism are also absent in both genomes, suggesting that both bacteria may use other alternative energy-generating pathways under anaerobic conditions. Indeed, genome analyses identified a gene encoding ATP/ADP translocase of the NttA family in both bacteria (27,58). Functional analysis confirmed that this transport protein was able to import ATP directly from the host cell's cytosol and translocate it into the bacterial cell. Thus, it can be concluded that NttA provides the 'Ca. Liberibacter sp.' cell with energy (105). This unique feature has been found in other obligate intracellular bacteria such as the intracellular bacteria Rickettsia prowazekii, Chlamydia psittaci, and Lawsonia intracellularis (40,97). Surprisingly, orthologs of the NttA transport family are absent from all other Rhizobiaceae. Given the reduced size of the 'Ca. Liberibacter sp.' (1.2 Mb) genome, it appears that this transport was acquired through horizontal gene transfer early in the evolution of 'Ca. Liberibacter' spp. Although the biochemical function of the translocase was confirmed, the biological role has yet to be determined. The capability of acquiring ATP from the host could play an important role in this obligate intracellular bacterium in planta and this feature may be exploited to facilitate a means for growing an axenic culture of this bacterium in vitro (105). The sequencing of bacterial genomes has given insight into the genomic inventory that enables prediction of putative biochemical information about 'Ca. L. solanacearum'. The precise substrate specificity for some significant pathways could not be determined with certainty without functional confirmation. However, this type of analysis provides a good overview of the biochemical and metabolic capability of this organism, which is particularly important for unculturable bacteria where conventional bacteriological approaches are often limited or not applicable.

Aspects of the etiology and epidemiology of ZC

The etiology and epidemiology of any plant disease is likely to be complex when it involves an insect vector, and ZC of potato is no different. Several important aspects of the disease will be reiterated here for clarity. The putative pathogen of ZC is 'Ca. L. solanacearum' (21,52,53,55) and it is vectored by the potato psyllid, B. cockerelli Šulc. The situation is made more complicated by the implication that another bacterium, 'Ca. L. psyllaurous', is the cause of a similar, yet distinct potato disease, psyllid yellows (PY) (Fig. 1D to F) (39,110). It was suggested that 'Ca. L. psyllaurous' is a secondary endosymbiont of the potato psyllid because it was demonstrated to be both vertically and horizontally transmitted through several life stages of the insect, raising additional questions regarding the ancestry of this bacterium (39). It is not a universally held belief, however, that PY is caused by 'Ca. L. psyllaurous' (101); instead, it may be caused by a toxin injected by the potato psyllid (17,29). It is interesting to note that the association of the potato psyllid in the development of ZC (71,74) predated the identification of 'Ca. L. solanacearum'/'Ca. L. psyllaurous' as putative pathogens of potato (39,52,53,55). Molecular characterization and phylogenetic analysis of the full-length 16S rRNA from the new species of 'Ca. Liberibacter' further confirmed the association of ZC disease of potato and the potato psyllid vector (56).

The etiological cause of ZC was further confirmed by grafting ZC-affected plant tissue to healthy plants, which developed typical symptoms of ZC, and the etiological agent was visualized using electron microscopy, as previously indicated (20,99). Using the available PCR technology, 'Ca. L. solanacearum' has been confirmed in ZC-affected potato and tomato plants from Texas (1,99,110), California (20), and Mexico (76,77,110). Using a wide array of PCR primers, 'Ca. L. solanacearum' was found to be associated with ZC-affected plants from throughout Texas in archived ZC-affected potato samples collected in 2005 to 2007 (110). The similarity among all of these 'isolates' of 'Ca. L. solanacearum'/'Ca. L. psyllaurous' was >99% (110). These studies also confirmed the presence of 'Ca. L. solanacearum'/'Ca. L. psyllaurous' in Colorado, Kansas, and Nebraska in 2008 and in seed potato tubers produced from Wyoming in 2007 (110).

Of much greater concern is the outbreak of ZC in the major potato-producing states of Idaho, Oregon, and Washington in 2011 (18,22). Although economic losses in these later states were confined, it does appear that ZC is spreading and is likely to be present wherever the insect vector B. cockerelli is found. The vector of ZC-associated 'Ca. L. solanacearum', B. cockerelli, is known to be a spatially and temporally variable insect in the United States and New Zealand (15,61,70).

Research has been ongoing to determine the phenology and seasonal migration of the psyllid throughout the native range, especially where it is known to be associated with ZC. It has been clearly demonstrated that psyllid populations differ in their ability to transmit the ZC-associated 'Ca. L. solanacearum' in potato (69). Although the precise explanation for this variability is not clearly understood, it may be due to genetic differences among psyllid populations, haplotypes or biotypes of the ZC-associated bacterium, or both. Previously, genetic variation of potato psyllids associated with ZC was determined at both the biotype level and the population level using either cytochrome oxidase I or interSSR analysis (45). The goal of this type of analysis has been to compare and contrast psyllids collected at different locations and at different times of the year and ultimately determine the movement patterns of this insect. It is epidemiologically important to determine whether there are biotypes of the potato psyllid that are more efficient vectors of 'Ca. L. solanacearum'. This information would further our understanding of insect genotypic diversity and possibly help predict the annual severity of ZC expression. It has been known for some time that there are at least two biotypes of B. cockerelli, a western and a central U.S. biotype (61). More recently, a "northwestern" biotype of B. cockerelli in the United States has been identified (101) and implicated in the ZC epidemic in the Pacific Northwest (18,22).

The impact that these different biotypes have on the epidemiology of ZC is not yet known but studies to determine the impact that psyllid populations have on disease development are ongoing. Initially, nothing was known about acquisition and transmission of 'Ca. L. solanacearum' and environmental effects. The successful transmission of an insect-vectored plant pathogen is largely dependent on vector transmission efficiency, and important pieces of the host–pathogen–vector relationship have been established. For example, adult psyllids have been shown to be more efficient than nymphs in transmitting 'Ca. L. solanacearum' (12). Acquisition success is influenced by feeding site and is highest when psyllids have access to the whole plant (64). Interestingly, inoculation access period had virtually no influence on the overall incidence of ZC, severity of disease development, and yield loss (11); however, the rate of successful inoculation increased with higher vector numbers, and plants inoculated with multiple psyllids had higher 'Ca. L. solanacearum' titers (64). Disease progress and tuber yield is clearly independent of psyllid density (12,64). A single adult potato psyllid is capable of successfully transmitting 'Ca. L. solanacearum' to potato in as little as 6 h (12). Air temperatures immediately following inoculation with Liberibacter-infective psyllids have been reported to affect ZC development (75), although more recent studies suggest that air temperatures following infection have no effect on latent period or symptom development (A. Rashed and C. Rush, personal communication). The stark difference in these results may be due to differences in the haplotype of the 'Ca. L. solanacearum' or the psyllid vector used in the respective experiments; however, that information is not currently known.

Temperatures as low as 17°C have been reported also to slow development of ‘Ca. L. solanacearum’ as do temperatures >32°C. Interestingly, very similar effects of temperature have been observed on the insect vector (17,60,109). The optimum temperature for psyllid development is estimated to be ≈27°C. Egg laying, egg hatch, and nymph development also slow at 32°C and cease at 35°C (60). Therefore, vector dynamics and pathogen development appear to be similarly reduced at temperatures of ≥32°C. The source or sources of ‘Ca. L. solanacearum’ from which B. cockerelli acquires the bacterium is still a source of speculation. Potato psyllids have long been thought to overwinter in the Lower Rio Grande Valley on wild Lycium spp. (36,94). This solanaceous host also has been found to be infected with ‘Ca. L. solanacearum’, albeit at very low frequencies (110). Therefore, it is suspected but not empirically proven that the potato psyllid can acquire the ZC bacterium while overwintering, which may explain why frequencies of ‘Ca. L. solanacearum’ are traditionally higher early in the growing season than later (19). Titers of ‘Ca. L. solanacearum’ in potato psyllids have been demonstrated to be substantially reduced in starved potato psyllids (95), which may explain why migrating populations of B. cockerelli have lower ‘Ca. L. solanacearum’ frequencies than those coming off overwintering hosts. Additionally, ‘Ca. L. solanacearum’ is known to have a negative effect on the population growth rate of the potato psyllid by reducing fecundity and survival of nymphs (80).

It is apparent that more detailed information on pathogen–vector relationships and interactions is necessary if risk assessment models are to be effective in predicting ZC development. Because the potato psyllid migrates from overwintering sites, it makes epidemiological sense that the development of ZC is more intense along the edges of an affected field rather than further infield (112). Disease progress was best explained by a seconddegree polynomial (quadratic) equation. Perhaps more interesting was the determination that ZC infections in potato fields did not develop spatially in a completely random pattern but, rather, in clusters, with ZC-infected plants situated in close proximity to each other (41). However, analysis of the infection clustering did not support secondary spread of the pathogen from an initial infection point. This is in direct contrast to earlier observations of ZC development in the field in Texas, where the disease developed in “holes” or pockets (Fig. 4). It is not known whether these holes are the result of secondary spread of the pathogen from individual focal points (71) or whether these clusters of infection have formed from plant-to-plant movement of competent vectors. A very fortunate attribute of ZC development is that seedborne ‘Ca. L. solanacearum’ does not appear to be epidemiologically important (42). Although the vegetative propagule used as “seed” to plant potato can harbor ‘Ca. L. solanacearum’ (42,86), it has not yet been demonstrated to be a source of inoculum in development of an epidemic. Seedborne ZC does have a detrimental effect on seed quality, however, largely through the reduction of germination rates and subsequent seedling vigor (42,86).

Figure 4 Development of zebra chip (ZC) in a cluster of infections in the field, which initially suggested secondary spread of the ZC pathogen from an initial focal point.

Host responses

ZC causes significant losses in tuber yield (11,42). The development of ZC symptoms in tubers occurs ≈3 weeks after infection, after which tuber development ceases (10). Concomitantly, tuber solids began to decrease as soon as ZC symptoms began to appear. The rapid development of ZC tuber symptoms in the aforementioned studies supports previous studies on movement of 'Ca. L. solanacearum' in potato and tomato (49). In these studies, the 'Ca. L. solanacearum' bacterium moved in the phloem and was more readily detected in the upper- and middle-tier leaves, before it was detected in fully developed leaves, 2 to 3 weeks after initial exposure to 'Ca. L. solanacearum'-inoculative psyllids (49). It was concluded that 'Ca. L. solanacearum' translocation within potato and tomato plants is similar to what has been previously described for phloem-limited viruses and phytoplasmas. A number of studies have investigated the physiological response of the host to infection at the cellular and molecular level. For example, there is evidence of cell death in potato tubers affected by ZC (66) and a build-up of host phenolics and plant defense-related proteins (3,81,108). Total protein content increases threefold, which includes increases in gene expression regulatory factors such as cyclophilin and tuber storage proteins such as patatin (3). Tyrosine has been suggested to be the primary cause of enzymatic browning in potato tubers (81). Glucose and sucrose levels also are elevated in ZC-diseased tubers, reducing process quality in potato chips due to the Maillard reaction that occurs during high-temperature frying (12,31). The characteristic development of this potato chip defect was the ultimate source of the name given to this potato disease, "zebra chip" (Fig. 5).

Figure 5 Range of symptoms of the zebra chip defect in potato chips. The development of these symptoms is due to a build-up of glucose in medullary ray tissue that caramelizes during high-temperature frying due to the Maillard Reaction.

Initial investigations have indicated that potato plants are more susceptible to the pathogen phenologically early in the plant's life cycle than later (31), although temperatures at the time of infection may affect the infection process, development of disease symptoms, and the bacterium–vector relationship (75). More recent studies indicate that plant age at the time of infection, over a broad range of temperatures, has no effect on the development of initial ZC symptoms or on the rate of symptom severity (64) (C. Rush, personal communication). These conflicting reports may be due to the specific interaction of 'Ca. L. solanacearum' haplotypes and potato cultivar and the conditions with which the studies were performed. Although all of the impacts proteobacterial pathogens have on affected plants are not understood, it is likely that the presence of the ZC-associated 'Ca. L. solanacearum' in the phloem is similar to another proteobacterium, a 'Ca. Phlomobacter sp.', and its effect on the roots of sugar beet (100). We know that 'Ca. L. solanacearum' infection of potato phloem disrupts carbohydrate flow to belowground portions of the plant, causing a number of the symptoms associated with ZC (11,31). The lack of carbohydrate flow to the roots will likely cause them to starve and decline in function, thereby causing the aboveground portions of the plant to also slowly decline and eventually die (Fig. 1A and B).

Host range

From the initial suggestion of 'Ca. L. solanacearum' as the putative pathogenic cause of ZC, it has been known that 'Ca. L. solanacearum' is also associated with a number of other solanaceous hosts in addition to potato and tomato (52,53,55). In New Zealand, 'Ca. L. solanacearum' has been associated with a new disease of Capsicum sp. and tomato while also causing a potato disease very similar to ZC in the United States (54,55). In addition, 'Ca. L. solanacearum' also has been found to infect S. betaceum (tamarillo) and

Physallis peruviana (Cape gooseberry) in New Zealand (53,54). 'Ca. L. solanacearum' has been associated with a psyllid-associated disorder of carrots in several countries in Europe (2,72,73,78,79). These are the first reports of a nonsolanaceous host for 'Ca. L. solanacearum'. The association of 'Ca. L. solanacearum' in carrot and with two other psyllid species raises interesting questions about the diversity of 'Ca. L. solanacearum' and its specific relationship with insect vectors, which is discussed elsewhere in this review. In the United States, a survey of 'Ca. L. solanacearum' in solanaceous hosts in Texas found that silverleaf nightshade (S. elaeagnifolium Cav.), wolfberry (Lycium barbarum L.), black nightshade (S. ptychanthum Dun.), and jalapeno pepper (Capsicum annuum L.) were hosts for the ZC-associated bacterium (110). The 'Ca. Liberibacter' spp. detected in all of these solanaceous plant hosts had similarity indices of 99.8% to the partial 16S rRNA gene sequences compared with the sequences of the same bacterium detected in ZC-affected potato plants (110). The role that these alternate hosts such as perennial solanaceous plants wolfberry and silverleaf nightshade have on the epidemiology of 'Ca. L. solanacearum' is not known but further studies are warranted and should be performed, as they have been for HLB caused by 'Ca. L. asiaticus' (23,25). For any of these plant hosts to be identified as epidemiologically significant reservoirs of 'Ca. L. solanacearum' inoculum, they will have to be demonstrated to retain the ZC-associated 'Ca. L. solanacearum' in an infectious state over an extended period of time and also be an important colonization host for B. cockerelli.

Molecular diagnosis

Detection of 'Ca. L. solanacearum' in its insect vector is essential for studying its etiology and epidemiology and in developing effective disease management strategies (36). Because ZC is associated with a prokaryotic bacterium, most of the PCR methods developed for 'Ca. L. solanacearum' detection have utilized a conventional PCR format, with primers to amplify 16S and 23S ribosomal DNA sequences (39,50,52, 54,91,99,110). Although a number of forward primers have been empirically designed based on the 16S rRNA sequence of 'Ca. L. solanacearum' (NCBI GenBank accession number EU980389 (50,55,110), the reverse primer most commonly used for the PCR reaction is the Liberibacter-universal primer OI2c (47). As a result, it is not surprising that some of the primers developed to detect 'Ca. L. solanacearum' also amplify some strains of 'Ca. L. asiaticus' and 'Ca. L. africanus' (92,110). Fortunately, all of the PCR primers developed to date are specific for 'Ca. Liberibacter' spp. associated with the ZC complex and do not detect other prokaryotic plant pathogens that affect potato (110). In addition to PCR primers using 16s and 23s rDNA genes, more recent primers have utilized genomic DNA such as a conserved bacterial housekeeping gene, adenylate kinase (91). These primers have been demonstrated to a detection limit of ≈0.65 ng (91). Genomic-based PCR primers also have been developed to facilitate detection of 'Ca. L. solanacearum' in the potato psyllid and have been found to be valuable in determining the frequency of the bacterium in its vector (19). Few studies have investigated the effectiveness of detecting 'Ca. L. solanacearum' in infected potato plant parts. In one study, the overall efficiency of the current PCR technology associated with the detection of true positives was demonstrated to be as low as 60% (110). Detection efficiencies have been generally determined to be higher for belowground portions of plants, such as stolons, roots, and tubers, as opposed to leaves, leaf petioles, and aboveground stems (50,110). However, these results might be contradicted by more recent studies which found 'Ca. L. solanacearum' detection frequencies higher in upper- and middletier leaves than in lower aboveground tissues; however, no attempt was made to assay belowground tissues (49). The presence of PCR inhibitors in tuber tissues can be problematic and caution needs to be exercised when using tuber tissue to detect 'Ca. L. solanacearum' infections (110). There has been interest in quantifying the development of 'Ca. L. solanacearum' in hosts once the plant becomes infected. In order to accomplish this, the expected amplicon for the PCR reaction was cloned in a plasmid and the number of copies of DNA estimated using the number of base pairs in the sequence and the average molecular mass of a base pair in double-stranded DNA (50). This method was useful in determining that 20 copies of the target 16S rDNA of 'Ca. L. solanacearum' could be detected in the DNA extract of a ZC-affected plant using a quantitative (q)PCR format. Using this methodology and applying it to naturally infected potato plants from the field, it was determined that the majority of plants harbored an 'Ca. L. solanacearum' population of 105 to 106 cells/g of tissue (50). As expected for a phloem-restricted pathogen, 'Ca. L. solanacearum' populations were three times higher in belowground than aboveground tissue, which likely explains why detection efficiencies are much higher when belowground plant tissue is used (50,110). In this study, 92% of all field samples tested positive for 'Ca. L. solanacearum' using qPCR whereas only 40 to 46% tested positive using conventional PCR, depending on the primers used (50). Molecular methods of pathogen detection can only be efficient when using high-quality DNA. Thus, it is extremely important to verify that the DNA is of sufficient quality to be amplified in a PCR reaction. When PCR assays utilize only a single set of primers to detect a microbe, a negative result does not distinguish between the absence of the target DNA sequences, the presence of PCR inhibitors, or poor DNA quality.

As a result, a multiplex PCR platform that combines pathogen-specific primers with an internal control that utilizes a plant-based DNA sequence target could be beneficial. Two such multiplex PCR methods have been developed to detect 'Ca. L. solanacearum' in plants, one utilizing a potato β-tubulin gene sequence (110) in a conventional PCR format, the other using a plant mitochondrial cytochrome oxidase gene sequence in a qPCR format (50). The use of such an internal control determined that 19% of the DNA samples tested gave a negative 'Ca. L. solanacearum' result due to poor DNA quality, clearly demonstrating the usefulness of such an approach (110). Despite recent improvements in PCR detection technology (19,91), there is still a need for improvement. For example, the development of a multiplex PCR format that would distinguish 'Ca. L. solanacearum' haplotypes (57) while simultaneously determining potato psyllid biotypes (45,103) could be very useful in epidemiological studies. Further improvements in qPCR also could facilitate epidemiological studies of ZC, as demonstrated with HLB-associated Liberibacter spp. in citrus (43,51,104). For example, qPCR has been found to be extremely useful in determining infestation rates of 'Ca. L. asiaticus' in the Asian citrus psyllid in Florida (51), which were 10 to 100%. PCR is currently being used to detect 'Ca. L. solanacearum' in potato psyllids collected throughout the United States as part of a ZC disease management plan (19,21). In contrast to the HLB situation, only ≈2% of these potato psyllids have tested positive for 'Ca. L. solanacearum' (21).

Disease management

In the absence of obvious genetic resistance, it is predictable that an agricultural industry will seek the use of insecticides in an attempt to effectively manage an insect-vectored disease. ZC is a relatively new disease in the United States (21,37) and, in the absence of other effective disease management strategies and tactics, efficacious insecticides have become the primary means by which the vector and, therefore, the disease are managed (32,36,62,63). Most of these studies have concentrated on the efficacy of various insecticides on the life stages of the potato psyllid. For example, thiamethoxam and abamectin were found to be effective in significantly reducing adult populations of B. cockerelli, although residual effects on adult mortality where relatively short (≈24 h) (32). Insecticides such as spiromesifen were found effective against nymphs (36). Unfortunately, there is a lack of evidence that the use of insecticides to manage psyllid populations will have any effect on reducing the incidence and severity of ZC disease development under conditions where potato psyllid populations harbor high frequencies of 'Ca. L. solanacearum' infection. For that reason, a few studies involving tomato and the potato psyllid have investigated the potential to improve host resistance and insecticides to effectively manage both the vector and the disease (62,63). Current pest management practices rely on the use of insecticides to increase psyllid mortality but they can also be used to effectively manage potato psyllid populations by impacting each life stage in ways that do not affect mortality. For example, some insecticides such as imidacloprid, abamectin, and pymetrozine have been shown to reduce probing durations while concomitantly increasing the amount of time adult psyllids spend off the leaf surface, suggesting that these insecticides also act as repellents (13). Perhaps more importantly, two of the insecticides used to manage psyllid populations, imidacloprid and abamectin, also significantly lowered the transmission of 'Ca. L. solanacearum', which may help reduce further infections (13). However, the development of insecticide resistance compromises the ability of any chemical treatment to effectively limit the rate of disease development, and some populations of potato psyllid have already been documented with resistance to imidacloprid (64). Growers also use a variety of other management tactics to limit the economic impact of ZC as it affects their potato crop. Laboratory testing of psyllids obtained from a trapping network has confirmed the suspicions of growers that the highest frequencies of 'Ca. L. solanacearum' in migrating populations occur early in the growing season (19). This may be because the potato psyllid can acquire the ZC-associated 'Ca. L. solanacearum' from an overwintering host, such as Lycium spp. or other solanaceous hosts, that are capable of harboring the bacterium (110). It has also recently been demonstrated that the potato psyllid can overwinter on S. dulcamara L. in Idaho (A. Jensen, personal communication). The role that this Solanum sp. may have on the epidemiology of 'Ca. L. solanacearum' is also warranted. As previously stated, it has also been demonstrated that colonies of B. cockerelli lose infectivity when starved (95), which may occur when psyllids migrate, thereby reducing the risk of bacteriliferous psyllids initiating a serious epidemic. As a direct result, potato growers in a number of potato-production areas of Texas where

ZC is endemic have resorted to delaying planting so as to avoid psyllid populations less likely to be vectoring 'Ca. L. solanacearum' (N. C. Gudmestad, unpublished data). Because psyllids move into fields from the periphery (112), potato growers also treat the edges of fields with insecticides so as to reduce the opportunity of the insect vector to become established in the entire field. Other disease management tactics being investigated include the development of traps that attract male and female adult psyllids through the use of volatile chemical attractants (38). Entomopathogenic fungi may also provide additional components useful in an integrated pest management strategy (48) in addition to the use of kaolin particle films and other biorational insecticides which appear to be useful in repelling all developmental stages of the potato psyllid (85,114). These potential alternative methods of ZC management are a priority and are being intensively investigated. The future for disease management likely lies with diseaseresistant potato cultivars (13). Although all potato cultivars currently grown in the United States are susceptible to ZC, a number of potato genotypes have been identified in which there was a significant reduction in 'Ca. L. solanacearum' transmission compared with susceptible controls. However, the exact nature of the host–parasite interaction contributing to the reduced transmission is still unclear (13). The development of ZC-resistant germplasm is a priority and is ongoing.

Conclusion and future perspectives

Since the first identification of ZC as a new disease of potato in the United States and New Zealand, a significant effort to determine the cause and, ultimately, the management of the disease has taken place. In addition to determining the cause of ZC, we now have a much deeper understanding of the putative proteobacterium involved. Although 'Ca. L. solanacearum' remains a relatively primitive pathogen in a genomic sense, it possesses a number of unique attributes that potentially allow it to be a successful plant pathogen. Although 'Ca. L. solanacearum' lacks substantial enzymatic capability and a T3SS (58), its genome does have other orthologous virulence-related means at its disposal. These need to be explored further because some hosts infected with 'Ca. L. solanacearum' become severely diseased, such as potato and tomato, while others such as eggplant and pepper can harbor dense populations of 'Ca. L. solanacearum'-infected psyllids with little or no ill effect (N. C. Gudmestad, unpublished). Furthermore, studies have shown that phosphonate fungicides induce SAR in potato (6), which has been demonstrated to significantly reduce 'Ca. L. solanacearum' titers when plants are treated prior to infection (93). It is not known whether this is related to the functional suppression or defect in the salicylate hydrolase gene in 'Ca. L. solanacearum'. Nevertheless, there are genomic data available that may provide the basis for the development of disease management tactics in the future. The 'Ca. L. solanacearum' vector, B. cockerelli, adds an additional level of complexity to the ZC disease of potato. Although it has been long recognized as an insect pest of potato in certain regions (17,29,94,109), its recent association with a putative bacterial pathogen is as perplexing as it is agriculturally problematic. If 'Ca. L. psyllaurous' is truly a secondary endosymbiont of B. cockerelli (39), then the question remains whether or not there are avirulent wild-type ancestors that still exist, perhaps in some other solanaceous plant host. Recent studies suggest there may be 'Ca. L. solanacearum' haplotype–B. cockerelli haplotype interrelationships that could affect ZC disease development. For example, PCR testing of the western B. cockerelli haplotype revealed only infections of haplotype A 'Ca. L. solanacearum' whereas central U.S. haplotypes of the potato psyllid are predominately infected with haplotype B 'Ca. L. solanacearum' (A. Wen and N. C. Gudmestad, unpublished). Although we have substantial information on acquisition, transmission and effect of psyllid densities on ZC development (11,12), there are no data to suggest that this is true for all haplotypes of the vector and the ZC-associated 'Ca. L. solanacearum'. Although there has been a great deal of progress made in our understanding of the ZC disease complex, it is apparent that there is much more to be learned before this disease can be managed long term in a sustainable manner. There are still substantial gaps in our knowledge of disease epidemiology, particularly as it affects vector survival and overwintering, as well as pathogen dynamics. There are also substantial gaps in our understanding of disease resistance mechanisms, which ultimately will be the primary means by which this disease is effectively and sustainably managed. Despite that, the depth of our understanding, relative to the genetic diversity of the putative pathogen and its vector, the migration of the vector, and the frequency of 'Ca. L. solanacearum' infection, coupled with effective insect and disease management integrated pest management strategies in the short term, have been impressive for a recently developed potato disease. The expectation is that future studies will continue to improve our understanding of this pathogen–host–vector interaction, while concomitantly reducing the economic impact of ZC in the United States and elsewhere.

ACKNOWLEDGMENTS

Trade names or commercial products in this publication are mentioned solely for the purpose of providing specific information and does not imply recommendation or endorsement by the United States Department of Agriculture (USDA). USDA is an equal opportunity provider and employer.This work was supported by a grant from the United States Department of Agriculture NIFA-SCRI (Project number 2009-51181-20176). We thank E. Civerolo for critical reading of this review article.

REFERENCES

1. Abad, J. A., Bandla, M., French-Monar, R. D., Liefting, L. W., and Clover, G. R. G. 2009. First report of the detection of 'Candidatus Liberibacter' species in zebra chip disease-infected potato plants in the United States. Plant Dis. 93:108.
2. Alfaro-Fernández, A., Cebrian, M. C., Villaescusa, F. J., Hermoso de Mendoza, A., Ferrandiz, J. C., Sanjuan, S., and Font, M. I. 2012. First report of 'Candidatus Liberibacter solanacearum' in carrot in mainland Spain. Plant Dis. 96:582.
3. Alvarado, V. Y., Odokonyero, D., Duncan, O., Mirkov, T. E., and Scholthof, H. B. 2012. Molecular and physiological properties associated with zebra complex disease in potatoes and its relation with 'Candidatus Liberibacter' contents in psyllid vectors. PLoS ONE 7:e37345.
4. Ammendola, S., Pasquali, P., Pistoia, C., Petrucci, P., Petrarca, P., Rotilio, G., and Battistoni, A. 2007. High-affinity Zn2+ uptake system ZnuABC is required for

bacterial zinc homeostasis in intracellular environments and contributes to the virulence of Salmonella enterica. Infect. Immun. 75:5867-5876.

5. Andersson, S. G. E., and Kurland, C. G. 1998. Reductive evolution of resident genomes. Trends Microbiol. 6:263-268.

6. Andreu, A. B., Guevara, M. G., Wolski, E. A., Daleo, G. R., and Caldiz, D. O. 2006. Enhancement of natural disease resistance in potatoes by chemicals. Pest. Manage. Sci. 62:162-170.

7. Bartscha, M., Gobbatoa, E., Bednareka, P., Debeyb, S., Schultzeb, J. L., Bautora, J., and Parkera, J. E. 2006. Salicylic acid-independent enhanced disease susceptibility1 signaling in Arabidopsis immunity and cell death is regulated by the Monooxygenase FMO1 and the nudix hydrolase NUDT7. Plant Cell 18:1038-1051.

8. Bové, J. M. 2006. Huanglongbing: A destructive, newly emerging, century-old disease of citrus. J. Plant Pathol. 88:7-37.

9. Brodersen, P., Malinovsky, F. G., Hématy, K., Newman, M. A., and Mundy, J. 2005. The role of salicylic acid in the induction of cell death in Arabidopsis acd11. Plant Physiol. 138:1037-1045.

10. Buchman, J. L., Fisher, T. W., Sengoda, V. G., and Munyaneza, J. E. 2012. Zebra chip progression: from inoculation of potato plants with Liberibacter to development of disease symptoms in tubers. Am. J. Potato Res. 89:159-168.

11. Buchman, J. L., Heilman, B. E., and Munyaneza, J. E. 2011. Effects of Liberibacter-infective Bactericera cockerelli (Hemiptera: Triozidae) density on zebra chip potato disease incidence, potato yield, and tuber processing quality. J. Econ. Entomol. 104:1783-1792.

12. Buchman, J. L., Sengoda, V. G., and Munyaneza, J. E. 2011. Vector transmission efficiency of Liberibacter by Bactericera cockerelli (Hemiptera: Triozidae) in zebra chip potato disease: Effects of psyllid life stage and inoculation period. J. Econ. Entomol. 104:1486-1495.

13. Butler, C. D., Gonzalez, B., Keremane, M. L., Lee, R. F., Novy, R. G., Miller, J. C., and Trumble, J. T. 2011. Behavioral responses of adult potato psyllid, Bactericera cockerelli (Hemiptera: Triozidae), to potato germplasm and transmission of 'Candidatus Liberibacter psyllaurous'. Crop Prot. 30:1233-1238.

14. Büttner, D., and He, S. Y. 2009. Type III protein secretion in plant pathogenic bacteria. Plant Physiol. 150:1656-1664.

15. Cameron, P. J., Surrey, M. R., Wigley, P. J., Anderson, J. A. D., Hartnett, D. E., and Wallace, A. R. 2009. Seasonality of Bactericera cockerelli in potatoes (Solanum tuberosum) in South Auckland, New Zealand. N. Z. J. Crop Hortic. Sci. 37:295-301.

16. Capela, D., Barloy-Hubler, F., Gouzy, J., Bothe, G., Ampe, F., Batut, J., Boistard, P., Becker, A., Boutry, M., Cadieu, E., Dréano, S., Gloux, S., Godrie, T., Goffeau, A., Kahn, D., Kiss, E., Lelaure, V., Masuy, D., Pohl, T., Portetelle, D., Pühler, A., Purnelle, B., Ramsperger, U., Renard, C., Thébault, P., Vandenbol, M.,

Weidner, S., and Galibert, F. 2001. Analysis of the chromosome sequence of the legume symbiont Sinorhizobium meliloti strain 1021. Proc. Natl. Acad. Sci. USA 98:9877-9882.

17. Cranshaw, W. S. 2001. Diseases caused by an insect toxin. Pages 73-74 in: Compendium of Potato Diseases. W. R. Stevenson, R. Loria, G. D. Franc, and D. P. Weingartner, eds. American Phytopathological Society, St. Paul, MN.
18. Crosslin, J. M., Hamm, P. B., Eggers, J. E., Rondon, S. I., Sengoda, V. G., and Munyaneza, J. E. 2012. First report of zebra chip disease and 'Candidatus Liberibacter solanacearum' on potatoes in Oregon and Washington State. Plant Dis. 96:452-453.
19. Crosslin, J. M., Lin, H., and Munyaneza, J. E. 2011. Detection of 'Candidatus Liberibacter solanacearum' in the potato psyllid, Bactericera cockerelli (Šulc), by conventional and real-time PCR. Southwest. Entomol. 36:125-135.
20. Crosslin, J. M., and Munyaneza, J. E. 2009. Evidence that the zebra chip disease and the putative causal agent can be maintained in potatoes by grafting and in vitro. Am. J. Potato Res. 86:183-187.
21. Crosslin, J. M., Munyaneza, J. E., Brown, J. K., and Liefting, L. W. 2010. Potato zebra chip disease: a phytopathological tale. Plant Health Progress. Online publication. doi:10.1094/PHP-2010-0317-01-RV
22. Crosslin, J. M., Olsen, N., and Nolte, P. 2012. First report of zebra chip disease and 'Candidatus Liberibacter solanacearum' on potatoes in Idaho. Plant Dis. 96:453.
23. Damsteegt, V. D., Postnikova, E. N., Stone, A. L., Kuhlmann, M., Wilson, C., Sechler, A., Schaad, N. W., Brlansky, R. H., and Schneider, W. L. 2010. Murrya paniculata and related species as potential hosts and inoculums reservoirs of 'Candidatus Liberibacter asiaticus', causal agent of Huanglongbing. Plant Dis. 94:528-533.
24. De Boer, S. H., Secor, G., Li, X., Gourley, J., Ross, P., and Rivera. V. 2007. Preliminary characterization of the etiologic agent causing zebra chip symptoms in potato. In: New and Old Pathogens of Potato in Changing Climate. Proc. EAPR Pathol. Sect. Hattula, Finland. A. Hannukkala and M. Segerstedt, eds. Agrifood Res. Working Pap. 142:28.
25. Deng, X., Zhou, G., Li, H., Chen, J., and Civerolo, E. L. 2007. Detection of 'Candidatus Liberibacter asiaticus' from wampee (Clausena lansium Skeels) by nested PCR. Plant Health Progress. Online publication. doi:10.1094/PHP-2007-0419-01-BR
26. Doddapaneni, H., Liao, H., Lin, H., Bai, X., Zhao, X., Civerolo, E. L., Irey, M., Coletta-Filho, H., and Pietersen. G. 2008. Comparative phylogenomics and multi-gene cluster analyses of the Citrus Huanglongbing (HLB)-associated bacterium 'Candidatus Liberibacter'. BMC Res. Notes 1:72. Online publication. doi:10.1186/1756-0500-1-72
27. Duan, Y., Zhou, L., Hall, D. G., Li, W., Doddapaneni, H., H. Lin, Liu, L., Vahling, C. M., Gabriel, D. W., Williams, K. P., Dickerman, A., Sun, Y., and Gottwald,

T. 2009. Complete genome sequence of citrus Huanglongbing bacterium, 'Candidatus Liberibacter asiaticus' obtained through metagenomics. Mol. Plant-Microbe Interact. 22:1011-1020.

28. Eichhorn, H., Lessing, F., Winterberg, B., Schirawski, J., Kamper, J., Muller, P., and Kahmann, R. 2006. A Ferroxidation/Permeation Iron Uptake System Is Required for Virulence in Ustilago maydis. Plant Cell 18:3332-3345.

29. Eyer, J. R., and Crawford, R. F. 1933. Observations on the feeding habits of the potato psyllid (Paratrioza cockerelli Šulc) and the pathological history of the "psyllid yellows" which it produces. J. Econ. Entomol. 26- 846-850.

30. Gabbianelli, R., Scotti, R., Ammendola, S., Petrarca, P., Nicolini, L., and Andrea, B. 2011. Role of ZnuABC and ZinT in Escherichia coli O157:H7 zinc acquisition and interaction with epithelial cells. BMC Microbiol. 11:36. Online publication. doi:10.1186/1471-2180-11-36

31. Gao, F., Jifon, J., Yang, X., and Liu, T.-X. 2009. Zebra ship disease incidence on potato is influenced by timing of potato psyllid infestation, but not by the host plants on which they were reared. Insect Sci. 16:399- 408.

32. Gharalari, A. H., Nansen, C., Lawson, D. S., Gilley, J., Munyaneza, J. E., and Vaughn, K. 2009. Knockdown mortality, repellency, and residual effects of insecticides for control of adult Bactericera cockerelli (Hemiptera: Psyllidae). J. Econ. Entomol. 102:1032-1038.

33. Gil, R., Sabater-Muñoz, B., Latorre, A., Silva, F. J., and Moya, A. 2002. Extreme genome reduction in Buchnera spp.: toward the minimal genome needed for symbiotic life. Proc. Natl. Acad. Sci. USA 99:4454-4458.

34. Glynn, J. M., Islam, M. S., Bai, Y., Wen, A., Gudmestad, N. C., Civerolo, E. L., and Lin, H. 2012. Multilocus sequence typing of 'Candidatus Liberibacter solanacearum' isolates from North America and New Zealand. J. Plant Pathol. 94:223-238.

35. González, V., Santamaría, R. I., Bustos, P., Hernández-González, I., Medrano-Soto, A., Moreno-Hagelsieb, G., Janga, S. C., Ramírez, M. A., Jiménez-Jacinto, V., Collado-Vides, J., and Dávila, G. 2006. The partitioned Rhizobium etli genome: genetic and metabolic redundancy in seven interacting replicons. Proc. Natl. Acad. Sci. USA 103:3834-3839.

36. Goolsby, J. A., Adamczyk, J. J., Bextine, B. E., Munyaneza, J. E., and Bester, G. 2007. Development of an IPM program for management of the potato psyllid to reduce incidence of zebra chip disorder in potatoes. Subtrop. Plant Sci. J. 59:85-94.

37. Gudmestad, N. C., and Secor, G. A. 2007. Zebra chip: a new disease of potato. Nebraska Potato Eyes 19:1-4.

38. Guedot, C., Horton, D. R., and Landolt, P. J. 2010. Sex attraction in Bactericera cockerelli (Hemiptera:Triozidae). Chem. Ecol. 39:1302-1308.

39. Hansen, A. K., Trumble, J. T., Stouthamer, R., and Paine, T. D. 2008. A new Huanglongbing (HLB) species, 'Candidatus Liberibacter psyllaurous', found to

infect tomato and potato, is vectored by the psyllid Bactericeca cockerelli (Sulc). Appl. Environ. Microbiol. 74:5862-5865.

40. Hatch, T. P., Al-Hossainy, E., and Silverman, J. A. 1982. Adenine nucleotide and lysine transport in Chlamydia psittaci. J. Bacteriol. 150:662-670.

41. Henne, D. C., Workneh, F., and Rush, C. M. 2012. Spatial patterns and spread of potato zebra chip disease in the Texas Panhandle. Plant Dis. 96:948-956.

42. Henne, D. C., Workneh, F., Wen, A., Price, J. A., Pasche, J. S., Gudmestad, N. C., and Rush, C. M. 2010. Characterization and epidemiological significance of plants grown from seed tubers affected by zebra chip disease. Plant Dis. 94:659-665.

43. Hung, T. H., Hung, S. C., Chen, C. N., Hsu, M. H., and Su, H. J. 2004. Detection by PCR of 'Candidatus Liberibacter asiaticus', the bacterium causing citrus huanglongbing in vector psyllids: Application to the study of vector-pathogen relationships. Plant Pathol. 53:96-102.

44. Ibarz Pavón, A. B., and Maiden, M. C. J. 2009. Multilocus sequence typing. Pages 129-140 in: Epidemiology of Microorganisms: Methods and Protocols. D. A. Caugant, ed. Humana Press, New York.

45. Jackson, B. C., Goolsby, J., Wyzykowski, A., Vitovsky, N., and Bextine, B. 2009. Analysis of genetic relationships between potato psyllid (Bactericera cockerelli) populations in the United States, Mexico and Guatemala using ITS2 and inter simple sequence repeat (ISSR) data. Subtrop. Plant Sci. 61:1-5.

46. Jagoueix, S., Bové, J. M., and Garnier, M. 1994. The phloem-limited bacterium of greening disease of citrus is a member of the alpha subdivision of the Proteobacteria. Int. J. Syst. Bacteriol. 44:379-386.

47. Jagoueix, S., Bové, J. M., and Garnier, M. 1996. PCR Detection of the two 'Candidatus Liberibacter' species associated with greening disease of citrus. Mol. Cell. Probes 10:43-50.

48. Lacey, L. A., Liu, T.-X., Buchman, J. L., Munyaneza, J. E., Goolsby, J. A., and Horton, D. R. 2011. Entomopathogenic fungi (Hypocreales) for control of potato psyllid, Bactericera cockerelli (Šulc.) (Hemiptera: Triozidae) in an area endemic for zebra chip disease of potato. Biol. Control 56:271-278.

49. Levy, J., Raviddran, A., Gross, D., Tamborindeguy, and Pierson, E. 2011. Translocation of 'Candidatus Liberibacter solanacearum', the zebra chip pathogen, in potato and tomato. Phytopathology 101:1285-1291.

50. Li, W, Abad, J. A., French-Monar, R. D., Rascoe, J., Wen, A., Gudmestad, N. C., Secor, G. A., Lee, I. M., and Levy, L. 2009. Multiplex real-time PCR for detection, identification and quantification of 'Candidatus Liberibacter solanacearum' in potato plants with zebra chip. J. Microbiol. Methods 78:59-65.

51. Li, W., Hartung, J. S., and Levy, L. 2006. Quantitative real-time PCR for detection and identification of 'Candidatus Liberibacter' species associated with citrus huanglongbing. J. Microbiol. Methods 66:104-115.

52. Liefting, L. W., Perez-Egusquiza, Z. C., Clover, G. R. G., and Anderson, J. A. D.

2008. A new 'Candidatus Liberibacter' species in Solanum tuberosum in New Zealand. Plant Dis. 92:1474.

53. Liefting, L. W., Sutherland, P. W., Ward, L. I., Paice, K. L., Weir, B. S., and Clover, G. R. G. 2009. A new 'Candidatus Liberibacter' species associated with diseases of solanaceous crops. Plant Dis. 93:208-214.

54. Liefting, L. W., Ward, L. I., Shiller, J. B., and Clover, G. R. G. 2008. A New 'Candidatus Liberibacter' species in Solanum betaceum (Tamarillo) and Physallis peruviana (Cape gooseberry) in New Zealand. Plant Dis. 92:1588.

55. Liefting, L. W., Weir, B. S., Pennycook, S. R., and Clover, G. R. G. 2009. 'Candidatus Liberibacter solanacearum', a Liberibacter associated with plants in the family Solanaceae. Int. J. Syst. Evol. Microbiol. 59:2274- 2276.

56. Lin, H., Doddapaneni, H., Munyaneza, J. E., Civerolo, E. L., Sengoda, V. G., Buchman, J. L., and Stenger, D. C. 2009. Molecular characterization and phylogenetic analysis of 16S rRNA from a new 'Candidatus Liberibacter' strain associated with zebra chip disease of potato (Solanum tuberosum L.) and the potato psyllid (Bactericera cockerelli Šulc). J. Plant Pathol. 91:215-219.

57. Lin, H., Islam, M. S., Bai, Y., Wen, A., Lan, S., Gudmestad, N. C., and Civerolo, E. L. 2012. Genetic diversity of 'Candidatus Liberibacter solanacearum' strains in the United States and Mexico revealed by simple sequence repeat markers. Eur. J. Plant Pathol. 132:297-308.

58. Lin, H., Lou, B., Glynn, J. M., Doddapaneni, H., Civerolo, E. L., Chen, C., Duan, Y., Zhou, L., and Vahling, C. M. 2011. The complete genome sequence of 'Candidatus Liberibacter solanacearum', the bacterium associated with potato zebra chip disease. PLoS ONE 6:e19135.

59. Lindgren, P. B., Peet, R. C., and Panopoulos, N. J. 1986. Gene cluster of Pseudomonas syringae pv. phaseolicola controls pathogenicity of bean plants and hypersensitivity on nonhost plants. J. Bacteriol. 168:512-522.

60. List, G. M. 1939. The effect of temperature upon egg deposition, egg hatch, and nymphal development of Paratrioza cockerelli (Šulc.). J. Econ. Entomol. 32:30-36.

61. Liu, D., Trumble, J. T., and Stouthamer, R. 2006. Genetic differentiation between eastern populations and recent introductions of potato psyllid (Bactericera cockerelli) into western North America. Entomol. Exp. Appl. 118:177-183.

62. Liu, D. G, and Trumble, J. T. 2004. Tomato psyllid behavioral responses to tomato plant lines and interactions of plant lines with insecticides. J. Econ. Entomol. 97:1078-1085.

63. Liu, D. G., and Trumble, J. T. 2005. Interactions of plant resistance and insecticides on the development and survival of Bactericera cockerelli [Šulc] (Homoptera: Psyllidae). Crop Prot. 24:111-117.

64. Liu, D. G., and Trumble, J. T. 2007. Comparative fitness of invasive and native populations of the potato psyllid (Bactericera cockerelli). Entomol. Exp. Appl. 123:35-42.

65. McDonald, B. A., and Linde, C. C. 2002. Pathogen population genetics, evolutionary potential, and durable resistance. Annu. Rev. Phytopathol. 40:349-379.

66. Miles, G. P., Samuel, M. A., Chen, J., Civerolo, E. L., and Munyaneza, J. E. 2010. Evidence of cell death is associated with zebra chip disease in potato tubers. Am. J. Potato Res. 87:337-349.

67. Moran, N. A., McCutcheon, J. P., and Nakabachi, A. 2008. Genomics and evolution of heritable bacterial symbionts. Annu. Rev. Genet. 42:165-190.

68. Morana, N. A., and Wernegreen, J. J. 2000. Lifestyle evolution in symbiotic bacteria: Insights from genomics. Trends Ecol. Evol. 15:321-326.

69. Munyaneza, J. E., Buchman, J. L., Upton, J. E., Goolsby, J. A., Crosslin, J. M., Bester, G., Miles, G. P., and Sengoda, V. G. 2008. Impact of different potato psyllid populations on zebra chip disease incidence, severity, and potato yield. Subtrop. Plant Sci. J. 60:27-37.

70. Munyaneza, J. E., Crosslin, J. M., and Buckman, J. L. 2009. Seasonal occurrence and abundance of the potato psyllid, Bactericera cockerelli, in South Central Washington. Am. J. Potato Res. 86:513-518.

71. Munyaneza, J. E., Crosslin, J. M., and Upton, J. E. 2007. Association of Bactericera cockerelli (Homoptera: Psyllidae) with "Zebra chip", a new potato disease in Southwestern United States and Mexico. J. Econ. Entomol. 100:656-663.

72. Munyaneza, J. E., Fisher, T. W., Sengoda, V. G., and Garczynski, S. F. 2010. First report of 'Candidatus Liberibacter solanacearum' associated with psyllid-affected carrots in Europe. Plant Dis. 94:639.

73. Munyaneza, J. E., Fisher, T. W., Sengoda, V. G., and Garczynski, S. F., Nissinen, A., Lemmetty, A. 2010. Association of "Candidatus Liberibacter solanacearum" with the psyllid, Trioza apicalis (Hemiptera: Triozidae) in Europe. J. Econ. Entomol. 4:1060-1070.

74. Munyaneza, J. E., Goolsby, J. A., Crosslin, J. M., and Upton, J. E. 2007. Further evidence that zebra chip potato disease in the Lower Rio Grande Valley of Texas is associated with Bactericera cockerelli. Subtrop. Plant Sci. J. 59:30-37.

75. Munyaneza, J. E., Sengoda, V. G., Buchman, J. L., and Fisher, T. W. 2012. Effects of temperature on 'Candidatus Liberibacter solanacearum' and zebra chip potato disease symptom development. Plant Dis. 96:18-23.

76. Munyaneza, J. E., Sengoda, V. G., Crosslin, J. M., De la Rosa-Lozano, G., and Sanchez, A. 2009. First report of 'Candidatus Liberibacter psyllaurous' in potato tubers with zebra chip disease in Mexico. Plant Dis. 93:552.

77. Munyaneza, J. E., Sengoda, V. G., Crosslin, J. M., Garzon-Tiznado, J. A., and Cardenas-Valenzuela, O. G. 2009. First report of 'Candidatus Liberibacter solanacearum' in tomato plants in México. Plant Dis. 93:1076.

78. Munyaneza, J. E., Sengoda, V. G., Stegmark, R., Arvidsson, A. K., Anderbrant, O., Yuvaraj, J. K., Ramert, B., and Nissinen, A. 2012. First report of 'Candidatus Liberibacter solanacearum' associated with psyllidaffected carrots in Sweden.

Plant Dis. 96:453.

79. Munyaneza, J. E., Sengoda, V. G., Sundheim, L., and Meadow, R. 2012. First report of 'Candidatus Liberibacter solanacearum' associated with psyllid-affected carrots in Norway. Plant Dis. 96:454.

80. Nachappa, P., Shapiro, A. A., and Tamborindeguy, C. 2012. Effect of 'Candidatus Liberibacter solanacearum' on fitness of its insect vector Bactericera cockerelli (Hemiptera: Triozidae), on tomato. Phytopathology 102:41-46.

81. Navarre, D. A., Shakya, R., Holden, J., and Crosslin, J. M. 2009. LC-MS analysis of phenolic compounds in tubers showing zebra chip symptoms. Am. J. Pot. Res. 86:88-95.

82. Nelson, W. R., Fisher, T. W., and Munyaneza, J. E. 2011. Haplotypes of 'Candidatus Liberibacter solanacearum' suggest long-standing separation. Eur. J. Plant Pathol. 130:5-12.

83. Ochman, H. 2005. Genomes on the shrink. Proc. Natl. Acad. Sci. USA 102:11959-11960.

84. Paulsen, I. T., Kang, K. H., Hance, M. E., and Ren, Q. 2004. Genome analysis of membrane transport. Pages 113-126 in: Microbial Genomes. C. M. Fraser, T. D. Read, and K. E. Nelson, eds. Humana Press Inc., Totowa, NJ.

85. Peng, L., Trumble, J. T., Munyaneza, J. E., and Liu, T. X. 2011. Repellency of kaolin particle film to potato psyllid, Bactericera cockerelli (Hemiptera:psyllidae), on tomato under laboratory and field conditions. Pest Manage. Sci. 67:815-824.

86. Pitman, A. R., Drayton, G. M., Kraberger, S. J., Genet, R. A., and Scott, I. A. W. 2011. Tuber transmission of 'Candidatus Liberibacter solanacearum' and its association with zebra chip on potato in New Zealand. Eur. J. Plant Pathol. 129:389-398.

87. Primrose, S. B., and Twyman, R. M. 2003. Principles of Genome Analysis and Genomics, third ed. S. B. Primrose and R. M. Twyman, eds. Blackwell Publishing Company, Malden, MA.

88. Raddadi, N., Gonella, E., Camerota, C., Pizzinat, A., Tedeschi, R., Crotti, E., Mandrioli, M., Bianco, P. A., Daffonchio, D., and Alma, A. 2011. 'Candidatus Liberibacter europaeus' sp. nov. that is associated with and transmitted by the psyllid Cacopsylla pyri apparently behaves as an endophyte rather than a pathogen. Environ. Microbiol. 13:414- 426.

89. Ramanan, N., and Wang, Y. 2000. A high-affinity iron permease essential for Candida albicans virulence. Science 288:1062-1064.

90. Rashed, A., Nash, T. D., Paetzold, L., Workneh, F., and Rush, C. M. 2012. Transmission efficiency of 'Candidatus Liberibacter solanacearum' and potato zebra chip disease progress in relation to pathogen titer, vector numbers, and feeding sites. Phytopathology 102:1079-1085.

91. Ravindran, A., Levy, J., Pierson, E., and Gross, D. C. 2011. Development of primers for improved PCR detection of the potato zebra chip pathogen, 'Candidatus Liberibacter solanacearum'. Plant Dis. 95:1542-1546.

92. Ravindran, A., Levy, J., Pierson, E., and Gross, D. C. 2012. Development of a loop-mediated isothermal amplification procedure as a sensitive and rapid method for the detection of 'Candidatus Liberibacter solanacearum' in potatoes and psyllids. Phytopathology 102:899-907.

93. Rodriguez La Torre, J. J. 2012. Movement and accumulation of 'Candidatus Liberibacter solanacearum' in potato plant. Ph.D. dissertation, North Dakota State University, Fargo, ND.

94. Romney, V. E. 1939. Breeding areas of the tomato psyllid, Paratrioza cockerelli (Šulc.). J. Econ. Entomol. 32:150-151.

95. Rush, C. M., Henne, D. C., Workneh, F., and Paetzold, L. 2010. Investigating titer variation of 'Candidatus Liberibacter solanacearum' in individual potato psyllids. In: Proc. 10th Annu. Zebra Chip Rep. Session. F. Workneh and C. M. Rush, eds. Dallas.

96. Saurabh, B., Sneha, S., Suvidya, R., Pramod, K., and B. Shailesh. 2011. Analysis of distribution and significance of simple sequence repeats in enteric bacteria Shigella dysenteriae SD197. Bioinformation 6:348-351.

97. Schmitz-Esser, S., Haferkamp, I., Knab, S., Penz, T., Ast, M., Kohl, C., Wagner, M., and Horn, M. 2008. Lawsonia intracellularis contains a gene encoding a functional rickettsia-like ATP/ADP translocase for host exploitation. J. Bacteriol. 190:5746-5752.

98. Secor, G. A., and Rivera-Varas, V. V. 2004. Emerging diseases of cultivated potato and their impact on Latin America. Rev. Latinoam. Papa (Suppl.) 1:1-8.

99. Secor, G. A., Rivera, V. V., Abad, J. A., Lee, I. M., Clover, G. R. G., Liefting, L. W., Li, X., and De Boer, S. H. 2009. Association of 'Candidatus Liberibacter solanacearum' with the zebra chip disease of potato established by graft and psyllid transmission, electron microscopy, and PCR. Plant Dis. 93:574-583.

100. Semetey, O, Gatineau, F, Bressan, A, and Boudon-Padieu, E. 2007. Characterization of a γ-3 proteobacteria responsible for the syndrome basses-richesses of sugar beet transmitted by Penastiridius sp. (Hemiptera, Cixiidae). Phytopathology 97:72-78.

101. Sengoda, V. G., Munyaneza, J. E., Crosslin, J. M., Buchman, J. L., and Pappu, H. R. 2010. Phenotypic and etiological differences between psyllid yellows and zebra chip diseases of potato. Am. J. Potato Res. 87:41-49.

102. Simpson, A. J., Reinach, F. C., Arruda, P., Abreu, F. A., Acencio, M., Alvarenga, R. Alves, L. M., Araya, J. E., Baia, G. S., Baptista, C. S., Barros, M. H., Bonaccorsi, E. D., Bordin, S., Bové, J. M., Briones, M. R., Bueno, M. R., Camargo, A. A., Camargo, L. E., Carraro, D. M., Carrer, H., Colauto, N. B., Colombo, C., Costa, F. F., Costa, M. C., Costa-Neto, C. M., Coutinho, L. L., Cristofani, M., Dias-Neto, E., Docena, C., El-Dorry, H., Facincani, A. P., Ferreira, A. J., Ferreira, V. C., Ferro, J. A., Fraga, J. S., Franca, S. C., Franco, M. C., Frohme, M., Furlan, L. R., Garnier, M., Goldman, G. H., Goldman, M. H., Gomes, S. L., Gruber, A., Ho, P. L., Hoheisel, J. D., Junqueira, M. L., Kemper, E. L., Kitajima, J. P., Krieger, J. E., Kuramae, E. E., Laigret, F., Lambais, M. R., Leite, L. C., Lemos, E. G., Lemos, M. V., Lopes, S. A., Lopes, C. R., Machado, J. A., Machado, M. A.,

Madeira, A. M., Madeira, H. M., Marino, C. L., Marques, M. V., Martins, E. A., Martins, E. M., Matsukuma, A. Y., Menck, C. F., Miracca, E. C., Miyaki, C. Y., Monteriro-Vitorello, C. B., Moon, D. H., Nagai, M. A., Nascimento, A. L., Netto, L. E., Nhani, A., Nobrega, F. G., Jr., Nunes, L. R., Oliveira, M. A., de Oliveira, M. C., de Oliveira, R. C., Palmieri, D. A., Paris, A., Peixoto, B. R., Pereira, G. A., Pereira, H. A., Pesquero, J. B., Jr., Quaggio, R. B., Roberto, P. G., Rodrigues, V., de M Rosa, A. J., de Rosa, V. E., de Sá, R. G., Santelli, R. V., Sawasaki, H. E., da Silva, A. C., da Silva, A. M., da Silva, F. R., da Silva, W. A., da Silveira, J. F., Jr., Silvestri, M. L., Siqueira, W. J., de Souza, A. A., de Souza, A. P., Terenzi, M. F., Truffi, D., Tsai, S. M., Tsuhako, M. H., Vallada, H., Van Sluys, M. A., Verjovski-Almeida, S., Vettore, A. L., Zago, M. A., Zatz, M., Meidanis, J., and Setubal, J. C. 2000. The genome sequence of the plant pathogen Xylella fastidiosa. The Xylella fastidiosa consortium of the organization for nucleotide sequencing and analysis. Nature 406:151-159.

103. Swisher, K. D., Munyaneza, J. E., and Crosslin, J. M. 2012. High resolution melting analysis of the cytochrome oxidase I gene identifies three haplotypes of the potato psyllid in the United States. Environ. Entomol. 41:1019-1028.

104. Teixeira, D. C., Saillard, C., Couture, C., Martins, E. C., Wulff, N. A., Eveillard-Jagoueix, S., Yamamoto, P. T., Ayres, A. J., and Bové, J. M. 2008. Distribution and quantification of 'Candidatus Liberibacter americanus', agent of huanglongbing disease of citrus in São Paulo State, Brazil, in leaves of an infected sweet orange tree as determined by PCR. Mol. Cell. Probes 22:139-150.

105. Vahling, C. M., Duan, Y., and Lin, H. 2010. Characterization of an ATP translocase identified in the destructive plant pathogen 'Candidatus Liberibacter asiaticus'. J. Bacteriol. 192:834-840.

106. Vahling-Armstrong, C. M., Zhou, H., Benyon, L., Morgan, J. K., and Duan, Y. 2012. Two Plant Bacteria, S. meliloti and 'Ca. Liberibacter asiaticus', Share Functional znuABC homologues that encode for a high affinity zinc uptake system. PLoS ONE 7:e37340.

107. Van Sluys, M. A., de Oliveira, M. C., Monteiro-Vitorello, C. B., Miyaki, C. Y., Furlan, L. R., Camargo, L. E. A., da Silva, A. C. R., Moon, D. H.,. Takita, M. A, Lemos, E. G., Machado, M. A., Ferro, M. I., da Silva, F. R., Goldman, M.H., Goldman, G.H., Lemos, M.V., El-Dorry, H., Tsai, S. M., Carrer, H., Carraro, D. M., de Oliveira, R. C., Nunes, L. R., Siqueira, W. J., Coutinho, L. L., Kimura, E. T., Ferro, E. S., Harakava, R., Kuramae, E. E., Marino, C. L., Giglioti, E., Abreu, I. L., Alves, L. M., do Amaral, A. M., Baia, G. S., Blanco, S. R., Brito, M. S., Cannavan, F. S., Celestino, A. V., da Cunha, A. F., Fenille, R. C., Ferro, J. A., Formighieri, E. F., Kishi, L. T., Leoni, S. G., Oliveira, A. R., Rosa, V. E., Sassaki, F. T., Jr., Sena, J. A., de Souza, A. A., Truffi, D., Tsukumo, F., Yanai, G. M., Zaros, L. G., Civerolo, E. L., Simpson, A. J., Almeida, N. F., Setubal, J. C., Jr., and Kitajima, J. P. 2003. Comparative analyses of the complete genome sequences of Pierce's disease and citrus variegated chlorosis strains of Xylella fastidiosa. J Bacteriol. 185:1018-1026.

108. Wallis, C. M., Chen, J., and Civerolo, E. L. 2012. Zebra chip-diseased potato

tubers are characterized by increased levels of host phenolics, amino acids, and defense-related proteins. Physiol. Mol. Plant Pathol. 78:66-72.

109. Wallis, R. L. 1955. Ecological studies on the potato psyllid as a pest of potatoes. U.S. Dep. Agric. Tech. Bull. 1107:1-24.

110. Wen, A., Mallik, I., Alvarado, V. Y., Pasche, J. S., Wang, X., Li, W., Levy, L., Lin, H., Scholthof, H., Mirkov, E., Rush, C. M., and Gudmestad, N. C. 2009. Detection, distribution, and genetic variability of 'Candidatus Liberibacter' species associated with the zebra complex disease of potato in the North America. Plant Dis. 93:1102-115.

111. Wood, D. W., Setubal, J. C., Kaul, R., Monks, D. E., Kitajima, J. P., Okura. V. K., Zhou, Y., Chen, L., Wood, G. E., Almeida, N. F., Woo, L., Jr., Chen, Y., Paulsen, I. T., Eisen, J. A., Karp, P. D., Bovee, D., Sr., Chapman, P., Clendenning, J., Deatherage, G., Gillet, W., Grant, C., Kutyavin, T., Levy, R., Li, M. J., McClelland, E., Palmieri, A., Raymond, C., Rouse, G., Saenphimmachak, C., Wu, Z., Romero, P., Gordon, D., Zhang, S., Yoo, H., Tao, Y., Biddle, P., Jung, M., Krespan, W., Perry, M., Gordon-Kamm, B., Liao, L., Kim, S., Hendrick, C., Zhao, Z. Y., Dolan, M., Chumley, F., Tingey, S. V., Tomb, J. F., Gordon, M. P., Olson, M. V., and Nester, E. W. 2001. The Genome of the natural genetic engineer Agrobacterium tumefaciens C58. Science 294:2317- 2323.

112. Workneh, F., Henne, D. C., Childers, A. C., Paetzold, L., and Rush, C. M. 2012. Assessments of the edge effect in intensity of potato zebra chip disease. Plant Dis. 96:943-947.

113. Yalpani, N., Silverman, P., Wilson, T. M., Kleier, D. A., and Raskin, I. 1991. Salicylic acid is a systemic signal and an inducer of pathogenesisrelated proteins in virus-infected tobacco. Plant Cell 3:809-818.

114. Yang, X.-B., Zhang, Y.-M., Hua, L., Peng, L.-N., Munyaneza, J. E., Trumble, J. T., and Liu, T.-X. 2010. Repellency of selected biorational insecticides to potato psyllid, Bactericera cockerelli (Hemiptera: Psyllidae). Crop Prot. 29:1320-1324.

115. Zhang, S., Flores-Cruz, Z., Zhou, L., Kang, B. H., Fleites, L. A., Gooch, M. D., Wulff, N. A., Davis, M. J., Duan, Y. P., and Gabriel, D. W. 2011. 'Ca. Liberibacter asiaticus' carries an excision plasmid prophage and a chromosomally integrated prophage that becomes lytic in plant infections. Mol. Plant-Microbe Interact. 24:458-468.

12
CHAPTER

REPLACEMENT IN PERENNIAL CROPS: CONDITIONS FOR SUCCESSFUL DISEASE MANAGEMENT

Mark S. Sisterson and Drake C. Stenger

Removing infected plants (roguing) is often used to manage spread of plant pathogens (1,5,13,16,19,46,52,56). In the case of perennial crops, rogued plants are often replaced with healthy plants. Roguing followed by replanting has two potential benefits. First, removing an infected plant eliminates a source of inoculum, potentially slowing pathogen spread. Second, infected plants may die or have reduced yield. Thus, replacing a diseased plant with a healthy plant may offset yield losses. While the potential benefits are enticing, whether or not roguing reduces losses depends on multiple factors. In this study, factors affecting successful implementation of a rogue and replace disease control strategy in a perennial cropping system are outlined and subsequent effects on pathogen spread and crop yield evaluated using spatially explicit simulation models.

A primary reason for removing infected plants is to eliminate inoculum sources. In this case, it is implicitly assumed that there is a risk of secondary pathogen spread. For vector-transmitted pathogens, risk of secondary pathogen spread depends on vector movement patterns, vector density within the crop, and role of the crop plant in the life cycle of the vector. For example, secondary spread of Cocoa swollen shoot virus (CSSV) may be common as cocoa is a primary host of mealybug vectors that may move between plants via an interlocking canopy (56). In contrast, some plant pathogens are spread by vectors that make trivial movements into the crop and then leave. For example, in the Central Valley of California, the primary vector of Xylella fastidiosa (causal agent of Pierce's disease of grape and almond leaf scorch disease) is the green sharpshooter, Draeculacephala minerva (12,48). This vector prefers grasses over grape and almond and is predominately found in permanent pastures and weedy alfalfa fields

Citation: Sisterson, M. S., and Stenger, D. C. 2013. Roguing with replacement in perennial crops: Conditions for successful disease management. Phyto-pathology 103:117-128.

(12,39,48), habitats that harbor plant species that serve as hosts of X. fastidiosa (41,48). As a result, movement of green sharpshooters into almond or grape is transient and risk of secondary pathogen spread appears to be low (12,29,47). Presuming secondary pathogen spread occurs, the extent to which roguing slows disease progress depends on how rapidly infected plants are removed. However, there are a number of logistical issues associated with efficient identification of infected plants in large-scale agricultural operations. For example, disease surveys may be hampered by a lack of appropriate and/or costeffective diagnostic tests or personnel trained to recognize symptoms.

Further, infected plants may continue to produce a useable yield before decline, making growers hesitant to remove diseased plants quickly. For example, farmers in Ghana are reluctant to remove CSSV-infected cocoa trees that are bearing pods and appear healthy (56). Consequently, there is often a considerable lag between infection and removal. Decisions to rogue and replant may be made by growers independently or growers may act in concert over a large area. The degree of coordination among growers is likely to affect the extent to which pathogen spread is slowed. For example, management of huanglongbing (HLB) disease in Brazil was more successful on large farms than on small farms (6). Similarly, DzahiniObiatey et al. (17) hypothesized that roguing failed to slow spread of CSSV in Ghana due to a lack of coordinated effort among farmers, as cocoa farms are highly fragmented. Replacing diseased plants with healthy plants may offset yield losses. In cases where plants die or produce no useable yield, the decision to replace diseased plants of a long-lived perennial crop is simple, as such plants have no value. However, some plant pathogens do not kill the host and have limited effects on yield (14,46,47). For example, yields of almond trees infected with X. fastidiosa are reduced by only 20 to 40% relative to uninfected almond trees (46,47). For some perennial crops, it may take years for a replacement plant to become fully productive (18,32,33). As a result, a mature infected plant may be more productive over the remaining lifetime of the crop than an uninfected immature replant (46). Thus, yield benefits associated with replanting are likely to be related to time required for plants to achieve full production and extent of yield loss due to disease.

Roguing as a disease control strategy has received some theoretical attention. Previous models largely used differential equations to describe disease dynamics (10,21,58). Here, a spatially explicit simulation model that captured key processes associated with primary and secondary pathogen spread was developed. Model assumptions were purposefully simple so that results would provide broad recommendations applicable to a wide array of systems. Farms were modeled as a grid of plants on an x/y coordinate system. Primary and secondary pathogen spread followed transition rules and plant yields were assigned based on plant age and infection status. Simulations were run at two scales. First, effects of roguing and replanting were evaluated on a single farm assuming that inoculum sources for primary pathogen spread were located outside of the simulated farm but not otherwise explicitly designated. Subsequently, the model was expanded to consider a crop growing region of 100 farms arranged in a square grid (10 farms × 10 farms) where rates of primary pathogen spread onto each farm were a function of disease incidence on neighboring farms. In this case, inoculum sources for primary pathogen spread were explicitly designated and replacement of infected plants on individual farms affected pathogen spread to neighboring farms.

The simulation approach taken here was similar to that used to develop guidelines for managing the evolution of resistance to insecticides by insects (9,23,28,36,44,45,54,55) and discussed further by Peck (35). This approach entails simulating the management strategy under a wide range of conditions to identify the region of parameter space associated with effective disease control. Ultimately, simulation results cannot conclusively demonstrate the effectiveness of a management strategy without field testing. However, for plant pathogens that spread rapidly or are invasive, researchers often do not have the opportunity to conduct large-scale, long-term field studies that incorporate appropriate controls (i.e., treatments that are expected to have high levels of pathogen spread) prior to making management recommendations. Accordingly, the simulation results presented here should be viewed as a tool to aid discussion of potential disease management strategies and for prioritizing strategies for field testing.

THEORY AND APPROACHES

Single farm model

The model was written in C++ using Microsoft Visual C++ (code available on request). The model was spatially explicit, with a farm defined as a square grid of plants with 75 rows and 75 plants per row (75 × 75 = 5,625 plants; Table 1). For each plant, the model tracked location on an x/y coordinate system, age, infection status, and yield based on age and infection status. The model assumed a perennial crop that was newly planted at the start of a simulation and followed the crop over 25 years with a time step equal to 1 year. Maturation periods of perennial crops vary. For example, short maturation times (1 to 2 years) are representative of crops such as banana which produce yield quickly, moderate maturation periods (4 to 6 years) are representative of crops such as grapes (34,59), and long maturation periods are representative of some citrus, almond, and cocoa varieties (>10 years; 18,32,33). Accordingly, a parameter for crop maturation period (M) was incorporated into the model. At time of planting, plants produced no yield and yield increased linearly from a minimum value of 0 to a maximum value of 1 when the crop reached maturity. Once a plant matured, the yield parameter based on maturity was held constant at 1 for the remainder of the simulation. For simulations of the single farm model, results using a short (M = 1 year) and long (M = 12 years) plant maturation period were presented to illustrate effects of plant maturation period on simulation results. Each year, each plant had a fixed probability of becoming infected due to primary pathogen spread (P; Table 1). To model this, a random number between 0 and 1 was obtained for each plant. If the random number was less than the probability that a plant was infected due to primary pathogen spread, the plant was infected; otherwise it remained uninfected. Infected plants could act as sources of inoculum for secondary pathogen spread (S), beginning the next year after becoming infected. Accordingly, the latent period (time between infection by a pathogen and onset of infectivity; 25) was fixed at 1 year for all simulations. As with primary pathogen spread, secondary pathogen spread occurred with a fixed probability (Table 1). As plants were arranged in a square grid, infected plants typically had eight neighbors (except for plants on farm edges). For each plant that bordered an infected plant, a random number between 0 and 1 was obtained. If this number was less than the probability of secondary pathogen spread, the plant was infected. For healthy plants that bordered more than one infected plant, risk of infection due to secondary spread was multiplicative. For example, if a healthy plant bordered two infected plants, there were two opportunities for the healthy plant to become infected, each with probability S. When considering values to use for rates of primary and secondary pathogen spread in simulations it is important to recognize that rates of primary and secondary pathogen spread within any given pathosystem are inherently variable and context dependent. For example, for vector-transmitted pathogens rates of primary spread onto a farm are a function of proximity to vector and inocula sources which vary over space and time. Similarly, rates of secondary pathogen spread are a function of a variety of factors including vector density, which varies based on the extent to which management tactics are applied to suppress vector populations. Accordingly, simulations of the single farm model used a wide range of values for primary (P = 0.01 to 0.10) and secondary pathogen spread (S = 0.01 to 0.20). Annual yield of each plant was calculated based on plant age and infection status. Once a plant was infected, yield was proportionately reduced relative to an uninfected plant of the same age.

Thus, yield loss due to infection (Y) was described by a parameter that had a value between 0 (infected plant produced no yield) and 1 (plants were tolerant with yield unaffected by infection) that described relative yield of an infected plant compared with an uninfected plant. For simplicity, yields of infected plants were assumed to decline immediately after infection. Accordingly, the model assumed no incubation period (time between infection by a pathogen and onset of symptom expression; 25). For simulations, two values for yield loss associated with infection were used: Y = 0.00 and 0.75. With Y = 0.0, infected plants produced no useable yield and were assumed to act as potential inoculum sources until removed. With Y = 0.75, infected plants were assumed to produce some useable yield similar to what was observed for almond trees infected with X. fastidiosa (46,47). After yields were calculated, the last step of each year was to determine which infected plants were replaced with healthy plants. For each set of simulations, the probability of identifying and replacing an infected plant had a fixed probability (R). Thus, for each infected plant a random number between 0 and 1 was obtained; if the random number was less than the probability that an infected plant was replaced, the plant was replaced. Four values for plant replacement rate were used in simulations: R = 0.00, R = 0.25, R = 0.50, and R = 1.0 (Table 1). As plant replacement was modeled as a binomial probability, rates of plant replacement can be expressed as the average length of time (after infection) an infected plant remained on the farm before it was removed. Thus, with R = 0.00, infected plants were never removed, with R = 0.25, infected plants remained on the farm for an average of 4 years, with R = 0.50, infected plants remained on the farm for an average of 2 years, and with R = 1.00, infected plants were removed the year the plant was infected. If a diseased plant was replaced, it was assumed to be replaced by a healthy plant of age zero that produced no yield at the time of planting. As with plants present at the start of a simulation, yields of replants increased linearly, until maturity.

Multi-farm model

Effects of grower cooperation/compliance for a disease management strategy involving roguing and replacement were evaluated by extending the single farm model to consider a square grid of 100 farms (10 × 10; Table 1). Rules governing plant growth and secondary pathogen spread for each farm in the multi-farm model were identical to those for the single farm model. The main difference between the single farm and multi-farm model was on how primary pathogen spread was modeled. In the single farm model, probability of a plant becoming infected via primary pathogen spread was a fixed value designated at the start of a simulation. In the multi-farm model, probability of primary infection within each farm was a function of disease incidence on neighboring farms (as farms were arranged in a square grid, farms typically had eight neighbors; except for farms on the edge of the grid). Thus, in the multi-farm model, probability of primary pathogen spread varied among farms and over time. In each year, the probability that a plant on a farm located at coordinates x,y was infected due to primary pathogen spread was determined by where $P_{x,y}$ was the probability that a plant on a farm located at coordinates x,y was infected via primary pathogen spread; $I_{x,y}$ was the number of infected plants on a farm located at coordinates x, y; $T_{x,y}$ was the total number of plants on a farm located at coordinates x, y, and w was a weighting factor that was varied in sensitivity analyses. For presentation, two values of the primary spread weighting factor (w) were used (w = 0.15 and 0.45) in combination with two values for rate of secondary pathogen spread (S = 0.01 and 0.10). Accordingly, this allowed for comparison of model results with relatively slow primary and secondary pathogen spread (w = 0.15, S = 0.01) to model results with relatively rapid primary and secondary pathogen spread (w = 0.45, S = 0.10). At the start of each simulation, all farms were assumed to be newly planted and a fixed proportion of farms were designated to engage in roguing/replacement and referred to as "managed farms." Farms that were not designated as managed farms, did not remove or replace infected plants and were referred to as "unmanaged farms." The proportion of farms engaged in roguing/replacement was varied from 0 to 1 in sensitivity analyses (Table 1). For all simulations of the multi-farm model, initial disease incidence was 0.1% with infected plants randomly distributed among all 100 farms.

Table 1. Parameters and values used in simulations of the single farm and multi-farm models

Model	Parameter (symbol)	Values presented
Single farm	Number of farms	1
	Number of plants per farm	5,625 (75 × 75)
	Probability of primary spread per plant per year (*P*)	0.0 to 0.1
	Probability of secondary spread per infected neighbor per year (*S*)	0.0 to 0.2
	Latent period (time between infection and onset of infectivity)	1 year
	Infectious period	Until removed
	Year maximum yield produced (*M*)	1, 12
	Yield of infected plants relative to uninfected plants (*Y*)	0.0, 0.75
	Incubation period (time between infection and yield decrease)	Immediate
	Probability of replacing an infected plant per year (*R*)	0.0, 0.25, 0.5, 1.0
Multi-farm	Number of farms	100 (10 × 10)
	Number of plants per farm	5,625 (75 × 75)
	Number of plants in region	562, 500
	Initial percentage of plants infected	0.1
	Probability of primary spread per plant per year ($P_{x,y}$)	Equation 1
	Primary spread weighting factor (*w*)	0.15, 0.45
	Probability of secondary spread per infected neighbor per year (*S*)	0.01, 0.1
	Latent period (time between infection and onset of infectivity)	1 year
	Infectious period	Until removed
	Year maximum yield produced (*M*)	1–16
	Yield of infected plants relative to uninfected plants (*Y*)	0.0, 0.75
	Incubation period (time between infection and yield decrease)	Immediate
	Probability of replacing an infected plant per year (*R*)	0.0 to 1.0
	Percentage of managed farms[a]	20, 80

[a] For each simulation a percentage of farms were designated as managed or unmanaged. Managed farms replaced infected plants and unmanaged farms did replace infected plants.

Simulation output

Output for the single farm model consisted of the percentage of extant plants infected in year 25 and the percentage of plants removed from the farm over all 25 years. As the percentage of plants removed was a cumulative total over all 25 years of production, values greater than 100% suggested that all plants on the farm were replaced and/or that some replants were repeatedly replaced. For the multi-farm model, output was the same as that for the single farm model except that the percentage of plants infected was tracked separately for managed and unmanaged farms. To examine effects of assumptions concerning time to plant maturity (M) and yield loss due to infection (Y), lifetime farm yields

$$P_{x,y} = \left(\frac{I_{x-1,y-1} + I_{x-1,y} + I_{x-1,y+1} + I_{x,y-1} + I_{x,y+1} + I_{x+1,y-1} + I_{x+1,y} + I_{x+1,y+1}}{T_{x-1,y-1} + T_{x-1,y} + T_{x-1,y+1} + T_{x,y-1} + T_{x,y+1} + T_{x+1,y-1} + T_{x+1,y} + T_{x+1,y+1}}\right) \times w$$

were output. As the model determined yield of each plant in each year based on plant age and infection status, lifetime farm yields were determined by summing yields of all plants on the farm over all years of the 25 year crop cycle. Lifetime farm yields decreased as plant maturation time (M) increased. To standardize yield results, yields in absence of disease over the 25 year crop cycle were determined for each plant maturation time evaluated. Yield from each simulation was expressed as a percentage of the maximum possible yield (relative yield) observed in the absence of disease based on plant maturation time. For the multi-farm model, relative yields were tracked separately on managed and unmanaged farms. Three sets of simulations were completed, each set building off the previous. The first set of simulations examined effects of plant replacement on disease incidence and yield in the single farm model with primary pathogen spread only. The second set examined effects of plant replacement on disease incidence and yield in the single farm model with both primary and secondary pathogen spread. The final set of simulations examined effects of plant replacement and percentage of farms replacing infected plants on disease incidence and yield in the multi-farm model. As the models were stochastic, 16 replicates for each parameter combination were completed and means were presented. Due to space limitations, only a subset of all simulations conducted could be presented. As a result, effects presented for some parameters on simulation outcome were limited to a single representative value or the comparative effects of a high and low value (Table 1). Nonetheless, results from a total of 8,192 simulations representing 512 unique parameter combinations are presented.

RESULTS

Single farm, primary pathogen spread only

Disease incidence after 25 years decreased as plant replacement rate increased (Fig. 1A). As there was no secondary pathogen spread, replacing infected plants did not alter the rate of infection from year to year. Accordingly, the percentage of plants replaced over the life of the crop was similar for all plant replacement rates (R > 0; Fig. 1B). With a high rate of plant replacement (R = 1.0), the percentage of plants infected equaled the rate of primary pathogen spread (Fig. 1A). Disease incidence increased as plant replacement rate decreased because infected plants remained on farms for longer periods before they were removed (Fig. 1A). With primary pathogen spread only, increasing plant replacement rate increased yields provided that infected plants produced no useable yield (Y = 0.00; Fig. 1C and D) or had short maturation periods (M = 1; Fig. 1E). However, with infected plants producing 75% of the yield of uninfected plants (Y = 0.75) and long plant maturation periods (M = 12), replacing infected plants (R ≥ 0.25) reduced yields relative to simulations without plant replacement (R = 0.00; Fig. 1F).

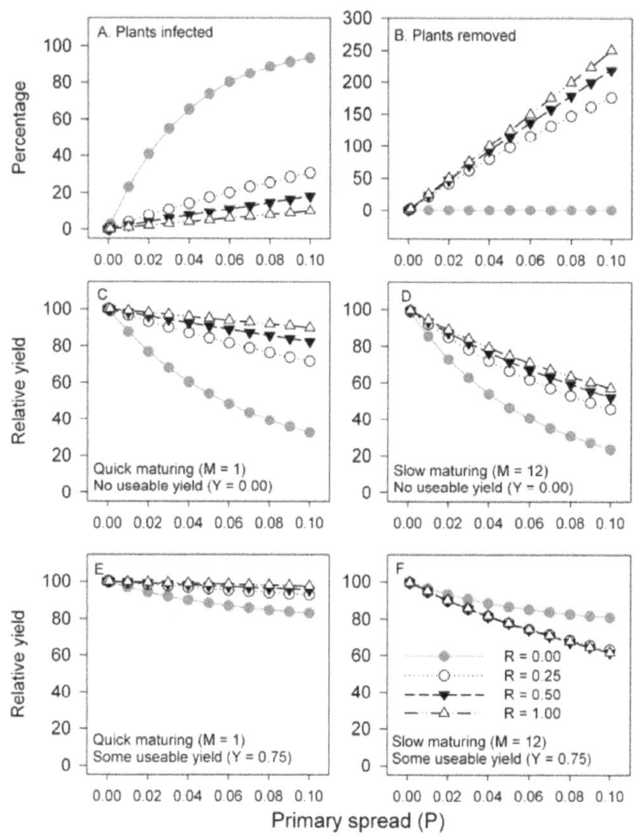

Figure 1. Effects of plant replacement on disease incidence and yield in the single farm model with primary pathogen spread only. A, Percentage of plants infected in the 25th year of production. B, Percentage of plants removed over all 25 years of production. Values >100% indicate that all plants on the farm were replaced or that some replants were repeatedly replaced. C to F, Relative yield with infected plants producing no useable yield (Y = 0.00; C and D) or 75% the yield of uninfected plants (Y = 0.75; E and F) and plants maturing in one (M = 1; C and E) or 12 years (M = 12; D and F). Four values for the probability of replacing an infected plant are shown (R).

Single farm, primary and secondary pathogen spread.

Disease incidence at the end of the 25 year crop cycle decreased as plant replacement rate increased (Fig. 2A). With a high plant replacement rate (R = 1.0), infected plants were removed the same year as becoming infected, thereby preventing secondary spread of the pathogen. Accordingly, with a high plant replacement rate (R = 1.0), disease incidence equaled the rate of primary pathogen spread (0.01); regardless of the rate of secondary pathogen spread used for a simulation (Fig. 2A). With a high plant replacement rate (R = 1.0), disease incidence not only remained low (Fig. 2A), but the percentage of plants replaced also was low as secondary pathogen spread was effectively halted (Fig. 2B). With moderate (R = 0.50) and low (R = 0.25) rates of plant replacement, infected plants remained in the field for an average of 2 or 4 years, respectively. As a result, with moderate (R = 0.50) and low (R = 0.25) rates of plant replacement, plants infected via primary pathogen spread were not removed before acting as inoculum sources for secondary pathogen spread. Accordingly, with moderate (R = 0.50) and low (R = 0.25) rates of plant replacement, disease incidence increased with the rate of secondary pathogen spread (Fig. 2A).

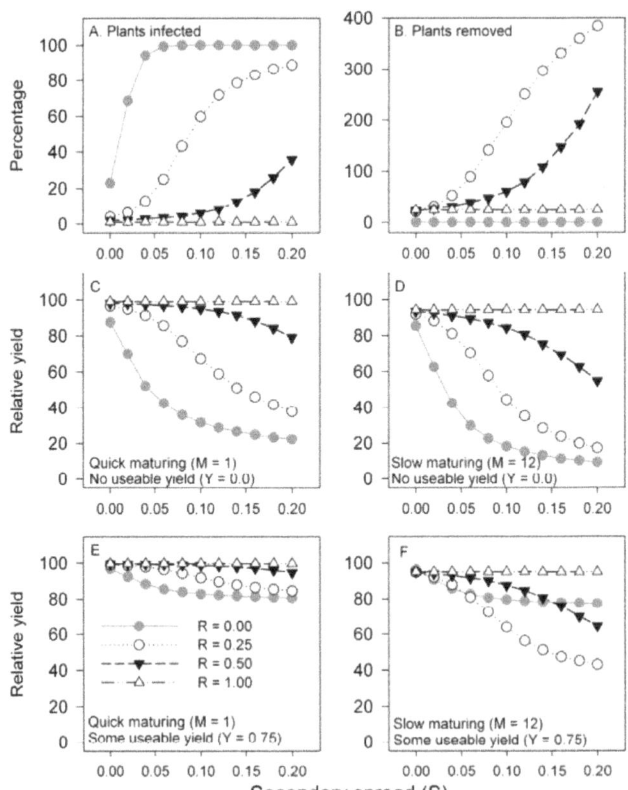

Figure 2. Effects of plant replacement on disease incidence and yield in the single farm model with primary and secondary pathogen spread. A, The percentage of plants infected in the 25th

year of production. B, The percentage of plants removed over all 25 years of production. Values >100% indicate that all plants on the farm were replaced or that some replants were repeatedly replaced. C to F, Relative yield with infected plants producing no useable yield (Y = 0.00; C and D) or 75% the yield of uninfected plants (Y = 0.75; E and F) and plants maturing in one (M = 1; C and E) or 12 years (M = 12; D and F). For all simulations primary spread (P) was assigned a value of 0.01. Four values for the probability of replacing an infected plant are shown (R).

Failure to slow secondary pathogen spread with moderate (R = 0.50) and low (R = 0.25) rates of plant replacement resulted in a high percentage of plants replaced during the 25 year crop cycle (Fig. 2B). With both primary and secondary pathogen spread, effects of plant replacement on yield (Fig. 2C to F) reflected effects of plant replacement on disease incidence (Fig. 2A) and percentage of plants replaced (Fig. 2B). Under such conditions, high yields were associated with parameter combinations that resulted in low disease incidence and replacement of few plants. With infected plants producing no useable yield (Y = 0.00; Fig. 2C and D) or short plant maturation times (M = 1; Fig. 2C and E), yields with plant replacement (R ≥ 0.25) were always greater than yields without plant replacement (R = 0.00; Fig. 2C to E). However, with infected plants producing 75% of the yield of uninfected plants (Y = 0.75) and long plant maturation times (M = 12), yields with plant replacement (R ≥ 0.25) were not always greater than yields without plant replacement (R = 0.00; Fig. 2F). Specifically, with infected plants producing 75% of the yield of uninfected plants (Y = 0.75) and a high plant replacement rate (R = 1.0), yields approached the maximum possible value (Fig. 2F) as secondary pathogen spread was effectively halted (Fig. 2A) and few plants were replaced (Fig. 2B). In contrast, with moderate (R = 0.50) or low (R = 0.25) plant replacement rates, secondary pathogen spread was not halted and the percentage of plants replaced was often high (Fig. 2B). As a result, with infected plants producing 75% of the yield of uninfected plants (Y = 0.75) and a long plant maturation time (M = 12), lifetime farm yields were lower with moderate (R = 0.50) and low (R = 0.25) rates of plant replacement compared with simulations without plant replacement (R = 0.00) for some levels of secondary pathogen spread (Fig. 2F).

Multi-farm, effects of plant replacement on disease incidence

In the multi-farm model, rate of primary pathogen spread onto any single farm was a function of disease incidence on neighboring farms. Accordingly, in the multi-farm model, replacing infected plants had two effects. First, replacing infected plants reduced secondary pathogen spread on managed farms as seen in the single farm model (Fig. 2A), provided that the probability of replacing an infected plant was sufficiently high. Second, replacing infected plants decreased rates of primary pathogen spread to neighboring farms potentially benefiting both managed and unmanaged farms, although the extent of this effect depended on the rate of plant replacement and the percentage of farms replacing infected plants (compare Fig. 3A, C, and E to Fig. 3B, D, and F). With 20% of farms replacing infected plants, most farms were unmanaged and had high disease incidence (Fig. 3C), contributing to high rates of primary pathogen spread among farms. Thus, with a minority of farms replacing infected plants, replacing infected plants reduced rates of secondary pathogen spread on managed farms, but had little effect on rates of primary pathogen spread among farms. As a result, with 20% of farms replacing infected plants, plant replacement was more effective at reducing disease incidence on managed farms with a low value for primary pathogen spread weighting factor (w = 0.15) than with a high value for primary pathogen spread weighting factor (w = 0.45; Fig. 3A). With only 20% of farms replacing infected plants, rates of primary pathogen spread were high such that on managed farms replants were often quickly infected, resulting in replacement of large numbers of plants (Fig. 3E). With 80% of farms replacing infected plants and high rates of plant replacement, secondary pathogen spread on managed farms was reduced and primary pathogen spread among all farms was regionally suppressed. As a result, with 80% of farms replacing infected plants, disease incidence on managed farms decreased to low levels as plant replacement rate increased (Fig. 3B). Regional suppression of primary pathogen spread among farms also resulted in lower disease incidence on unmanaged farms, provided that rates of secondary pathogen spread were low (S = 0.01; Fig. 3D). With a majority of farms replacing infected plants and high rates of plant replacement, pathogen spread was not only effectively suppressed (Fig. 3B), but this degree of suppression was accomplished with replacement of a relatively small number of plants (Fig. 3F). A high plant replacement rate was critical to observing the aforementioned effect as inefficient removal of infected plants, even with a majority of farms engaged in management, resulted in replacement of large numbers of plants (Fig. 3F), with little reduction in disease incidence (Fig. 3B) particularly with high rates of secondary pathogen spread (S = 0.10)

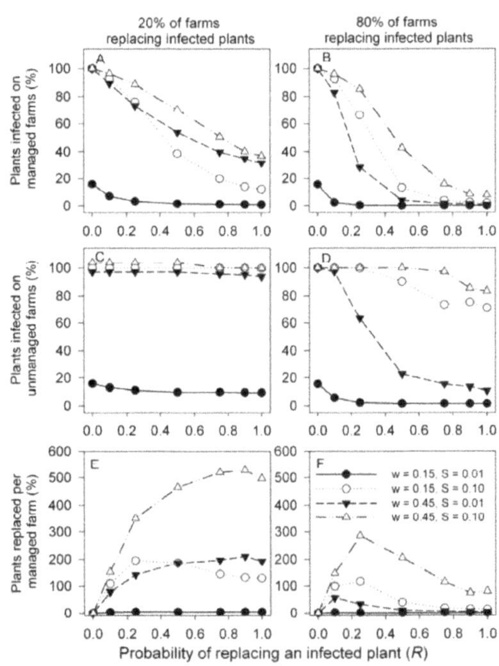

Figure 3. Disease incidence on managed (A and B) and unmanaged (C and D) farms in the 25th year of production as a function of primary spread weighting factor (w), rate of secondary spread

(S), and rate of plant replacement (R) with 20% of farms replacing infected plants (A and C) or 80% of farms replacing infected plants (B and D). Percentage of plant replaced over all 25 years of production per managed farm with 20% of farms replacing infected plants (E) or 80% of farms replacing infected plant (F). Values of the percentage of plants replaced per managed farm that exceed 100% indicate that all plants on managed farms were replaced or that some replants were repeatedly replaced. With R = 0.00, no farms replaced infected plants.

Multi-farm, effects of plant replacement on yield

Effects of plant replacement and percentage of farms replacing infected plants on managed farm yields (Fig. 4) were directly related to effects of plant replacement on disease incidence (Fig. 3A and B) and the percentage of plants replaced (Fig. 3E and F). In general, yields on managed farms were higher with parameter combinations that resulted in low disease incidence and replacement of few plants (compare Fig. 3A, B, E, and F to Fig. 4). With infected plants producing no useable yield (Y = 0.00), replacing infected plants (R > 0.00) always increased yield on managed farms relative to simulations without plant replacement (R = 0.00), although yield benefits increased as the plant replacement rate increased and as the percentage of farms replacing infected plants increased (Fig. 4A and B). With infected plants producing 75% of the yield of uninfected plants (Y = 0.75), replacing infected plants (R > 0.00) did not always increase yields on managed farms relative to simulations without plant replacement (R = 0.00; Figs. 4C and D and 5). With infected plants producing 75% of the yield of uninfected plants (Y = 0.75) and 20% of farms replacing infected plants, replacing infected plants often resulted in decreased yields on managed farms relative to simulations without plant replacement (Fig. 4C), with yield reductions increasing as plant maturation period increased (Fig. 5). This occurred due to the large number of plants replaced on managed farms when only 20% of farms replaced infected plants (Fig. 3E). In contrast, with infected plants producing 75% of the yield of uninfected plants (Y = 0.75) and 80% of farms replacing infected plants, plant replacement generally increased yields provided that the probability of replacing an infected plant was high (Figs. 4D and 5). This occurred because with a high percentage of managed farms and high rates of plant replacement, primary pathogen spread was suppressed at the regional scale and secondary pathogen spread was suppressed on managed farms, resulting in replacement of few plants (Fig. 3F). Effects of plant replacement on unmanaged farm yields reflected effects of plant replacement on disease incidence on unmanaged farms. In such cases, higher yields were associated with parameter combinations that resulted in lower disease incidence (compare Fig. 3C and D to Fig. 6). With 20% of farms replacing infected plants, plant replacement had little effect on disease incidence (Fig. 3C) or yield on unmanaged farms (Fig. 6A and C), regardless of the effects of infection on yield. In contrast, with 80% of farms replacing infected plants and high rates of plant replacement, primary pathogen spread was regionally suppressed (Fig. 3B and D). Accordingly, with 80% of farms replacing infected plants and high rates of plant replacement, plant replacement on managed farms also increased yields on unmanaged farms, particularly if secondary pathogen spread was low (S = 0.01; Fig. 6B and D).

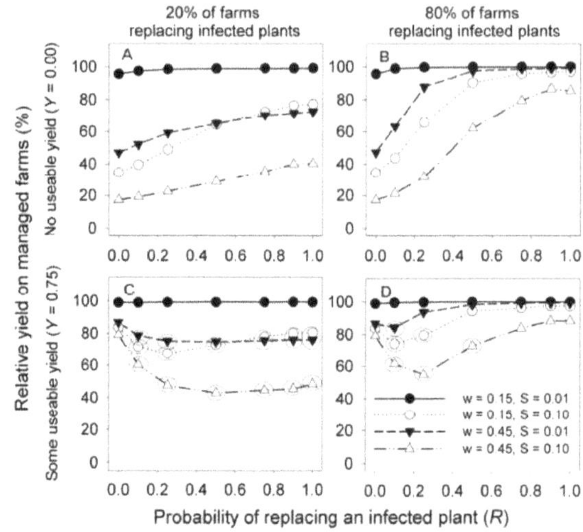

Figure 4. Effects of plant replacement on managed farm yields with infected plants producing no useable yield (Y = 0.00; A and B) or infected plants producing 75% of the yield of an uninfected plant (Y = 0.75; C and D) and with 20% of farms replacing infected plants (A and C) or 80% of farms replacing infected plants (B and D). In all panels, primary spread weighting factor (w), rate of secondary spread (S), and plant replacement rate (R) were varied. For all simulations plant maturation time (M) was 12. Circled values indicate simulations where relative yields with plant replacement (R > 0.00) were less than relative yields without plant replacement (R = 0.00). With R = 0.00, no farms replaced infected plants.

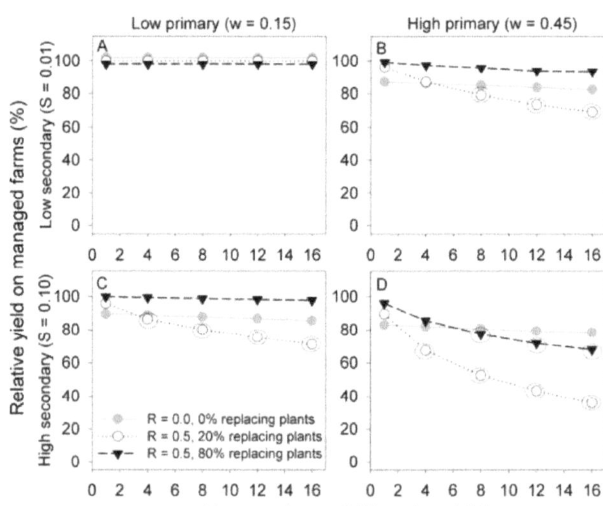

Figure 5. Effects of plant maturation period on managed farm yields with infected plants producing 75% of the yield of uninfected plants (Y = 0.75). Relative yields on managed farms with low

primary (w = 0.15) and secondary (S = 0.01) spread (A), high primary (w = 0.45) and low secondary (S = 0.01) spread (B), low primary (w = 0.15) and high secondary (S = 0.10) spread (C), and high primary (w = 0.45) and secondary (S = 0.10) spread (D). For all panels, results without plant replacement (R = 0.0 on all farms) and with moderate plant replacement (R = 0.5) on 20 and 80% of farms are shown. Circled values indicate simulations where relative yields with moderate plant replacement (R = 0.50) were less than relative yields without plant replacement (R = 0.00).

DISCUSSION

Model results and field observations

Results of this simulation study provide insight into conditions associated with successful disease suppression and yield optimization due to roguing and replacement of infected plants. Specifically, efficient replacement of infected plants (Figs. 2 and 3) combined with a high degree of compliance among farms (Fig. 3B) effectively slowed pathogen spread, which in turn resulted in replacement of fewer plants (Figs. 2B and 3F) and increased yields (Fig. 2C to F and Fig. 4B and D). In contrast, inefficient replacement of infected plants (Figs. 2 and 3) and/or limited compliance among farms (Fig. 3A) failed to slow pathogen spread, which in turn resulted in replacing large numbers of plants on managed farms (Figs. 2B and 3E) with little yield benefit (Figs. 2C to F and 4A and C). Furthermore, with inefficient plant replacement and infected plants producing 75% of the yield of uninfected plants, plant replacement often resulted in lower yields than when infected plants were never replaced (Figs. 4C and D and 5), a situation where growers actively engaged in disease control faired more poorly than their neighbors making no effort to control disease (compare Fig. 4C and D to Fig. 6C and D). To contextualize simulation results, perennial cropping systems that used roguing as a management strategy are discussed below. These comparisons serve to illustrate that parameters identified as critical to success of roguing with replacement in simulations appear to be important in real world systems and that application of the overarching conclusions from the model are relevant. Estimates of biological parameter values for each of the systems discussed are available online as Supplementary Table S1.

As the models made many generalized assumptions, some discussion of perspective is warranted. The strength of using generalized assumptions lies in the ability to make broad conclusions and look for commonalities across systems. The weakness of this approach is that it easy to identify model assumptions that are an oversimplification or to question the appropriateness of a particular parameter value relative to a specific system. Recognizing the trade-off between broad applicability and specificity is important. Accordingly, interpretation of model results should focus on factors that are likely to affect the success of roguing across a wide range of conditions, rather than subtle differences in assumptions or parameter values that are unlikely to affect the overall conclusions regarding the success or failure of roguing. In California, key vectors of X. fastidiosa vary by region, as do characteristics of X. fastidiosa spread (24,29,49). In the Central Valley of California, X. fastidiosa infection of grapevines and almonds is largely due to movement of green sharpshooters from irrigated pastures or weedy alfalfa fields (12,39,48), habitats that harbor plant species that are hosts of X. fastidiosa (41,48), into vineyards or almond orchards (8,12,29).

While green sharpshooters are routinely caught on traps placed on vineyard and orchard borders, populations are generally not found within vineyards or orchards (8,12) as grapevines and almond trees are considered occasional hosts (39). Similar to simulations assuming primary pathogen spread only (Fig. 1A), roguing of Pierce's disease affected vines in the Central Valley of California had no effect on spread of X. fastidiosa within vineyards (29). Although both crops are long-lived and require multiple years to reach maturity, effects of X. fastidiosa infection on almond trees and grapevines differ. Pierce's disease may be lethal (24) whereas almond leaf scorch-affected trees continue to produce a useable yield for years after infection (46,47). As a result, in areas where the green sharpshooter is the primary vector for both crops, recommendations for plant replacement differ. As seen in simulations, there is economic incentive to remove infected (dead) grapevines from vineyards even in the absence of secondary pathogen spread, as replacement vines will contribute to future yields (Fig. 1D). In contrast, decisions regarding removal of X. fastidiosa infected almond trees depend largely on whether the orchard will remain in place for a sufficient period of time for replacement trees to mature to a point where they will out-produce infected trees (Fig. 1F; 46,47). Beginning in the 1930s, citrus in California was affected by quick decline disease caused by Citrus tristeza virus (CTV) infection of sweet orange scions grafted onto sour orange rootstock (5).

In response, growers adopted use of a CTV-tolerant rootstock and, in cooperation with the state of California, instituted an eradication policy in which CTV-infected trees were identified and removed. However, most resident CTV strains in California cause mild or no disease symptoms on citrus grafted on tolerant rootstocks (5,37) making growers reluctant to remove CTV-infected trees. As a consequence, participation in eradication efforts by five contiguous pest control districts in Central California was not uniform. Specifically, two of five pest control districts, representing 45% of the citrus acreage in Central California, opted to not participate in eradication efforts (60). The three participating districts engaged in roguing/replacement from 1995 until 2007 and 2008 when the program was abandoned as disease incidence within eradication zones had risen to levels where eradication was considered impractical (5). Simulations suggest two factors may have contributed to failure of roguing/replacement to suppress CTV incidence within eradication zones.

First, eradication zones bordered citrus growing areas not engaged in eradication, creating a situation analogous to simulations with a minority of growers complying with a roguing/replacement program (Fig. 3A, C, and E). Further, early diagnosis of CTV infection was inefficient. Studies indicated that CTV infection may be undetectable via ELISA (the primary detection method during the eradication program) for up to 2 years postinfection, during which vectors may be able to acquire and transmit the virus (26). Accordingly, plant replacement rates were at best moderate ($R \leq 0.50$). In response, a new management strategy has been advocated (5): targeted removal of trees infected with virulent (but uncommon) CTV strains causing significant disease regardless of rootstock. The model indicates that success of the targeted approach to eliminate virulent strains will depend largely on a sufficiently high level of grower compliance and prompt identification/removal of trees infected with virulent strains. HLB disease is a worldwide threat to citrus production. The disease was reported to occur in Brazil in 2004 and Florida in 2005 (25). In both locations, the presumptive causal agent(s) ('Candidatus Liberibacter asiaticus' and 'Ca. L. americanus' in Brazil; 'Ca. L. asiaticus' only in Florida) has spread to most citrus production areas (6,51). Infected groves appear to be important sources of inoculum for primary pathogen spread into neigh boring groves and within groves there is evidence of secondary pathogen spread (25). Farm level experiments in Brazil found that successful management of HLB via roguing/replacement and vector control was positively associated with grove size and distance from unmanaged groves (6). Increasing grove size and distance from unmanaged groves is analogous to increasing the percentage of managed farms in the multi-farm model (Fig. 3).

Despite evidence supporting regional management of HLB in Brazil, only 35% of citrus acreage is actively engaged in management (6). In Florida, management of HLB also may be impacted by a lack of large-scale coordination/compliance. Two factors contribute to a lack of large-scale coordination/compliance in Florida. First, as of 2009 there were 57,650 abandoned hectares of citrus in Florida which have been documented to serve as a source of vectors and Ca. L. asiaticus (7,57). Second, some growers are attempting to keep HLB-affected groves productive via fertilizer treatments (50,51), although recent evidence suggests that fertilizer treatments may be ineffective (27). From a regional and long-term perspective, maintaining highly affected groves via fertilizer treatments is likely to reduce efficacy of management on neighboring farms, although the extent of this effect depends on the percentage of unmanaged farms and the degree to which primary pathogen spread among farms can be suppressed by reducing vector populations (Fig. 3). Roguing followed by replacement has been used to manage spread of CSSV among cocoa plantings in Ghana. Specifically, CSSV was first reported in Ghana in 1936 and an eradication campaign was begun by 1946 (56). The eradication program is ongoing, although there have been periods of disruption and protocols occasionally have been modified (17,56).

Primary spread of CSSV is recognized to originate from older infected plantings and secondary pathogen spread occurs due to movement of mealybug vectors among plants within plantings (17). During eradication attempts, entire plantings were replaced often with improved germplasm. However, infected trees were often replaced slowly (e.g., R was low) as cocoa maintains some degree of productivity after infection (56). Further, replaced plantings often bordered plantings that contained diseased trees that were not replaced (16). For example, 12,447 CSSV outbreaks were identified in the eastern region of Ghana between 2006 and 2010. During this period, only 890 (7%) outbreaks were replanted leaving an estimated 11,957,300 CSSV-infected trees in the field (15). Despite removal of millions of CSSV infected trees in Ghana, success has been limited (17,56). This observation is reflected in simulations which indicate that huge numbers of trees can be replaced (Fig. 3E and F) with little impact on disease incidence (Fig. 3A and B) when a minority of growers replace trees or when plant replacement is inefficient. More recent eradication protocols used in the Western Region of Ghana advocate distancing new plantings from potential sources of infection via barriers of CSSV immune crops (16).

Consideration of model assumptions

The models used here were purposefully simple so that the results would be broadly applicable. Thus, it is important to consider effects of simplifying assumptions on interpretation of model results. In simulations, the latent period (time between infection by a pathogen and onset of infectivity; 25) was fixed at 1 year. Similarly, the model(s) assumed no incubation period (time between infection by a pathogen and onset of symptom expression; 25) as yields of infected plants declined immediately. Latent and incubation periods may vary from months to years (5,25,26) and the lengths of each period are likely to have important effects on pathogen spread and/or yield. For example, the time-scale over which removal of an infected plant is defined as "efficient" increases with latent period. Likewise, infected plants with a brief incubation period are likely to be removed sooner than infected plants with a protracted incubation period as infected plants with clear visual symptoms are more easily detected than asymptomatic plants. Latent periods for pathogens such as CTV, CSSV, and 'Ca. Liberibacter spp.' appear to be shorter than incubation periods (5,25,26,56). Accordingly, efficient removal of plants affected by CTV, HLB, and CSSV is hampered by the fact that plants may serve as an acquisition source before symptoms are apparent. In simulations with infected plants producing no useable yield (Y = 0.00), infected plants did not die and continued to act as inoculum sources for secondary pathogen spread until removed. In some cases, infected plants may die (4,22) or may be less attractive to vectors than healthy plants (20,42). For most vector transmitted pathogens, dead plants do not serve as a source of inocula. In such cases, plant death can be viewed as a form of roguing that does not require any action on the part of a grower. Simulations assuming that infected plants died within 4, 6, or 10 years after infection were conducted using the multi-farm model (data not shown). Results indicated that roguing was more effective at suppressing pathogen spread as the average lifespan of infected plants decreased. For all simulations, infected plants were replaced on a plant-byplant basis. Specifically, the model simulated a scenario where farms were surveyed at the end of each year and if an infected plant was detected it was removed. However, alternative plant replacement rules could be considered. For example, Thresh and Owusu (56) advocated removing cocoa plants infected with CSSV and all plants immediately adjacent to an infected plant as plants neighboring infected plants are likely already infected but not symptomatic. In France, decisions concerning plant replacement in peach orchards affected by Plum pox virus strain M depend on disease incidence (13). Specifically, with incidence 10%, entire orchards were removed. The model presented here is sufficiently flexible, that such plant replacement rules could be simulated and it is likely that evaluation of different plant replacement rules may find that different approaches are more suitable under different conditions. Assumptions concerning inoculum sources for primary pathogen spread differed for the single farm and multi-farm models. The single farm model did not explicitly designate inoculum sources contributing to primary pathogen spread. Consequently, the single-farm model is suitable for considering spread of plant pathogens regardless of primary inoculum sources. In contrast, the multi-farm model explicitly designated inoculum sources as other farms producing the same crop. Accordingly, application of the results from the multi-farm model is more limited than those of the single farm model. However, the observation that efficacy of plant replacement decreased as percentage of unmanaged farms increased (Fig. 3) is likely to extend to situations where inoculum sources exist in other crops or noncrop habitats. For example, on the North Coast of California (e.g., Napa and Sonoma counties), native plants in riparian habitats are important sources of primary inocula for spread of X. fastidiosa to grapevines by the blue-green sharpshooter (11,38,40). As roguing within riparian habitats is impractical, such habitats are equivalent to unmanaged farms. Accordingly, it is reasonable to expect the risk of primary pathogen spread into a vineyard to increase as the percentage of surrounding riparian areas increase.

CONCLUSIONS

One of the challenges associated with managing the spread of invasive and/or rapidly spreading plant pathogens is that management strategies must often be implemented prior to or with limited field testing. For example, epidemics of Pierce's disease in southern California in the late 1990s due to introduction of an exotic vector required immediate action (49,53). Similarly, detection of the Asian citrus psyllid (primary vector of 'Ca. Liberibacter spp.') in 2008 (2) and an HLB-affected tree in 2012 (3) in California required immediate action. For situations where immediate action is required, simulation models provide a transparent and rigorous method for evaluating a wide range of management tactics to gain insight into which strategies may be most effective at suppressing pathogen spread. The models presented here were purposefully simple so that they could be applied to a broad array of systems, with a specific goal of emphasizing the effects of rate of tree removal on the quantity of trees removed and ultimately yield. While roguing as a control strategy was emphasized, management is most likely to be successful when diverse tactics are applied. For example, roguing was more effective at suppressing pathogen spread with low rates of primary and secondary pathogen spread compared with simulations with high rates of primary and secondary pathogen spread (Figs. 2 and 3). For vector-transmitted pathogens, rates of pathogen spread are dependent on factors affecting vector behavior (i.e., factors affecting vector movement among plants and transmission efficiency) and vector density (number of vectors per plant) (30,31,42,43). Thus, for vector-transmitted pathogens roguing/plant replacement is likely to be most successful when combined with vector management strategies that reduce rates of primary and secondary pathogen spread.

ACKNOWLEDGMENTS

Mention of trade names or commercial products in this publication is solely for the purpose of providing specific information and does not imply recommendation or endorsement by the U.S. Department of Agriculture. USDA is an equal opportunity provider and employer.

REFERENCES

1. Allen, R. N. 1978. Epidemiological factors influencing the success of roguing for the control of bunch top disease of bananas in New South Wales. Austral. J. Agric. Res. 29:535-544.

2. Anonymous. 2008. Single Asian citrus psyllid detected in San Diego. California Department of Food and Agriculture. Sacramento, CA. Press Release #08-057.

3. Anonymous. 2012. Citrus disease huanglongbing detected in Hacienda Heights area of Los Angeles County. California Department of Food and Agriculture. Sacramento, CA. Press Release #12-012.

4. Bar-Joseph, M., Marcus, R. and Lee, R. F. 1989. The continuous challenge of citrus tristeza virus control. Ann. Rev. Phytopathology 27:291- 316.

5. Barnier, J., Grafton-Cardwell, B., and Polek, M. 2010. Citrus tristeza virus (CTV): Then and now. Citrograph Nov./Dec. 16–23.

6. Belasque, J., Bassanezi, R. B., Yamamoto, P. T., Ayres, A. J., Tachibana, A., Violante, A. R., Tank, A., Di Giorgi, F., Tersi, F. E. A., Menezes, G. M., Dragone, J., Jank, R. H., and Bove, J. M. 2010. Lessons from huanglongbing management in São Paulo State, Brazil. J. Plant Pathol. 92:285-302.

7. Boina, D. R., Meyer, W. L., Onagbola, E. O., and Stelinski, L. L. 2009. Quantifying dispersal of Diaphorina citri (Hemiptera: Psyllidae) by immunomarking and potential impact of unmanaged groves on commercial citrus management. Environ. Entomol. 38:1250-1258.

8. Cabrera-La Rosa, J. C., Johnson, M. W., Civerolo, E. L., Chen, J., and Groves, R. L. 2008. Seasonal population dynamics of Draeculacephala minerva (Hemiptera: Cicadellidae) and transmission of Xylella fastidiosa. J. Econ. Entomol. 101:1105-1113.

9. Caprio, M. A., Storer, N., Sisterson, M. S., Peck, S., and de Holanda Nunes Maia, A. 2008. Assessing the risk of the evolution of resistance to pesticides using spatially complex simulation models. Pages 90-117 in: Global Pesticide Resistance in Arthropods. M. E. Whalon, D. MotaSanchez, and R. M. Hollingworth, eds. CAB International, Wallingford, UK.

10. Chan, M.-S., and Jeger, M. J. 1994. An analytical model of plant virus disease dynamics with roguing and replanting. J. Appl. Ecol. 31:413-427.

11. Coletta-Filho, H. D., Bittleston, L. S., and Almeida, R. P. P. 2011. Spatial genetic structure of a vector-borne generalist pathogen. Appl. Environ. Microbiol. 77:2596-2601.

12. Daane, K. M., Wistrom, C. M., Shapland, E. B., and Sisterson, M. S. 2011. Seasonal abundance of Draeculacephala minerva and other Xylella fastidiosa vectors in California almond orchards and vineyards. J. Econ. Entomol. 104:367-374.

13. Dallot, S., Gottwald, T., Labonne, G., and Quiot, J.-B. 2004. Factors affecting the spread of Plum pox virus strain M in peach orchards subjected to roguing in France. Phytopathology 94:1390-1398.

14. Daniells, J. W., Geering, A. D. W., Bryde, M. J., and Thomas, J. E. 2001. The effect of Banana streak virus on the growth and yield of dessert bananas in tropical Australia. Ann. Appl. Biol. 139:51-60.

15. Domfeh, O., Dzahini-Obiatey, H., Ameyaw, G. A., Abaka-Ewusie, K., and Opoku, G. 2011. Cocoa swollen shoot virus disease situation in Ghana: A review of current trends. African J. Agric. Res. 6:5033-5039.

16. Dzahini-Obiatey, G., Ameyaw, A., and Ollennu, L. A. 2006. Control of cocoa swollen shoot disease by eradicating infected trees in Ghana: A survey of treated and replanted areas. Crop Prot. 25:647-652.

17. Dzahini-Obiatey, H., Domfeh, O., and Amoah, F. M. 2010. Over seventy years of a viral disease of cocoa in Ghana: From researchers' perspective. African J. Agric. Res. 5:476-485.

18. Edwin, J., and Masters, W. A. 2005. Genetic improvement and cocoa yields in Ghana. Exp. Agric. 41:491-503.

19. Fauquet, C., and Fargette, D. 1990. African cassava mosaic virus: Etiology, epidemiology, and control. Plant Dis. 74:404-411.

20. Fereres, A., and Moreno, A. 2009. Behavioral aspects influencing plant virus transmission by homopteran insects. Virus Res. 141:158-168.

21. Fishman, S., and Marcus, R. 1984. A model for spread of plant disease with periodic removals. J. Mathematical Biol. 21:149-158.

22. Folimonova, S. Y., Robertson, C. J., Garnsey, S. M., Gowda, S., and Dawson, W. O. 2009. Examination of the responses of different genotypes of citrus to huanglongbing (citrus greening) under different conditions. Phytopathology 99:1346-1354.

23. Georghiou, G. P., and Taylor, C. E. 1977. Operational influences in the evolution of insecticide resistance. J. Econ. Entomol. 70:653-658.

24. Goodwin, P., and Purcell, A. H. 1992. Pierce's disease. Pages 76-84 in: Grape Pest Management. D. L. Flaherty, L. P. Christensen, W. T. Lanini, J. J. Marois, P. A. Phillips, and L. T. Wilson, eds. University of California, Division of Agriculture and Natural Resource, Oakland.

25. Gottwald, T. R. 2010. Current epidemiological understanding of citrus huanglongbing. Annu. Rev. Phytopathol. 48:119-139.

26. Gottwald, T. R., Garnsey, S. M., and Riley, T. D. 2002. Latency of systemic infection in young field-grown sweet orange trees following graft-inoculation with Citrus tristeza virus. Pages 48-53 in: Proc. 15th Conf. Int. Organ. Citrus Virol. IOCV, Riverside, CA.

27. Gottwald, T. R., Graham, J. H., Isrey, M. S., McCollum, T. G., and Wood, B. W. 2012. Inconsequential effect of nutritional treatments on huanglongbing control, fruit quality, bacterial titer and disease progress. Crop Prot. 36:73-82.

28. Gould, F., Cohen, M. B., Bentur, J. S., Kennedy, G. G., and Van Duyn, J. 2006. Impact of small fitness costs on pest adaptation to crop varieties with multiple toxins: A heuristic model. J. Econ. Entomol. 99:2091-2099.

29. Hewitt, W. B., Frazier, N. W., Freitag, J. H., and Winkler, A. J. 1949. Pierce's disease investigations. Hilgardia 19:207-262.

30. Jeger, M. J., Holt, J., Van Den Bosch, F., and Madden, L. V. 2004. Epidemiology of insect-transmitted plant viruses: Modeling disease dynamics and control interventions. Physiol. Entomol. 29:291-304.

31. Jeger, M. J., Van Den Bosch, F., Madden, L. V., and Holt, J. 1998. A model for analysing plant-virus transmission characteristics and epidemic development. IMA J. Math. Appl. Medicine Biol. 15:1-18.

32. Lampinen, B. D., Connell, J. H., Verdegall, P., Viveros, M., Metcalf, S. G., Negron, C., Thorpe, M. A., Gradziel, T. M., and Micke, W. C. 2006. Progress report: Regional almond variety trials planted in 1993. University of California Progress Report 83806.

33. Muraro, R. P., and Castle, W. S. 1996. Resetting and replanting options in Florida citrus groves and the financial consequences. Proc. Fla. State Hort. Soc. 109:128-131.

34. Peacock, W. L., Vasquez, S. J., Hashim, J. M., Fidelibus, M. W., Leavitt, G. M., Klonsky, K. M., and De Moura, R. L. 2006. Sample costs to establish and a vineyard and produce grapes for raisins: Tray dried raisin San Joaquin Valley. University of California Cooperative Extension publication number GR-SJ-06-1.

35. Peck, S. L. 2004. Simulation as experiment: A philosophical reassessment for biological modeling. Trends Ecol. Evol. 19:530-534.

36. Peck, S. L., Gould, F., and Ellner, S. P. 1999. Spread of resistance in spatially extended regions of transgenic cotton: Implications for management of Heliothis virescens (Lepidoptera: Noctuidae). J. Econ. Entomol. 92:1-16.

37. Polek, M., Gumpf, D. J., Wallen, C. M., and Riley, K. M. 2005. Biological characterization of naturally occurring Citrus tristeza virus strains in California citrus. Pages 68-74 In: Proc. 16th Conf. Int. Organ. Citrus Virol. IOCV, Riverside, CA.

38. Purcell, A. 1974. Spatial patterns of Pierce's disease in the Napa Valley. Am. J. Enol. Viticult. 25:162-167.

39. Purcell, A. H., and Frazier, N. W. 1985. Habitats and dispersal of the principal leafhopper vectors of Pierce's disease in the San Joaquin Valley. Hilgardia 53:1-32.

40. Purcell, A. H., and Saunders, S. R. 1999. Fate of Pierce's disease strains of Xylella fastidiosa in common riparian plants in California. Plant Dis. 83:825-830.

41. Shapland, E. B., Daane, K. M., Yokota, G. Y., Wistrom, C., Connell, J. H., Duncan, R. A., and Viveros, M. A. 2006. Ground vegetation survey for Xylella fastidiosa in California almond orchards. Plant Dis. 90:905-909.

42. Sisterson, M. S. 2008. Effects of insect-vector preference for healthy or infected plants on pathogen spread: Insights from a model. J. Econ. Entomol. 101:1-8.

43. Sisterson, M. S. 2009. Transmission of insect-vectored pathogens: Effects of vector fitness as a function of infectivity status. Environ. Entomol. 38:345-355.

44. Sisterson, M. S., Antilla, L., Carrière, Y., Ellers-Kirk, C., and Tabashnik, B. E. 2004. Effects of insect population size on evolution of resistance to transgenic crops. J. Econ. Entomol. 97:1413-1424.

45. Sisterson, M. S., Carrière, Y. Dennehy, T. J., and Tabashnik, B. E. 2005. Evolution of resistance to transgenic crops: Interactions between insect movement and field distribution. J. Econ. Entomol. 98:1751-1762.

46. Sisterson, M. S., Chen, J., Viveros, M. A., Civerolo, E. L., Ledbetter, C., and Groves, R. L. 2008. Effects of almond leaf scorch disease on almond yield: Implications for management. Plant Dis. 92:409-414.

47. Sisterson, M. S., Ledbetter, C. A., Chen, J., Higbee, B. S., Groves, R. L., and Daane, K. M. 2012. Management of almond leaf scorch disease: Long term data on yield, tree vitality, and disease progress. Plant Dis. 96:1037-1044.

48. Sisterson, M. S., Thammiraju, S. R., Lynn-Patterson, K., Groves, R. L., and Daane, K. M. 2010. Epidemiology of diseases caused by Xylella fastidiosa in California: Evaluation of alfalfa as a source of vectors and inocula. Plant Dis. 94:827-834.

49. Sisterson, M. S., Yacoub, R., Montez, G., Grafton-Cardwell, E. E., and Groves, R. L. 2008. Distribution and management of citrus in California: Implications for management of glassy-winged sharpshooters. J. Econ. Entomol. 101:1041-1050.

50. Spann, T., and Schumann, A. 2012. Using good horticultural practices to maintain yield of HLB-affected groves. Citrus Industry June:6-11.

51. Spann, T. M., Atwood, R. A., Dewdney, M. M., Ebel, R. C., Ehsani, R., England, G., Futch, S. H., Gaver, T., Hurner, T., Oswalt, C., Rogers, M. E., Roka, F. M., Ritenour, M. A., Zekri, M., Boman, B. J., Chung, K.-R., Danyluk, D., Goodrich-Schneider, R., Morgan, K. T., Morris, R. A., Muraro, R. P., Roberts, P., Rouse, R. E., Schumann, A. W., Stansly, P. A., and Stelinski, L. L. 2010. IFAS Guidance for Huanglongbing (Greening) Management. IFAS Publication number HS1165. University of Florida Institute of Food and Agricultural Sciences, Gainesville, FL.

52. Spharim, I., and Shalhevet, S. 1988. Formulating policy to cope with virus disease in citrus orchards. Agric. Syst. 27:215-224.

53. Stenger, D. C., Sisterson, M. S., and French, R. 2010. Population genetics of Homalodisca vitripennis reovirus validates timing and limited introduction to California of its invasive insect host, the glassy-winged sharpshooter. Virology 407:53-59.

54. Tabashnik, B. E., Gassmann, A. J., Crowder, D. W., and Carrière, Y. 2008. Insect resistance to Bt crops: Evidence versus theory. Nat. Biotechnol. 26:199-202.

55. Tabashnik, B. E., Sisterson, M. S., Ellsworth, P. C., Dennehy, T. J., Antilla, L., Liesner, L., Whitlow, M., Staten, R. T., Fabrick, J. A., Unnithan, G. C., Yelich, A. J., Ellers-Kirk, C., Harpold, V. S., Li, X., and Carrière, Y. 2010. Suppressing resistance to Bt cotton with sterile insect releases. Nat. Biotechnol. 28:1304-1307.

56. Thresh, J. M., and Owusu, G. K. 1986. The control of cocoa swollen shoot disease

in Ghana: An evaluation of eradication procedures. Crop Prot. 5:41-52.

57. Tiwari, S., Lewis-Rosenblum, H., Pelz-Stelinski, K., and Stelinski, L. L. 2010. Incidence of Candidatus Liberibacter asiaticus infection in abandoned citrus occurring in proximity to commercially managed groves. J. Econ. Entomol. 103:1972-1978.
58. Van den Bosch, F., and de Roos, A. M. 1996. The dynamics of infectious disease in orchards with roguing and replanting as control strategy. J. Mathematical Biol. 35:129-157.
59. Vasquez, S. J. H., Buckey, J. M., Peacock, W. L., Klonsky, K. M., and De Moura, R. L. 2007. Samples costs to establish and produce table grapes: San Joaquin Valley South. University of California Cooperative Extension publication number GR-VS-07-1.
60. Yokomi, R. K., and DeBorde, R. L. 2005. Incidence, transmissibility, and genotype analysis of Citrus tristeza virus (CTV) isolates from CTV eradicative and noneradicative districts in central California. Plant Dis. 89:859-866.

13
CHAPTER

ROOT-KNOT NEMATODE MANAGEMENT

Judy A. Thies, Richard F. Davis, John D. Mueller, Richard L. Fery, David B. Langston and Gilbert Miller

The southern root-knot nematode, Meloidogyne incognita (Chitwood) Kofoid and White, is a major limiting factor to pepper and cucurbit production in the United States and worldwide (2,3,15,19,24). In the United States, preplant soil fumigation with methyl bromide is the primary method for controlling root-knot nematodes in high-value vegetable crops (22). Because of the ozone-depleting properties of methyl bromide, its production in the United States will cease by 1 January 2005 (23). Although other soil fumigants and nematicides have varying degrees of usefulness for controlling root-knot nematodes in vegetable crops, environmental concerns and high costs of reregistration may limit use of these pesticides as feasible alternatives to methyl bromide. Thus, there is a great deal of interest in development of alternative methods for managing root-knot nematodes in vegetable crops. One approach would be the utilization of root-knot nematode-resistant vegetable cultivars as rotational crops to limit damage caused by root-knot nematodes in subsequent susceptible vegetable crops. In Louisiana, 'Dasher II' cucumber doublecropped after the root-knot nematoderesistant 'Celebrity' tomato produced heavier yields of marketable fruit and had less severe root-galling and lower population densities of second-stage juveniles of M. incognita in the soil than the same cucumber cultivar grown after the susceptible 'Heatwave' tomato (1,7). Muskmelons double-cropped after 'Celebrity' tomato also exhibited less severe root-galling and produced heavier yields than muskmelons grown after 'Heatwave' tomato (6). The U. S. Vegetable Laboratory, United States Department of Agriculture–Agricultural Research Service (USDA-ARS), Charleston, SC, released the root-knot nematode-resistant bell pepper cv. Charleston Belle in 1997 (5). This open-pollinated cultivar is homozygous

Citation: Judy A. Thies, Richard F. Davis, John D. Mueller, Richard L. Fery, David B. Langston, and Gilbert Miller Root-Knot Nematode Management, Plant Dis. 88:589-593.

for the dominant N gene that controls resistance to M. incognita (5), M. arenaria races 1 and 2, and M. javanica (17). 'Keystone Resistant Giant' is the susceptible recurrent parental cultivar used in the backcross breeding procedure (six backcrosses) to develop 'Charleston Belle' (5). The reactions of 'Charleston Belle' and its near-isogenic parent, 'Keystone Resistant Giant', to rootknot nematodes have been characterized in greenhouse and growth-chamber environments (16,17), but there is little information available on the use of resistant bell pepper cultivars in double-cropping systems with susceptible vegetable crops. The objective of these studies was to determine the potential of using a resistant bell pepper cultivar as a rotation crop for managing root-knot nematodes in double-cropped cucumber and squash. A secondary objective was to evaluate the effectiveness of resistance to the southern root-knot nematode conditioned by the N gene in 'Charleston Belle' in field tests in South Carolina and Georgia.

MATERIALS AND METHODS

These studies were conducted at the Clemson University Edisto Research and Education Center, Blackville, SC and the Crop Protection and Management Research Unit, USDA-ARS, Tifton, GA. Two double-crop pepper and cucurbit experiments were conducted at each site. Bell peppers were planted in the spring, followed by a fall crop of either cucumber or yellow squash. The pepper genotypes used in these studies were a set of near-isogenic bell pepper cultivars, Charleston Belle (NN) and Keystone Resistant Giant (nn), that differ in resistance to root-knot nematodes controlled by the N gene. The cucumber and squash cultivars used in all experiments were Stonewall and Cougar, respectively.

Egg inocula

Egg inocula of M. incognita race 3 were maintained on tomato (Lycopersicon esculentum Mill. cv. Rutgers) and pole bean (Phaseolus vulgaris L. cv. Kentucky Wonder 191) in isolated greenhouse benches. M. incognita egg inocula were extracted from infected tomato and bean roots using 0.5% NaOCl (9).

Bell pepper–cucumber double-crop study at Blackville, SC

Seed of both bell pepper cultivars were planted in the greenhouse on 30 March 2002. On 3 May 2002, the field was tilled and beds were formed and covered with 3-ml black plastic. On 6 May 2002, the pepper seedlings were inoculated with approximately 5,000 M. incognita race 3 eggs per seedling and, on 8 May 2002, the seedlings were transplanted into single-row plots on beds 2 m apart. Each plot contained 25 plants spaced 30 cm apart. The experimental design was a randomized complete block with 10 replications. At planting, eight cores of soil 20 cm deep were collected from each plot and combined into a composite sample. Second-stage juveniles (J2) were extracted from 113 cm3 of soil from the composite sample using the sugar flotation method (11). Standard cultural and insect control practices were followed and plots were subsurface drip irrigated. Soil temperatures were recorded at 10-min intervals at 5-cm and 15-cm depths in the row from 4 June through 17 October 2002. Fruit were harvested from 10 plants in the center of each plot on 17 and 30 July 2002. On 30 July 2002, the root systems and surrounding soil of six pepper plants (fourth through ninth planting positions from the front of the plot) were dug from each plot. The roots were removed carefully from the soil and the soil samples were saved for extraction of J2. The roots were washed and rated for gall severity and egg mass production using a scale of 1 to 5, where 1 = 0 to 3% root system galled or covered with egg masses; 2 = 4 to 25%, 3 = 26 to 50%, 4 = 51 to 79%, and 5 = ≥80% root system galled or covered with egg masses (20). M. incognita eggs were extracted from a bulked 20-g subsample of roots from each plot using 1% NaOCl (9). J2 were extracted from 113 cm3 of soil as previously described. On 31 July 2002, the pepper plants remaining in the plots were sprayed with glyphosate. On 5 August 2002, the shoots of the pepper plants were clipped at the crown and removed, and the black plastic was sprayed with whitewash to increase the reflectivity of the plastic mulch. On 8 August 2002, two seed of 'Stonewall' cucumber were direct seeded into each hole that previously had been planted with a bell pepper plant. After seedling emergence, seedlings were thinned to one per hill. Standard cultural and pest control practices were followed and plots were subsurface drip irrigated. Fruit were harvested from 10 plants (plant positions 10 through 19 from the front of the plot) in the center of each plot on 26 September and 1, 3, 8, and 17 October 2002. On 17 October 2002, the root systems and surrounding soil of 10 cucumber plants were dug from the center of each plot. The roots were removed carefully from the soil and the soil samples were saved for extraction of J2. The roots were washed and rated for gall severity and egg mass production using a scale of 1 to 5, as previously described. M. incognita eggs were extracted from a bulked 20-g subsample of roots from each plot using 1% NaOCl (9). J2 of M. incognita were extracted from 113 cm3 of soil as previously described.

Bell pepper–squash double-crop study at Blackville, SC

The bell pepper and squash experiments were conducted as described for the bell pepper–cucumber study at Blackville, SC. Planting, harvesting, and sampling dates for bell pepper were the same as previously described. Seed of 'Cougar' yellow squash were sown on 8 August 2002. Squash fruit were harvested on 17, 23, and 26 September and 1 and 3 October 2002. On 3 October 2002, soil and root samples were collected, roots were evaluated for severity of galling, and M. incognita eggs and J2 were extracted from roots and soil, respectively, as previously described. Bell pepper–cucumber double-crop study at Tifton, GA. Field and bed preparation, cultural practices, and experimental design were as described for the Blackville study. Seed of peppers were sown in the greenhouse on 31 March 2002 and transplanted to the field on 29 April 2002. At planting and on 12 August, soil samples were collected and M. incognita J2 were extracted as previously described. Fruit were harvested from all plants in each plot on 24 July and 12 August 2002. 0n 14 August 2002, the root systems of six plants were dug from the center of each plot. Roots were rated for gall severity and eggs were extracted from 20 g of fibrous roots as previously described. The plots were treated as previously described in preparation for the fall cucumber crop. Sowing and culture of cucumber were as previously described. On 22 August 2002, seed of 'Stonewall' cucumber were sown. Fruit were harvested from 10 cucumber plants in the center of each plot on 9, 11, 15, 18, 21, 25, and 28 October and 1 and 4 November 2002. On 4

November 2002, soil samples were collected, root systems of 10 cucumber plants were dug, and roots were rated for gall severity. M. incognita J2 and eggs were extracted from soil and root samples, respectively, as previously described.

Table 1. Gall index, numbers of Meloidogyne incognita eggs per gram of fresh root, second-stage juveniles (J2) per 100 cm3 of soil, total fruit yield, and total fruit numbers of bell pepper cvs. Charleston Belle (CB) and Keystone Resistant Giant (KRG)a

Cultivar[b]	Gall index[c]	Eggs[d]	J2[d]	Yield (kg/plot)	No. of fruit
Blackville, SC					
CB	1.4****	725****	8*	2.6*	48*
KRG	4.6	9,515	20	1.7	31
Tifton, GA					
CB	1.3****	2,424****	16***	2.2	41
KRG	3.9	11,632	225	1.6	29
Combined[e]					
CB	1.3****	1,574****	12****	2.4**	44****
KRG	4.3	10,573	122	1.7	30

[a] According to the *F* test, *, **, ***, and **** = significant at $P < 0.05$, 0.01, 0.001, and 0.0001, respectively.
[b] 'Charleston Belle' is resistant and 'Keystone Resistant Giant' is susceptible to *M. incognita*.
[c] Gall index: 1 = 0 to 3%, 2 = 4 to 25%, 3 = 26 to 50%, 4 = 51 to 79%, and 5 = 80 to 100% root system galled.
[d] Data were $\log_{10}(x + 1)$ transformed before analysis.
[e] Combined analysis of both tests.

Table 2. Gall index, numbers of Meloidogyne incognita eggs per gram of fresh root, second-stage juveniles (J2) per 100 cm3 of soil, total fruit yield, and total fruit numbers of cucumber cv. Stonewall double-cropped after bell pepper cvs. Charleston Belle (CB) and Keystone Resistant Giant (KRG)[a]

Previous cultivar[b]	Gall index[c]	Eggs[d]	J2[d]	Yield (kg/plot)	No. of fruit
Blackville, SC					
CB	3.5***	3,970	645	5.8****	19***
KRG	4.9	7,346	1,190	0.7	2
Tifton, GA					
CB	4.9	6,651*	293	4.9***	22***
KRG	5.0	11,749	151	0.7	3
Combined[e]					
CB	4.2****	5,310***	469	5.3****	20****
KRG	4.9	9,548	671	0.7	3

[a] According to the *F* test, *, **, ***, and **** = significant at $P < 0.05$, 0.01, 0.001, and 0.0001, respectively.
[b] 'Charleston Belle' is resistant and 'Keystone Resistant Giant' is susceptible to *M. incognita*.
[c] Gall index: 1 = 0 to 3%, 2 = 4 to 25%, 3 = 26 to 50%, 4 = 51 to 79%, and 5 = 80 to 100% root system galled.
[d] Data were $\log_{10}(x + 1)$ transformed before analysis.
[e] Combined analysis of both tests.

Bell pepper–squash double-crop study at Tifton, GA.

The bell pepper– squash experiment was conducted as described for the bell pepper–cucumber study at Tifton, GA. Planting, harvesting, and sampling dates for bell pepper were the same. On 22 August 2002, seed of 'Cougar' yellow squash were sown. Fruit were harvested on 3, 9, 11, 15, and 18 October 2002. On 29 October 2002, root samples were collected and evaluated for root gall severity, and M. incognita eggs were extracted and counted. On 13 November 2002, 10 cores of soil were collected from each plot and J2 were extracted as previously described.

Data analysis

Nematode egg and J2 data were log10(x + 1) transformed to normalize the error variance before analysis (14). Data were analyzed using the GLM procedure of SAS for Windows (version 6.12; SAS Institute, Cary, NC).

RESULTS

Bell pepper–cucumber double-crop study at Blackville, SC

Density of M. incognita J2 was approximately 10 per 100 cm3 of soil at planting. Soil temperatures ranged from 13.4 to 43.0°C at a 5-cm depth and from 16.0 to 38.3°C at a 15-cm depth from 4 June to 17 October 2002. Because bell pepper data for the cucumber and squash experiments at Blackville, SC were not significantly different and there were no interactions between the two studies, the bell pepper data were combined for analysis (Table 1). The resistant cv. Charleston Belle had a root gall severity index of 1.4 and the root systems supported 725 M. incognita eggs per gram of fresh root (Table 1). In contrast, the susceptible cv. Keystone Resistant Giant had a root gall severity index of 4.6 and the root systems supported 9,515 M. incognita eggs per gram of fresh root. 'Charleston Belle' had 92% fewer ($P < 0.0001$) eggs per gram of fresh root than its susceptible isoline 'Keystone Resistant Giant'. There were 60% fewer ($P < 0.05$) J2 in the soil at the last harvest in plots where 'Charleston Belle' had been grown than in plots where 'Keystone Resistant Giant' had been grown. 'Charleston Belle' produced 1.5× greater ($P \leq 0.05$) numbers of fruit and fruit yield than 'Keystone Resistant Giant'. Cucumber plants grown in plots following the resistant cv. Charleston Belle had less severe root galling ($P \leq 0.001$) than those following the susceptible 'Keystone Resistant Giant' (gall indices were 3.5 versus 4.9, respectively; Table 2; Fig. 1A and B). Cucumber yields were 87% heavier ($P \leq 0.0001$) and numbers of fruit were 89% higher ($P \leq 0.001$) in plots previously planted to 'Charleston Belle' than to 'Keystone Resistant Giant'.

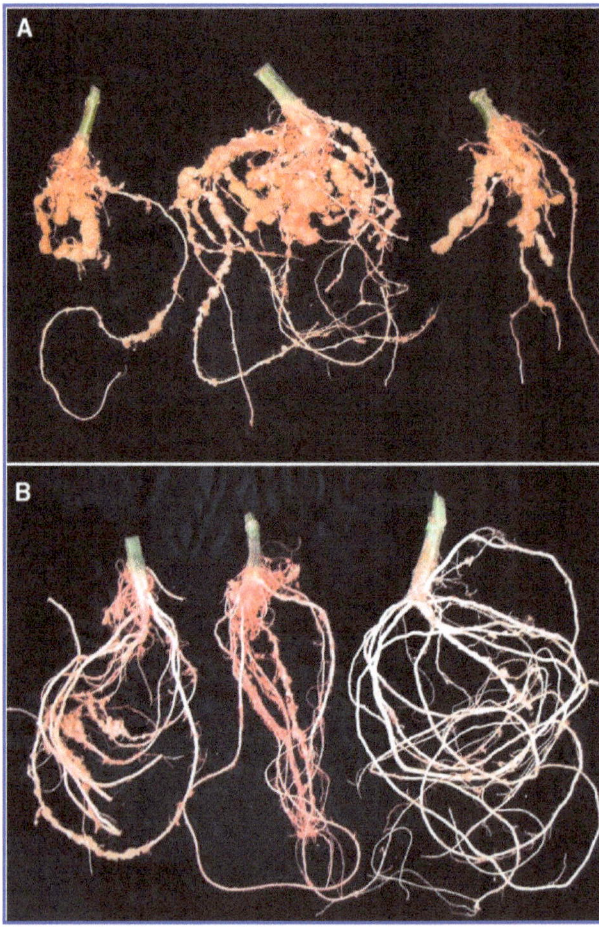

Figure 1. **A**, Roots of 'Stonewall' cucumber grown in a double-crop sequence following the root-knot nematode-susceptible bell pepper 'Keystone Resistant Giant', and **B,** roots of 'Stonewall' cucumber following the root-knot nematode-resistant bell pepper 'Charleston Belle', in Blackville, SC.

Bell pepper–cucumber double-crop study at Tifton, GA

Density of M. incognita J2 was approximately 119 per 100 cm3 of soil at planting. Bell pepper data for the cucumber and squash experiments at Tifton, GA were not significantly different and there were no interactions between the two studies; therefore, the bell pepper data were combined for analysis (Table 1). 'Charleston Belle' had a root gall severity index of 1.3 and the root systems supported 2,424 M. incognita eggs per gram of fresh root (Table 1). 'Keystone Resistant Giant' had a root gall severity index of 3.9 and root systems supported 11,632 M. incognita eggs per gram of fresh root. 'Charleston Belle' had 79% fewer ($P \leq 0.0001$) eggs per gram of fresh root and 92%

fewer (P ≤ 0.001) J2 in the soil at the final harvest than 'Keystone Resistant Giant'. Cucumber plants grown in plots previously planted to the resistant cv. Charleston Belle had 43% fewer (P ≤ 0.05) M. incognita eggs per gram of fresh root than those planted to 'Keystone Resistant Giant' (Table 2). Cucumber yields were 86% heavier (P ≤ 0.001) and numbers of fruit were 86% higher (P ≤ 0.001) in plots previously planted to 'Charleston Belle' than to 'Keystone Resistant Giant'.

Bell pepper–squash double-crop study at Blackville, SC

Bell pepper results were presented previously (see bell pepper–cucumber double-crop study at Blackville, SC). The root gall severity was less (P ≤ 0.001; Table 3; Fig. 2) and M. incognita egg production was 59% less (P ≤ 0.05) for squash following 'Charleston Belle' than those grown after 'Keystone Resistant Giant'. Likewise, numbers of J2 in the soil were 66% less (P ≤ 0.01) in plots where squash followed 'Charleston Belle' compared with plots where squash followed 'Keystone Resistant Giant'. Squash yields were 54% heavier (P ≤ 0.05) and numbers of squash fruit were 45% higher (P ≤ 0.01) when grown following 'Charleston Belle' compared with 'Keystone Resistant Giant'.

Bell pepper–squash double-crop study at Tifton, GA

Bell pepper results were presented previously (see bell pepper–cucumber double-crop study at Tifton, GA). The root gall severity index and M. incognita egg production on squash plants grown after 'Charleston Belle' were not significantly different from those grown after 'Keystone Resistant Giant' (Table 3). Squash yields were 59% heavier (P ≤ 0.05) and numbers of squash fruit were 60% higher (P ≤ 0.05) when grown in plots following 'Charleston Belle' compared with 'Keystone Resistant Giant'.

Table 3. Gall index, numbers of Meloidogyne incognita eggs per gram of fresh root, second-stage juveniles (J2) per 100 cm3 of soil, total fruit yield, and total fruit numbers of squash cv. Cougar double-cropped after bell pepper cvs. Charleston Belle (CB) and Keystone Resistant Giant (KRG)a

Previous cultivar[b]	Gall index[c]	Eggs[d]	J2[d]	Yield (kg/plot)	No. of fruit
Blackville, SC					
CB	3.3***	10,710*	686**	3.5*	22**
KRG	4.7	26,418	2,004	1.6	12
Tifton, GA					
CB	4.6	13,020	248	1.7*	10*
KRG	4.9	15,444	306	0.7	4
Combined[e]					
CB	4.0****	11,865*	467**	2.7**	16***
KRG	4.8	20,931	1,155	1.2	8

[a] According to the *F* test, *, **, ***, and **** = significant at $P < 0.05$, 0.01, 0.001, and 0.0001, respectively.

[b] 'Charleston Belle' is resistant and 'Keystone Resistant Giant' is susceptible to *M. incognita*.

[c] Gall index: 1 = 0 to 3%, 2 = 4 to 25%, 3 = 26 to 50%, 4 = 51 to 79%, and 5 = 80 to 100% root system galled.

[d] Data were $\log_{10}(x + 1)$ transformed before analysis.

[e] Combined analysis of both tests.

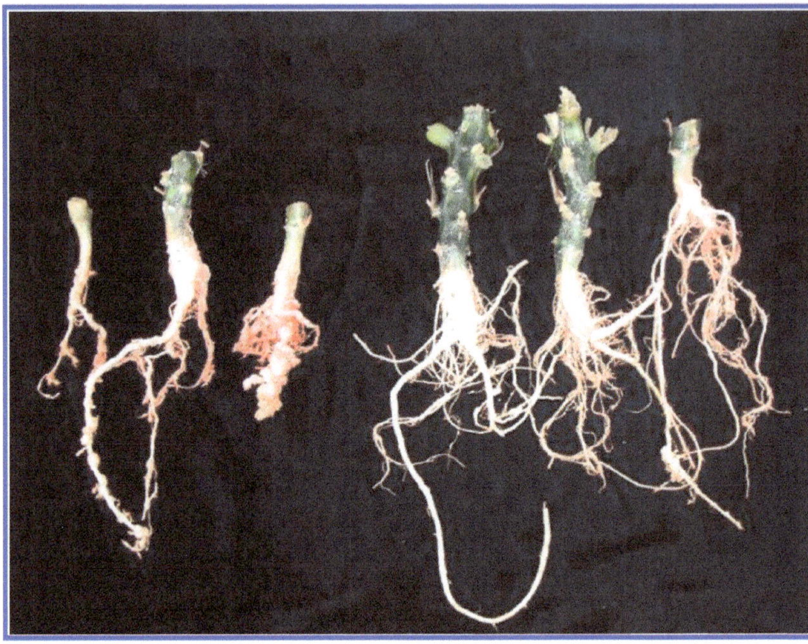

Figure 2. Roots of 'Cougar' yellow squash plants grown in a double-crop sequence following the rootknot nematode–susceptible bell pepper 'Keystone Resistant Giant' (left) and following the root-knot nematode–resistant bell pepper 'Charleston Belle'(right), in Blackville, SC.

DISCUSSION

'Charleston Belle' exhibited a high level of resistance and 'Keystone Resistant Giant' was susceptible in both the Blackville, SC and Tifton, GA studies, confirming results of previous experiments in which root galling and reproduction of M. incognita were minimal on 'Charleston Belle' and root galling was severe and reproduction of M. incognita was high on 'Keystone Resistant Giant' (16,17). 'Charleston Belle' (NN) produced significantly greater fruit yield than its susceptible near-isoline 'Keystone Resistant Giant' (nn) in the Blackville study, similar to results of another study conducted at Blackville, SC in 2001 (21). In prior trials, the resistant 'Carolina Cayenne' pepper, which also is homozygous for the N gene and carries an additional recessive gene that confers resistance to M. incognita, produced significantly greater yields than its susceptible near-isoline PA-136 when grown in M. incognita-infested fields (19). Results of the present and previous studies demonstrate the effectiveness of the N gene in controlling the southern root-knot nematode in field-grown pepper (19,21). Breakdown of resistance to M. incognita under high soil temperatures has been reported in several horticultural crop species, including tomato, bean, and sweet potato (4,8,10,13). Resistance breaking due to high soil temperatures has been reported in both bell pepper (C. annuum) and in Scotch Bonnet-type peppers (C. chinense) in controlled environment studies (16,18). Although the resistant bell cvs. Charleston Belle and

Carolina Wonder exhibited a partial loss of resistance at 28 and 32°C, reproduction of M. incognita on these cultivars was only 20% of that on the susceptible parental cvs. Keystone Resistant Giant and Yolo Wonder B, respectively, and root gall indices of the resistant cultivars were still within the moderately resistant range (16). Because the majority of bell peppers grown in the United States are produced in hot climates, knowledge of the heat stability of root-knot nematode resistance is critical both in the development of pepper cultivars and in making recommendations about nematode management to pepper growers. Results of the present study demonstrate that resistance exhibited by 'Charleston Belle' is effective even under the high soil temperatures recorded in the Blackville, SC study, in which maximum daily temperatures ranged from 25.7 to 38.3°C at a 15-cm soil depth. Thus, it appears that resistance conditioned by the N gene was stable at the high soil temperatures in the present study, even though resistance of 'Charleston Belle' was partially lost at constant soil temperatures of 28 and 32°C in controlled environment experiments (16). A possible explanation for the differences in response of 'Charleston Belle' under field and controlled environment conditions is that soil temperatures fluctuated during the course of 24 h in the plastic mulched pepper beds, whereas the soil temperature in the growth chamber was relatively stable at 32°C, which provided the conditions necessary for a partial loss of resistance in 'Charleston Belle'. 'Stonewall' cucumber plants grown in plots following 'Charleston Belle' had less severe root galling and 44% fewer eggs per gram of fresh root than cucumber plants grown after 'Keystone Resistant Giant' (averaged over both locations). Cucumber fruit yields (averaged over both locations) were 87% greater in plots following 'Charleston Belle' than in plots following 'Keystone Resistant Giant'. These results are similar to those reported from Louisiana in which 'Dasher II' cucumber doublecropped following the root-knot nematoderesistant 'Celebrity' tomato produced heavier yields, had reduced root-galling, and lower densities of M. incognita J2 in the soil than the same cultivar grown after the susceptible 'Heatwave' tomato (1,7). In the present studies, 'Cougar' squash plants grown following 'Charleston Belle' pepper had less severe root galling, 43% fewer eggs per gram of fresh root, and 60% fewer J2 present in the soil at the final harvest than squash plants grown following 'Keystone Resistant Giant' pepper (averaged over both locations). Fruit yields of squash grown following 'Charleston Belle' (averaged over both locations) were 55% greater than fruit yields of squash grown following 'Keystone Resistant Giant'. In another double-crop study conducted in Tifton, GA, soil populations of M. incognita recovered in 'Charleston Belle' plots were about 11% of those recovered in the susceptible bell pepper 'Camelot' plots (24). Fruit weights of 'Dixie' squash grown after 'Charleston Belle' and after 'Camelot' bell peppers were not different. In Florida, yield of 'Lemondrop' squash grown in the spring following the previous summer's rotation crop of 'Hale' castor, 'Deltapine 90' cotton, velvetbean, or crotalaria produced heavier yields than squash grown after root-knot nematode-susceptible 'Florunner' peanut (12). Root-gall severity ratings of squash grown after any of the abovementioned crops (root-gall severity ratings ranged from 3.7 to 4.8 using a 1-to-5 rating scale) were not different. Our results demonstrated the effectiveness of resistance to the southern root-knot nematode conditioned by the N gene in 'Charleston Belle' pepper under field conditions in South Carolina and Georgia. Furthermore, the present studies showed that root-knot nematode-resistant bell pepper cultivars, such as Charleston Belle, will be useful tools for managing M. incognita in double-cropping systems with cucurbit crops. The utilization of resistant vegetable cultivars will provide economical and environmentally compatible alternatives to methyl bromide and other nematicides for managing M. incognita in vegetable crops. An integrated pest management approach that utilizes other nematode management techniques, such as biocontrol organisms or resistant cover crops, in combination with root-knot nematode-resistant pepper may be necessary in order to achieve acceptable yields of highly susceptible cucurbits in a double-crop production scheme.

ACKNOWLEDGMENTS

We thank S. B. Merrill, P. Berland, J. Carey, T. Hilton, and A. K. Montfort for technical assistance in conducting the experiments.

REFERENCES

1. Colyer, P. D., Kirkpatrick, T. L, Vernon, P. R., Barham, J. D., and Bateman, R. J. 1998. Reducing Meloidogyne incognita injury to cucumber in a tomato-cucumber double-cropping system. J. Nematol. 30:226-231.
2. Di Vito, M., Cianciotta, V., and Zaccheo, G. 1992. Yield of susceptible and resistant pepper in microplots infested with Meloidogyne incognita. Nematropica 22:1-6.
3. Di Vito, M., Greco, N., and Carella, A. 1985. Population densities of Meloidogyne incognita and yield of Capsicum annuum. J. Nematol. 17:45-49.
4. Dropkin, V. H. 1969. The necrotic reaction of tomatoes and other hosts resistant to Meloidogyne: Reversal by temperature. Phytopathology 59:1632-1637.
5. Fery, R. L., Dukes, P. D., Sr., and Thies, J. A. 1998. 'Carolina Wonder' and 'Charleston Belle': Southern root-knot nematode resistant bell peppers. HortScience 33:900-902.
6. Hanna, H. Y. 2000. Double-cropping muskmelons with nematode-resistant tomatoes increases yield, but mulch color has no effect. HortScience 35:1213-1214.
7. Hanna, H. Y., Colyer, P. D., Kirkpatrick, T. L., Romaine, D. J., and Vernon, P. R. 1994. Feasibility of improving cucumber yield without chemical control in soils susceptible to nematode build-up. HortScience 29:1136-1138.
8. Holtzmann, O. V. 1965. Effect of soil temperature on resistance of tomato to root-knot nematode (Meloidogyne incognita). Phytopathology 55:990-992.
9. Hussey, R. S., and Barker, K. R. 1973. A comparison of methods of collecting inocula of Meloidogyne spp., including a new technique. Plant Dis. Rep. 57:1025-1028.
10. Jatala, P., and Russell, C. C. 1972. Nature of sweet potato resistance to Meloidogyne incognita and the effects of temperature on parasitism. J. Nematol. 4:1-7.
11. Jenkins, W. R. 1964. A rapid centrifugal-flotation technique for separating nematodes from soil. Plant Dis. Rep. 48:692.

12. McSorley, R., Dickson, D. W., de Brito, J. A., Hewlett, T. E., and Frederick, J. J. 1994. Effects of tropical rotation crops on Meloidogyne arenaria population densities and vegetable yields in microplots. J. Nematol. 26:175-181.

13. Mullin, B. A., Abawi, G. S., and Pastor-Corrales, M A. 1991. Modification of resistance expression of Phaseolus vulgaris to Meloidogyne incognita by elevated soil temperatures. J. Nematol. 23:182-187.

14. Noe, J. P. 1985. Analysis and interpretation of data from nematological experiments. Pages 187-196 in: An Advanced Treatise on Meloidogyne. Vol. II: Methodology. K. R. Barker, C. C. Carter, and J. N. Sasser, eds. North Carolina State University Graphics, Raleigh.

15. Sasser, J. N., and Freckman, D. W. 1987. A world perspective on nematology: The role of the society. Pages 7-14 in: Vistas on Nematology: A Commemoration of the Twenty-Fifth Anniversary of the Society of Nematologists. J. A. Veech and D. W. Dickson, eds. Society of Nematology, Hyattsville, MD.

16. Thies, J. A., and Fery, R. L. 1998. Modified expression of the N gene for southern rootknot nematode resistance in pepper at high soil temperatures. J. Am. Soc. Hortic. Sci. 123:1012-1015.

17. Thies, J. A., and Fery, R. L. 2000. Characterization of resistance conferred by the N gene to Meloidogyne arenaria races 1 and 2, M. hapla, and M. javanica in two sets of isogenic lines of Capsicum annuum L. J. Am. Soc. Hortic. Sci. 125:71-75.

18. Thies, J. A., and Fery, R. L. 2000. Heat stability of resistance to Meloidogyne incognita in Scotch Bonnet peppers (Capsicum chinense Jacq.). J. Nematol. 32:356-361.

19. Thies, J. A., Mueller, J. D., and Fery, R. L. 1997. Effectiveness of resistance to southern root-knot nematode in 'Carolina Cayenne' pepper in greenhouse, microplot, and field tests. J. Am. Soc. Hortic. Sci. 122:200-204.

20. Thies, J. A., Mueller, J. D., and Fery, R. L. 1998. Use of a resistant pepper as a rotational crop to manage southern root-knot nematode. HortScience 33:716-718.

21. Thies, J. A., Mueller, J. D., Fery, R. L., Miller, G., and Varn, J. E. 2003. Response of bell pepper cultivars near-isogenic for the N gene to Meloidogyne incognita in field trials. HortScience. 38:1394-1396.

22. U.S. Department of Agriculture. 1993. Alternatives to methyl bromide: Assessments of research needs and priorities. Proc. USDA Workshop on Alternatives to Methyl Bromide, Arlington, VA.

23. U.S. Environmental Protection Agency. 2000. Protection of Stratospheric Ozone: Incorporation of Clean Air Act Amendments for Reductions in Class I, Group VI Controlled Substances. Federal Register 65 (No. 229):70795-70804.

24. Webster, T. M., Csinos, A. S., Johnson, A. W., Dowler, C. C., Sumner, D. R., and Fery, R. L. 2001. Methyl bromide alternatives in a bell pepper-squash rotation. Crop Prot. 20:605-614.

14
CHAPTER

SOILBORNE PATHOGENS IN AN IRRIGATED CROPPING SYSTEM: EFFECTS OF TILLAGE, RESIDUE MANAGEMENT, AND CROP ROTATION

T. C. Paulitz, K. L. Schroeder, and W. F. Schillinger,

Irrigated crops are produced on 645,000 ha in east-central Washington and northcentral Oregon (1). Approximately 70% of this cropland is irrigated with water diverted from rivers, while water for the remaining hectares is pumped from deep wells. Much of the irrigated cropland is found in areas of less than 200 mm annual precipitation, where dryland farming is difficult or impossible. Wind erosion is the biggest environmental concern throughout the region (21,31). Major irrigated field crops include potatoes (Solanum tuberosum L.), corn (Zea mays L.), wheat (Triticum aestivum L.), barley (Hordeum vulgare L.), oat (Avena sativa L.), alfalfa (Medicago sativa L.), and dry bean (Phaseolus vulgaris L.). Winter canola (Brassica napus L.) is not yet widely grown but has received increased attention in recent years due to its reported benefits for subsequent crops in the rotation (10). Due to many available crop options, there are no "set" cropping systems. Farmers rotate crops to reduce pests and diseases, optimize water use, and take advantage of current market opportunities. Many deep-well irrigators plant winter wheat on the same field every year. Irrigated wheat grain yields range from 6,000 to 9,400 kg/ha with residue production of 11,000 kg/ha or more. After grain harvest in August, the traditional practice is to burn the stubble and invert

Citation: Paulitz, T. C., Schroeder, K. L., and Schillinger, W. F. 2010. Soilborne pathogens in an irrigated cropping system: Effects of tillage, residue management, and crop rotation. doi:10.1094 / PDIS-94-1-0061

the surface soil with moldboard plow tillage in preparation for planting in September. Generally, farmers feel they need to burn their fields because high residue levels hamper planting and to control the problematic grass weed, downy brome (Bromus tectorum L.). Alternatives to field burning are needed to reduce smoke emissions and maintain air quality. Therefore, new crop rotation and stubble management strategies are needed to make no-till (without burning) work. A group of deep-well irrigated wheat farmers approached Washington State University (WSU) and USDA-ARS scientists in 1998 concerning the future of their farming operations. The farmers were concerned about potential regulations to reduce or eliminate cereal stubble burning and desired research on how to farm profitably without field burning. Reduction or elimination of tillage could also reduce soil erosion. The experiment reported here was designed jointly by farmer advisors and WSU and USDA-ARS scientists. The experiment involved both crop rotation and stubble management factors. The objective of the experiment was to determine the feasibility of planting directly into high levels of residue (no-till) as a substitute for burning in irrigated cropping systems. Specific objectives of the larger project were to:

1. Test a 3-year crop rotation of winter wheat–spring barley–winter canola directly planted (no-till) (i) into standing stubble, (ii) after mechanical removal of stubble, and (iii) after burning the stubble. This was compared with the check treatment of annual winter wheat planted after stubble burning + moldboard plowing.
2. Evaluate and develop effective techniques for planting crops into heavy surface stubble using no-till methods.
3. Document cumulative effects of a diverse no-till crop rotation under three stubble management practices on diseases, soil physical and biological properties, water use efficiency, weed ecology, and farm economics, and to compare these effects to those under the check treatment.

The experiments reported here were part of objectives 1 and 3, and were designed to answer several root disease questions: (i) Will root diseases increase when tillage is eliminated? Previous work has shown that Rhizoctonia root rot and bare patch can increase in no-till rotations (17,24,27). (ii) Will residue management reduce disease severity? Gaeumannomyces graminis var. tritici (causal agent of take-all) and Fusarium pseudograminearum and F. culmorum (causal agents of Fusarium crown rot) survive in crown tissue, and several studies have shown increases in diseases in no-till systems where residue is left on the surface (2,32). In addition, irrigated winter wheat produces more residue compared to dryland winter wheat, as 1.3 to 2.0 kg of straw is produced for every 1.0 kg grain (9). (iii) Will crop rotation reduce root diseases? Take-all can be controlled by rotating with broadleaf crops such as canola or pea (Pisum sativum L.), which are resistant to take-all (12), but Rhizoctonia solani AG-8 (cause of Rhizoctonia root rot and bare patch) has a broad host range, and crop rotation does not reduce bare-patch (7). In this paper, we report on soilborne pathogens and root diseases in these cropping systems, specifically Rhizoctonia root rot and bare patch, take-all, Fusarium crown rot (35), and lesion nematodes (Pratylenchus neglectus and P. thornei)

MATERIALS AND METHODS

Experiment layout and treatments

A 6-year irrigated cropping study was conducted at the WSU Dryland Research Station near Lind during the 2001 to 2006 crop years. The experimental area had been in a 2-year cultivated dryland winter wheat–summer fallow rotation for at least 30 years prior to initiation of the experiment. To obtain uniform residue levels at the beginning of the experiment, the entire 4-ha experimental area was planted to 'Madsen' winter wheat in September 1999. The irrigated grain yield in August 2000 was 7,400 kg/ha, and straw production was 11,000 kg/ha. Beginning in the 2001 crop year, a 3-year crop rotation of winter wheat–spring barley–canola was practiced under the three stubble management methods. These were planted: (i) directly into standing stubble, (ii) after removal of stubble by swathing and baling of the straw, or (iii) after burning the stubble. All no-till treatments were planted and fertilized in one pass with a low disturbance Cross Slot no-till drill (3). A check treatment of continuous annual winter wheat sown with a conventional double-disc drill after stubble burning and moldboard plowing was also included. The experimental design was a split block with four replications.

Main plots (15 × 150 m) were crops and subplots (15 × 50 m) were residue management. Crops were planted perpendicular to the residue management treatments. Each portion of the 3-year no-till crop rotation in each residue management method was sown each year. Thus, there were 40 plots (3 crops × 3 residue management practices × 4 replications = 36 plots + 4 replications of the check treatment of continuous winter wheat after burning and plowing). The burn and plow check plots were 15 × 150 m in size. The timing of irrigation varied from year to year, but the entire experiment received 380 mm of irrigation water every year with 150 mm applied in the fall and 230 mm applied in the spring. Water was applied via hand-line sprinklers, and the same irrigation timing was used for all crops. Average annual precipitation at the site during the 6-year study period was 221 mm. All treatments received 190 kg N, 34 kg P, and 22 kg S per hectare each year. Soils at the site contain adequate potassium (K) for crop production with extractable soil K levels >450 kg/ha. The burn–plow continuous winter wheat received all fertilizer in granular form with 135 kg N, 34 kg P, and 22 kg S per hectare applied with a granular spreader in September prior to plowing and planting. The sources that made up this mix were urea, monoammonium phosphate, and ammonium sulfate.

The additional 55 kg/ha N was applied as granular urea top-dress to the growing wheat crop in April. The no-till winter wheat received 135 kg N, 34 kg P, and 22 kg S per hectare as liquid at time of planting with the Cross Slot no-till drill. The liquid fertilizer mix was Solution 32 (NH4NO3 + urea) + ammonium phosphate + thiosul. The additional 55 kg/ha N was applied as a granular urea top-dress to the growing wheat crop in April. Winter canola received the same fertilizer regime as the no-till winter wheat. However, winter canola was killed by cold and/or disease in 5 out of 6 years; therefore, 55 kg/ha of liquid Solution 32 N was applied at planting of spring canola in those 5 years. Spring barley received 190 kg N, 34 kg P, and 22 kg S per hectare as Solution 32 + ammonium phosphate + thiosul at time of planting applied with the Cross

Slot no-till drill. Winter wheat (cv. Madsen) and spring barley (cv. Baronesse) were sown at a rate of 112 kg/ha during all years. Winter canola and spring canola sowing rates varied from 4 to 8 kg/ha and averaged 5.6 kg/ha over the 6 years. Additional details on field operations and inputs for production of winter wheat, spring barley, and canola in the no-till rotation and for continuous winter wheat after burning and plowing are reported by Schillinger et al. (29) and Zaikin et al. (41).

Pathogen quantification from fungal DNA from soil

In March or April, three shovels of soil were randomly taken from each replicate block of each treatment, and the soil within each replicate was combined. One kilogram of soil from each sample was air-dried and frozen at –20°C for 2 weeks prior to sending to the South Australia Research and Development Institute (SARDI), Adelaide, SA. SARDI developed a suite of tests for the quantification of six fungal pathogens and three nematode pathogens of wheat (20). This system has been used commercially (over 16,000 samples processed) in the Australian grain belt for over a decade (20) and can help growers predict the risk of disease and select appropriate management strategies. DNA was extracted from the soil and quantified according to the methods of Ophel-Keller et al. (20). DNA of tested fungal and nematode pathogens was reported as pg/g soil. For statistical analysis, DNA concentrations were log transformed to maintain normality for the ANOVA analysis

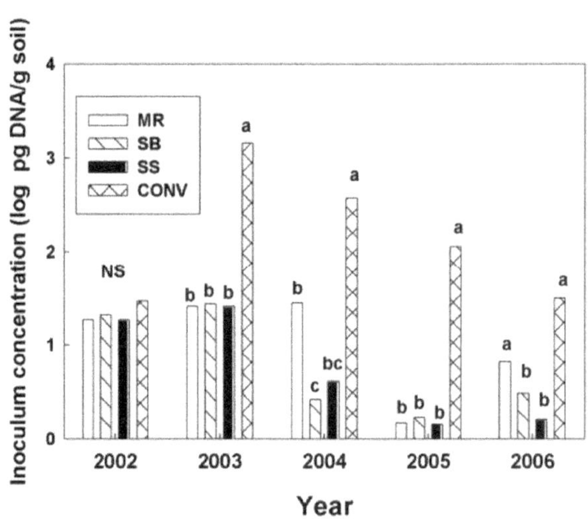

Figure 1. DNA concentration of Gaeumannomyces graminis var. tritici in March–April in soil of a winter wheat crop grown in four crop rotation/residue management treatments: winter wheat–spring barley– canola no-till rotation with mechanical stubble removal (MR), stubble burning (SB), or stubble left standing (SS), and continuous winter wheat with stubble burned and mold-board plowed (CONV). Treatments within each year with the same letters are not statistically different, P = 0.05, Tukey's mean separation test.

Plant and disease measurements

Each year, approximately in mid-April for winter wheat and mid-May for spring barley, three samples were taken within each plot approximately 15 m apart along a transect in the center of the plot. A sample consisted of a shovelful within the row, with approximately five plants in each sample.

Plants were kept at 4°C until evaluation. Three plants were randomly chosen from each sample, for a total of nine plants per replicate. Roots were washed, and the number of tillers, plant height, length of first true leaf, number of seminal and number of crown roots were recorded. The number of roots with typical Rhizoctonia root rot symptoms (spear-tipping, constricted cortex, brown root tips) were counted, and incidence was expressed as the percentage of seminal or crown roots that were diseased. Disease severity was also rated using a 0 to 8 scale (16). Takeall severity was measured on winter wheat using a 0 to 5 scale (19). Grain yield was obtained in late July or early August by harvesting a 50-m-long swath through each plot with a plot combine with a 1.5-m-wide cutting platform operated 20 cm above the ground. The plot combine was equipped with a custom-built blowing air system to uniformly distribute straw and chaff (28). Immediately thereafter, a commercial combine equipped with a straw chopper and chaff spreader was used to harvest the entire experimental area, cutting slightly lower than 20 cm above the ground

Statistical analysis

Analysis of variance was performed on all transformed DNA data and plant parameters, with year, residue/tillage treatment, and block as main effects. Means were separated with Tukey's test. For the disease ratings, data were analyzed with ordinal logistic analysis, and means were separated with the non-parametric Kruskal-Wallis test. Analyses were performed with JMP 4.0 (SAS Institute, Cary, NC).

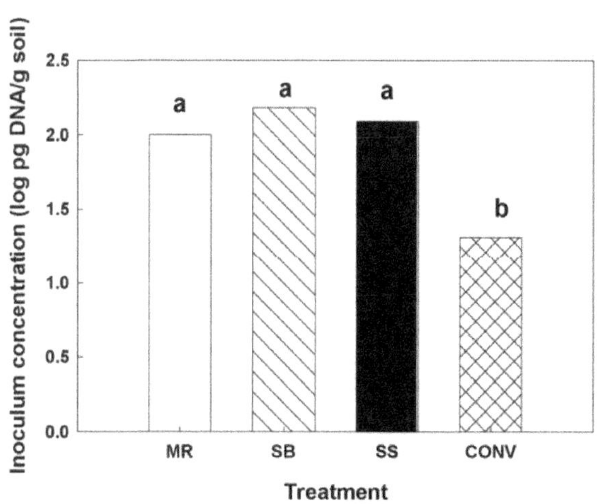

Figure 2. DNA concentration of Rhizoctonia solani AG-8 in March–April in soil of a winter wheat crop grown in four crop rotation/residue management treatments averaged

across all years: winter wheat–spring barley– canola no-till rotation with mechanical stubble removal (MR), stubble burning (SB), or stubble left standing (SS), and continuous winter wheat with stubble burned and moldboard plowed (CONV). Treatments with the same letters are not statistically different, P = 0.05, Tukey's mean separation test.

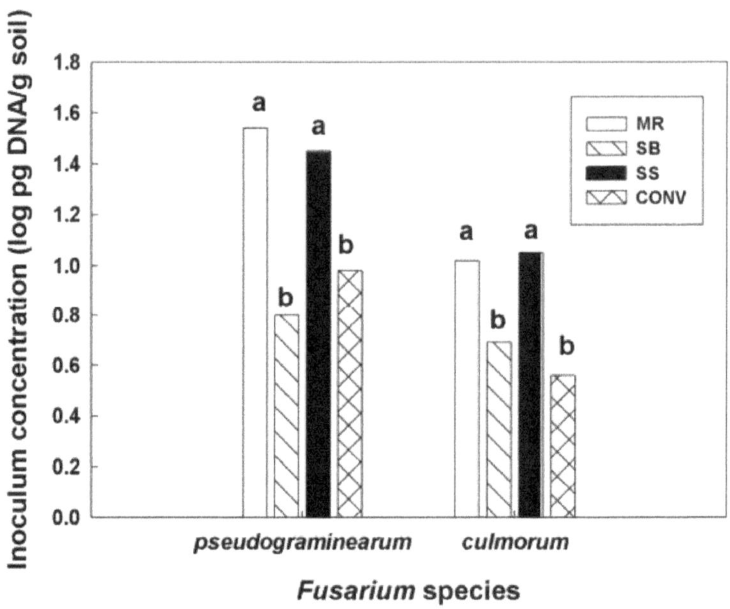

Figure 3. DNA concentration of Fusarium pseudograminearum and F. culmorum in March–April in soil of a winter wheat crop grown in four crop rotation/residue management treatments averaged across all years: winter wheat–spring barley–canola no-till rotation with mechanical stubble removal (MR), stubble burning (SB), or stubble left standing (SS), and continuous winter wheat with stubble burned and moldboard plowed (CONV). Treatments within each Fusarium species with the same letters are not statistically different, P = 0.05, Tukey's mean separation test.

RESULTS

Pathogen quantification from fungal DNA from soil

For the take-all pathogen G. graminis var. tritici, the year × treatment interaction was significant (P < 0.001). In the first year that DNA was measured, there was no significant treatment effect (Fig. 1). However, in the following year, continuous winter wheat had significantly higher DNA levels than in the winter wheat treatments with rotation crops. This difference was present in the three subsequent crop years, although the level of G. graminis var. tritici in the continuous winter wheat treatment declined over time. The DNA concentrations of R. solani AG-8 were significantly affected by both year (P < 0.001) and residue treatment (P < 0.001), but there was no significant year × treatment interaction (P = 0.13). Averaged over all years, continuous winter wheat had significantly less DNA of Rhizoctonia (1.31 log pg DNA/g) compared to the mechanical removal (2.00), stubble burned (2.18), and standing stubble (2.09) treatments (Fig. 2). F. pseudograminearum DNA concentrations in winter wheat were affected by culmorum showed similar trends (Fig. 3), although the overall DNA concentrations of F. culmorum were lower than those of F. pseudograminearum. The DNA levels of P. neglectus in winter wheat had a significant treatment × year interaction (P = 0.005). In 2 years (2002 and 2005), there was no treatment effect (Fig. 4). However, in the other 3 years, significantly higher levels were found in the continuous winter wheat treatment with burning and plowing, compared to the no-till treatments with standing stubble or mechanical stubble removal. The DNA of P. thornei was below the detection limit.

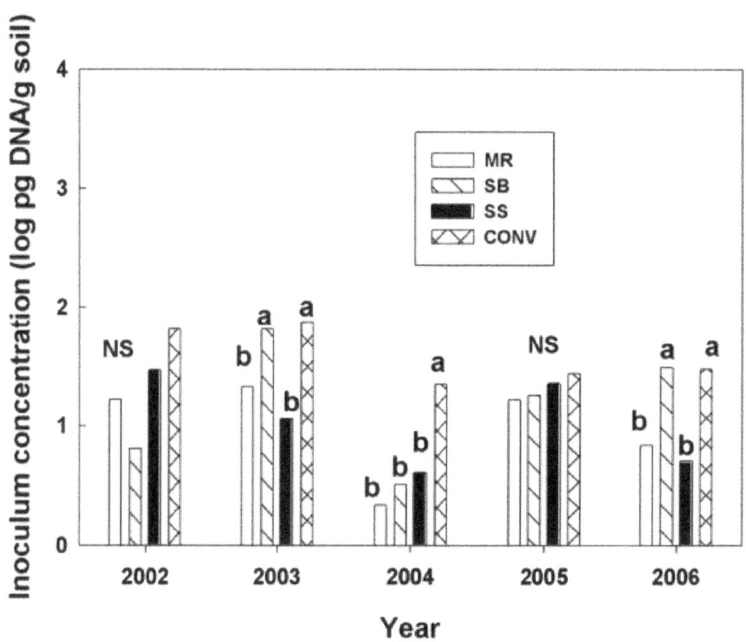

Figure 4. DNA concentration of Pratylenchus neglectus in March–April in soil of a winter wheat crop grown in four crop rotation/residue management treatments: winter wheat–spring barley–canola no-till rotation with mechanical stubble removal (MR), stubble burning (SB), or stubble left standing (SS), and continuous winter wheat with stubble burned and moldboard plowed (CONV). Treatments within each year with the same letters are not statistically different, P = 0.05, Tukey's mean separation test.

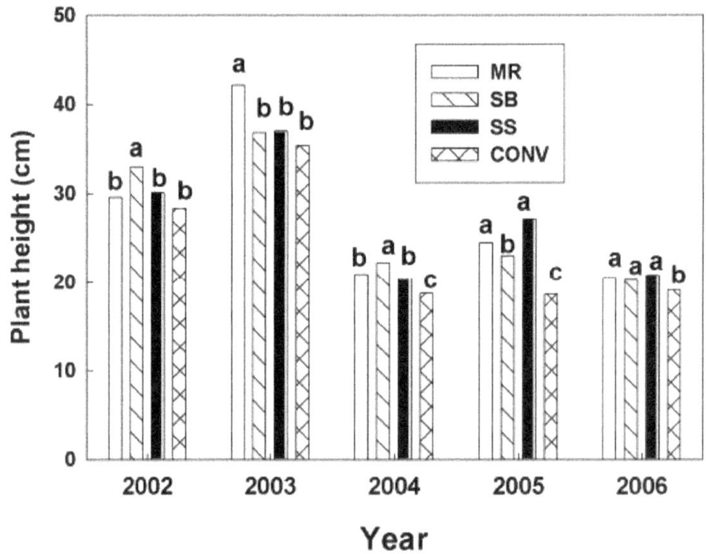

Figure 5. Plant height of winter wheat seedlings in April grown in four crop rotation/residue management treatments: winter wheat–spring barley–canola no-till rotation with mechanical stubble removal (MR), stubble burning (SB), or stubble left standing (SS), and continuous winter wheat with stubble burned and moldboard plowed (CONV). Treatments within each year with the same letters are not statistically different, P = 0.05, Tukey's mean separation test.

Plant parameters

Both tiller number and plant height of winter wheat were affected by residue treatment and interacted with year (P = 0.05 and < 0.001) and interacted with year (P < 0.006 and 0.0001, respectively). In general, the stubble burned treatment had the tallest plants with the greatest number of tillers (Figs. 7 and 8). Barley grown with the stubble burned treatment also had the greatest number of crown roots in 2003, 2004, and 2005 (Fig. 9).

Disease measurements

The highest take-all severity ratings on wheat were found in the continuous winter wheat with burning and plowing, where ratings were two to three units higher than in the diversified no-till rotation treatments (29). Takeall in the continuous wheat treatment increased from 2002 to 2004, and then declined in 2006. In 2003 the take-all rating was 3.8, but in 2006 it was only 2.0. The percentage of wheat crown roots and seminal roots exhibiting symptoms of Rhizoctonia root rot was variable across years (P < 0.0001), and year interacted with treatment (P < 0.001 and 0.0001, respectively, Figs. 10 and 11). In 2003 and 2006, the continuous winter wheat had the lowest level of crown and seminal root infection symptoms compared to the other treatments. The continuous winter wheat treatment also had the lowest Rhizoctonia root rot rating in 2003 and

2006 (Fig. 12). The method of residue management had no effect on the incidence of Rhizoctonia root rot on barley crown roots, and the effect on seminal roots was variable. The stubble burned treatment had numerically the lowest disease level in 2003, 2004, and 2005. Rhizoctonia root rot rating was also variable across years. In 2002, 2003, and 2004, method of residue management had no effect. But in 2005, the stubble burned treatment had the lowest rating (2.5 compared to 4.4 in the mechanical removal treatment and 3.7 in the standing stubble treatment), while in 2006, the standing stubble had the lowest rating (2.6 compared to 3.8 in the mechanical removal treatment and 3.1 in the stubble burned treatment).

DISCUSSION

Statistically significant year × treatment interactions for almost all of the variables measured suggest that variable environmental conditions each year play a strong role in disease expression. Another explanation is that disease and inoculum dynamics change over the course of a cropping system or rotation (29). For example, takeall increased over time in the continuous winter wheat treatment, and burning and plowing did not reduce this disease over the first 4 years of the rotation. This was verified by higher levels of G. graminis var. tritici DNA detected in the soil, which was accompanied by increased disease, significantly shorter plants with fewer tillers, and significantly less grain yield (29,41) compared to no-till winter wheat in a 3-year rotation with barley and canola. The effect of crop rotation on reducing G. graminis var. tritici is well documented (12).

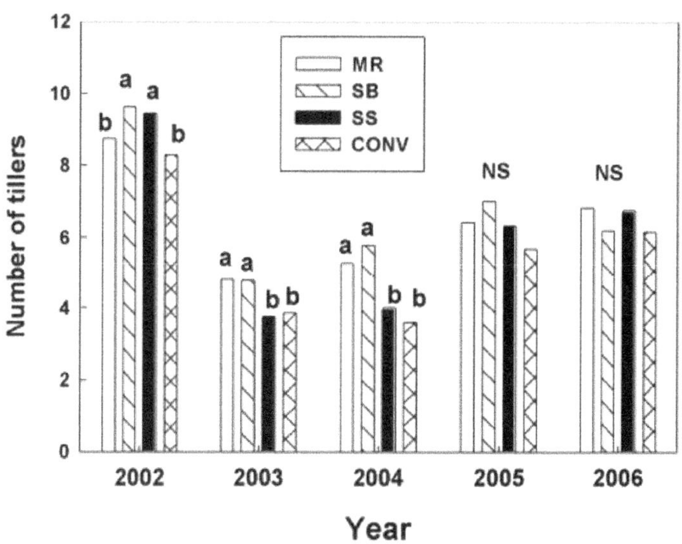

Figure 6. Number of tillers of winter wheat seedlings in April grown in four crop rotation/residue management treatments: winter wheat–spring barley–canola no-till rotation with mechanical stubble removal (MR), stubble burning (SB), or stubble left standing (SS), and continuous winter wheat with stubble burned and moldboard plowed (CONV). Treatments within each year with the same letters are not statistically different, P = 0.05, Tukey's mean separation test.

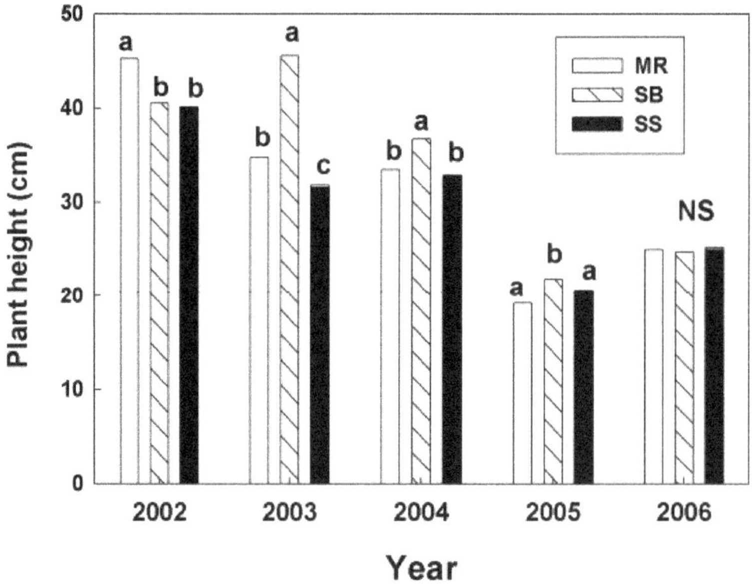

Figure 7. Plant height of spring barley seedlings in May grown in three crop rotation/residue management treatments: winter wheat–spring barley–canola no-till rotation with mechanical stubble removal (MR), stubble burning (SB), or stubble left standing (SS). Treatments within each year with the same letters are not statistically different, P = 0.05, Tukey's mean separation test.

Broadleaf crops such as canola are not hosts for G. graminis var. tritici, and although this pathogen will infect barley, less disease and inoculum are produced (8). To our knowledge, this is the first demonstration of the dynamics of inoculum increase of G. graminis var. tritici during the first few years of monocropping, using DNA-based techniques. We also observed take-all decline after 4 to 5 years of monocropping (29). This phenomenon, documented around the world, is characterized as a reduction in disease after a few years of disease increase in monoculture wheat (11) and is caused by increased populations of antagonistic Pseudomonas spp. that produce antifungal compounds such as 2,4-diacetylphloroglucinol and phenazine in the G. graminis var. tritici lesions on the root and in the rhizosphere (39). Take-all decline was first observed at the WSU Dryland Research Station at Lind in the late 1970s by Cook et al. (4). Take-all decline was correlated with high levels of 2,4-diacetylphloroglucinol producers at this location (18,25,26).

We also detected high levels of 2,4-diacetylphloroglucinol producers in the continuous winter wheat with burning and plowing treatment (R. Allende-Molar, unpublished). The only consistent effect of residue management method on plant growth of winter wheat and spring barley was fewer crown roots on wheat in the standing stubble treatment. This could be due to cooler soil temperatures from the mulching effect of the stubble and the retention of more soil moisture in the spring. Schroeder and Paulitz (30) observed a similar effect in wheat, comparing no-till to conventionally planted (i.e., after tillage) wheat. The number of tillers and plant height of barley were significantly highest in the stubble burned treatment. The development and heading of barley was also accelerated in this treatment, where flowering occurred about 1 week earlier than in the other treatments (data not shown). This effect is likely due to faster soil warming because of the lack of residue and the darker soil surface (5). Another major pathogen in this region is R. solani AG-8, which causes root rot, stunting, and bare patch.

The continuous winter wheat with burning and plowing had lower levels of Rhizoctonia in the soil, compared to the no-till treatments. The reduction of Rhizoctonia with tillage has been well documented in the Pacific Northwest of the United States and in Australia (17,24,27,38). The concentration of Rhizoctonia DNA in no-till soil was approximately 1 log unit higher than in the burn and plow treatment. The real comparison would have been with a healthy crop in the continuous annual burn and plow winter wheat that was not infected with the take-all pathogen. Among residue management treatments, there was no consistent effect on Rhizocto nia disease or inoculum. Rhizoctonia produces inoculum and survives in the roots rather than the crown, so mechanical removal of straw or burning will not reduce inoculum. In addition, burning does not heat the soil to lethal temperatures due to the thermal buffering capacity of the soil. For example, in a bluegrass burning trial in eastern Washington, the soil temperature 1 to 3 cm deep did not change, although the soil surface reached 240°C (14). We did not detect significant levels of Fusarium crown rot in winter wheat in these experiments. This disease is exacerbated by plant water stress (6,22), often triggered by early fall planting and excessive N. Irrigation probably masked the disease symptoms, even though the pathogen is present in soils in this region, and probably caused latent infections in our experiment. However, based on the soil DNA, the highest levels of F. pseudograminearum were observed in the treatments with standing stubble and mechanical removal treatments, compared to the continuous burn and plow and stubble burned treatments, especially by year 3. In the burn and plow treatment with a moldboard plow, the old crowns were inverted and mixed into the soil, where the residue quickly decays. This is consistent with literature showing higher levels of Fusarium crown rot in no-till systems with high levels of residue compared to conventionally tilled systems (2,32). Previous studies have shown that F. pseudograminearum survives primarily as mycelium in the crowns (13), in contrast to F. culmorum which survives as chlamydospores formed from macroconidia in the soil.

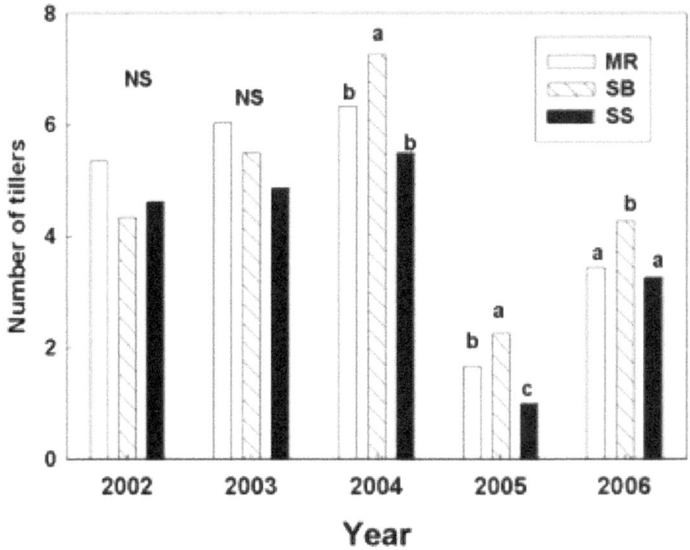

Figure 8. Number of tillers of spring barley seedlings in May grown in three crop rotation/residue management treatments: winter wheat–spring barley–canola no-till rotation with mechanical stubble removal (MR), stubble burning (SB), or stubble left standing (SS). Treatments within each year with the same letters are not statistically different, P = 0.05, Tukey's mean separation test.

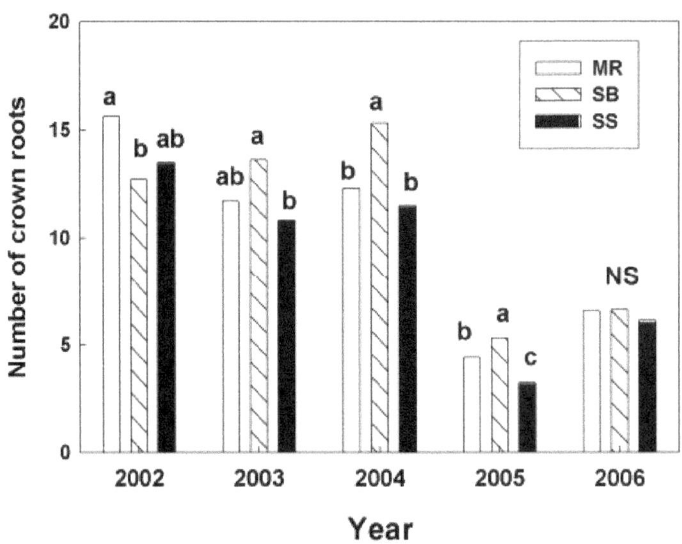

Figure 9. Number of crown roots of spring barley seedlings in May grown in three crop rotation/residue management treatments: winter wheat–spring barley–canola no-

till rotation with mechanical stubble removal (MR), stubble burning (SB), or stubble left standing (SS). Treatments within each year with the same letters are not statistically different, P = 0.05, Tukey's mean separation test.

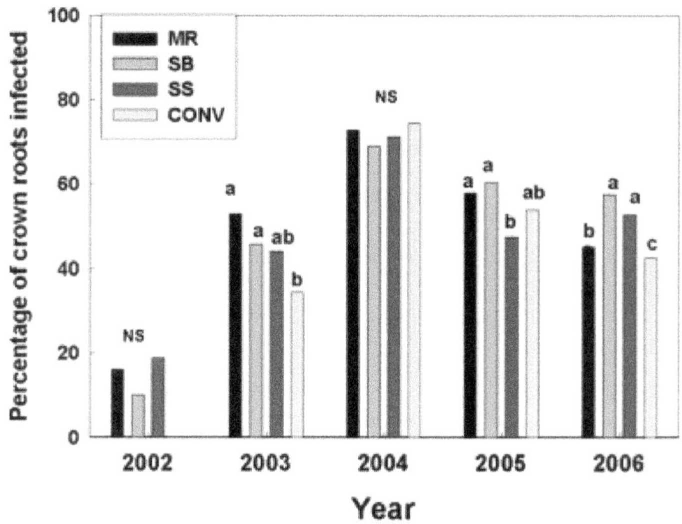

Figure 10. Percentage of crown roots with symptoms of Rhizoctonia in winter wheat seedlings in April grown in four crop rotation/residue management treatments: winter wheat–spring barley–canola no-till rotation with mechanical stubble removal (MR), stubble burning (SB), or stubble left standing (SS), and continuous winter wheat with stubble burned and moldboard plowed (CONV). Treatments within each year with the same letters are not statistically different, P = 0.05, Tukey's mean separation test.

We consistently found lower levels of DNA of F. culmorum in the soil compared to DNA of F. pseudograminearum. Previous surveys in the mid-1990s (32) showed that F. pseudograminearum was the predominant species in the lower precipitation winter wheat–summer fallow areas of eastern Washington and north-central Oregon, compared to F. culmorum, which was more predominant in the higher precipitation areas with continuous annual cropping. The root lesion nematode, P. neglectus, is widely distributed in the dryland areas of the Pacific Northwest, and can cause significant yield losses based on experiments with nematicides and resistant varieties (36). We found the highest concentration of DNA in the continuous winter wheat with burning and plowing, demonstrating that tillage had a minimal effect on these nematodes. Although P. neglectus has a wide host range, some crops, such as pea, lentil (Lens culinaris Medik.), and some cultivars of canola, result in less nematode reproduction (37). Barley is also a less conducive host for nematode reproduction (37), and is less impacted by nematode infection (33). The canola and barley in the rotation may have reduced the populations of P. neglectus compared to a monoculture of wheat, which is highly susceptible to the nematode infection and reproduction. Although canola was not a focus of this paper, we did observe major disease problems that could be a limiting factor in canola production. Winter canola was killed in the fall in 5 of 6 years due to a combination of cold air temperatures and soilborne plant pathogens in the fall. To survive the winter, the healthy plant must form a rosette, and hence must be planted as early as possible. In 2002, winter canola was planted on 27 August and irrigated.

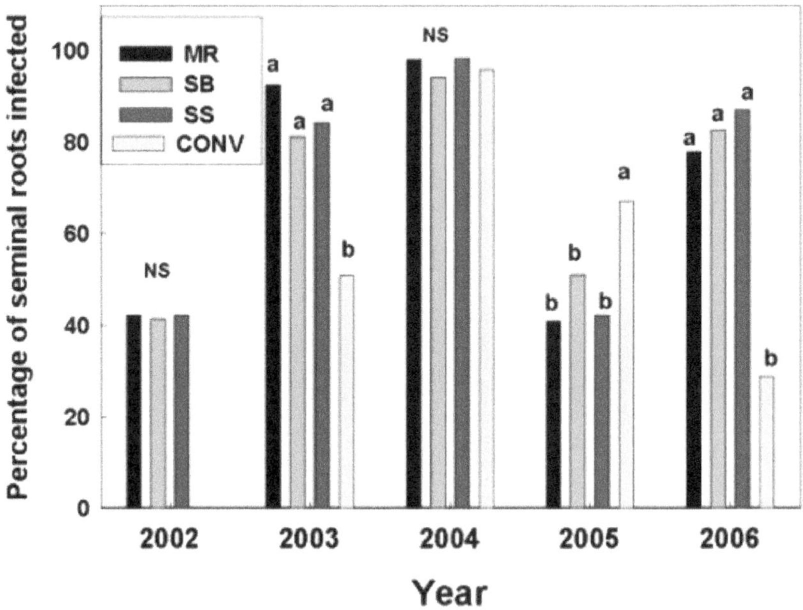

Figure 11. Percentage of seminal roots with symptoms of Rhizoctonia in winter wheat seedlings in April grown in four crop rotation/residue management treatments: winter wheat–spring barley–canola no-till rotation with mechanical stubble removal (MR), stubble burning (SB), or stubble left standing (SS), and continuous winter wheat with stubble burned and moldboard plowed (CONV). Treatments within each year with the same letters are not statistically different, P = 0.05, Tukey's mean separation test.

A good stand was established by mid-September, but barley volunteers from the previous crop were also established. The barley was controlled with the grass weed herbicide Assure II (quizalofop). After 2 weeks, the established canola began to die in the field. Isolation and greenhouse testing verified that the pathogen was R. solani AG-2-1 (23), which causes postemergence damping-off and wirestem of canola and other brassicas (15,40). This disease was probably exacerbated by the green-bridge effect of killing barley within the crop (34). Cropping systems have a changing dynamic of disease and soilborne pathogens that are influenced by tillage practices, rotation crops, and cultural methods such as residue management. This 6-year study has demonstrated the complex interactions that occur, where one disease may be of primary importance, even though other diseases are present. Although we expected Rhizoctonia root rot to be a major limitation in a no-till system, take-all in the burn and plow monoculture winter wheat system turned out to be the greatest constraint to plant health. Although

Rhizoctonia DNA concentrations in the soil were higher in the no-till systems, no major yield losses occurred (29,41), possibly because of the compensatory effect of adequate water or some unknown microbial suppression. However, economics will play the greatest role in farmers' acceptance of this system. During the years that this experiment was conducted, wheat prices were low, and barley and canola prices even lower, and all crops in the rotations had negative net economic returns (29,41).

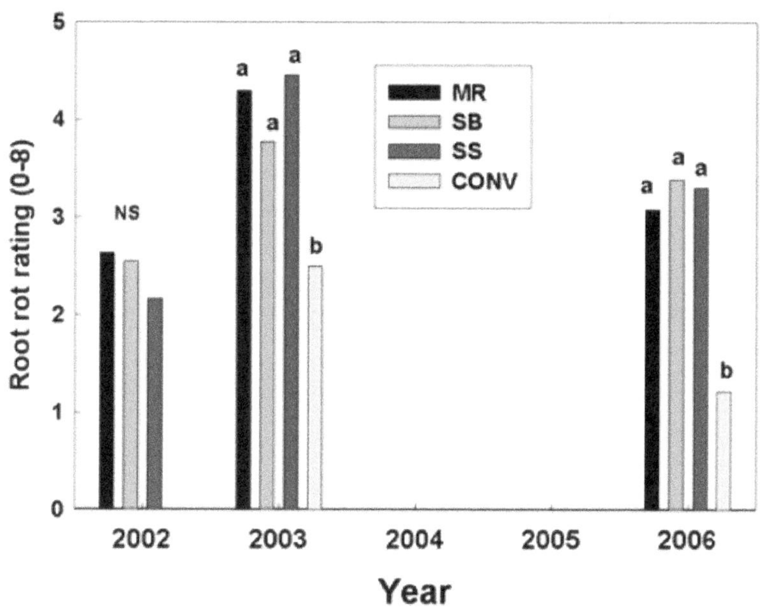

Figure 12. Rhizoctonia root rot rating of winter wheat seedlings in April grown in four crop rotation/residue management treatments: winter wheat–spring barley–canola no-till rotation with mechanical stubble removal (MR), stubble burning (SB), or stubble left standing (SS), and continuous winter wheat with stubble burned and moldboard plowed (CONV). Treatments within each year with the same letters are not statistically different, P = 0.05, Kruskal-Wallis mean separation test. Data were not taken in 2004 and 2005.

ACKNOWLEDGMENTS

We thank Bruce Sauer, Harry Schafer, Steve Schofstoll, Timothy Smith, Brian Fode, Karen Adams, and Rebecca Flanagan for their excellent technical assistance. Funding for the experiment was provided by the USDA-ARS, WSU, Washington Wheat Commission, the Columbia Plateau PM10 Project, the Solutions to Economic and Environmental Problems (STEEP) Project, and the Washington Department of Ecology.

REFERENCES

1. Anonymous. 2008. Washington statistics. USDA-National Agricultural Statistics Service, Washington, DC.

2. Bailey, K. L., Gossen, B. D., Lafond, G. P., Watson, P. R., and Derksen, D. A. 2001. Effect of tillage and crop rotation on root and foliar diseases of wheat and pea in Saskatchewan from 1991–1998: Univariate and multivariate analyses. Can. J. Plant Sci. 81:789-803.

3. Baker, C. J., Saxton, K. E., and Ritchie, W. R. 1996. No-tillage seeding: Science and practice. CAB International, Cambridge, UK.

4. Cook, R. J. 1981. The influence of rotation crops on take-all decline phenomenon caused by Gaeumannomyces graminis var. tritici, biological control. Phytopathology 71:189-192.

5. Cook, R. J., and Haglund, W. A. 1991. Wheat yield depression associated with conservation tillage caused by root pathogens in the soil not phytotoxins from straw. Soil Biol. Biochem. 23:1125-1132.

6. Cook, R. J., and Papendick, R. I. 1972. Influence of water potential of soils and plants on root disease (fungi). Annu. Rev. Phytopathol. 10:349-374.

7. Cook, R. J., Schillinger, W. F., and Christensen, N. W. 2002. Rhizoctonia root rot and wheat take-all in diverse direct seed cropping systems. Can. J. Plant Pathol. 24:349-358.

8. Cook, R. J., and Veseth, R. J. 1991. Wheat Health Management. American Phytopathological Society, St. Paul, MN.

9. Donaldson, E., Schillinger, W. F., and Dofing, S. M. 2001. Straw production and grain yield relationships in winter wheat. Crop Sci. 41:100-106.

10. Guy, S. O., and Gareau, R. M. 1998. Crop rotation, residue durability, and nitrogen fertilizer effects on winter wheat production. J. Prod. Agric. 11:457-461.

11. Hornby, D. 1983. Suppressive soils. Annu. Rev. Phytopathol. 21:65-85.

12. Hornby, D. 1998. Take-All of Cereals: A Regional Perspective. CAB International, Wallingford, UK.

13. Inglis, D. A., and Cook, R. J. 1986. Persistence of chlamydospores of Fusarium culmorum in wheat field soils of eastern Washington. Phytopathology 76:1205-1208.

14. Johnston, W. J., Golob, C. T., Sitton, J. W., and Schultz, T. R. 1996. Effect of temperature and postharvest field burning of Kentucky bluegrass on germination of sclerotia of Claviceps purpurea. Plant Dis. 80:766-768.

15. Kataria, H. R., and Verma, P. R. 1992. Rhizoctonia solani damping-off and root rot in oil seed rape and canola. Crop Prot. 11:8-13.

16. Kim, D.-S., Cook, R. J., and Weller, D. M. 1997. Bacillus sp. L324-92 for biological control of three root diseases of wheat grown with reduced tillage. Phytopathology 87:551-558.

17. MacNish, G. C. 1985 Methods of reducing rhizoctonia patch of cereals in Western Australia. Plant Pathol. 34:175-181.

18. McSpadden Gardener, B. B., and Weller, D. M. 2001. Changes in populations of rhizosphere bacteria associated with take all of wheat. Appl. Environ. Microbiol.

67:4414-4425.

19. Moore, K. J., and Cook, R. J. 1984. Increased take-all of wheat with direct drilling in the Pacific Northwest. Phytopathology 74:1044- 1049.
20. Ophel-Keller, K., McKay, A., Hartley, D., Herdina, and Curran, J. 2008. Development of a routine DNA-based testing service for soilborne diseases in Australia. Australas. Plant Pathol. 37:243-253.
21. Papendick, R. I. 2004. Farming with the Wind II: Wind Erosion and Air Quality Control on the Columbia Plateau and Columbia Basin. Special Report by the Columbia Plateau PM10 Project Rep. XB 1042. Washington Agric. Exp. Stn., Pullman.
22. Papendick, R. I., and Cook, R. J. 1974. Plant water stress and development of Fusarium [roseum] foot rot in wheat subjected to different cultural practices. Phytopathology 64:358- 363.
23. Paulitz, T. C., Okubara, P. A., and Schillinger, W. F. 2006. First report of damping-off of canola caused by Rhizoctonia solani AG 2-1 in Washington State. Plant Dis. 90:829.
24. Pumphrey, F. V., Wilkins, D. E., Hane, D. C., and Smiley, R. W. 1987. Influence of tillage and nitrogen fertilizer on Rhizoctonia root rot (bare patch) of winter wheat. Plant Dis. 71:125-127.
25. Raaijmakers, J. M., Bonsall, R. F., and Weller, D. M. 1999. Effect of population density of Pseudomonas fluorescens on production of 2,4-diacetylphloroglucinol in the rhizosphere of wheat. Phytopathology 89:470-475.
26. Raaijmakers, J. M., and Weller, D. M. 1998. Natural plant protection by 2,4-diacetylphloroglucinol-producing Pseudomonas spp. in take-all decline soils. Mol. PlantMicrobe Interact. 11:144-152.
27. Rovira, A. D. 1986. Influence of crop rotation and tillage on Rhizoctonia bare patch of wheat. Phytopathology 76:669-673.
28. Schillinger, W. F., Smith, T. A., and Schafer, H. L. 2008. Chaff and straw spreader for a plot combine. Agron. J. 100:398-399.
29. Schillinger, W. F., Young, D. L., Kennedy, A. C., and Paulitz, T. C. 2010. Diverse no-till irrigated crop rotations instead of burning and plowing continuous wheat. Field Crops Res. 115:39-49.
30. Schroeder, K. L., and Paulitz, T. C. 2006. Root diseases of wheat and barley during the transition from conventional tillage to direct seeding. Plant Dis. 90:1247-1253.
31. Schwartz, J., and Neas, L. M. 2000. Fine particles are more strongly associated than coarse particles with acute respiratory health effects in schoolchildren. Epidemiology 11:6- 10.
32. Smiley, R. W., Collins, H. P., and Rasmussen, P. E. 1996. Diseases of wheat in long-term agronomic experiments at Pendleton, Oregon. Plant Dis. 80:813-820.
33. Smiley, R. W., and Machado, S. 2009. Pratylenchus neglectus reduces yield of winter wheat in dryland cropping systems. Plant Dis. 93:263-271.

34. Smiley, R. W., Ogg, A. G., and Cook, R. J. 1992. Influence of glyphosate on Rhizoctonia root rot, growth, and yield of barley. Plant Dis. 76:937-942.

35. Smiley, R. W., and Patterson, L.-M. 1996. Pathogenic fungi associated with Fusarium foot rot of winter wheat in the semiarid Pacific Northwest. Plant Dis. 80:944-949.

36. Smiley, R. W., Whittaker, R. G., Gourlie, J. A., and Easley, S. A. 2005. Suppression of wheat growth and yield by Pratylenchus neglectus in the Pacific Northwest. Plant Dis. 89:958-968.

37. Taylor, S. P., Hollaway, G. J., and Hunt, C. H. 2000. Effect of field crops on population densities of Pratylenchus neglectus and P. thornei in southeastern Australia. 1. P. neglectus. J. Nematol. 32:591-599.

38. Weller, D. M., Cook, R. J., MacNish, G., Bassett, E. N., Powelson, R. L., and Petersen, R. R. 1986. Rhizoctonia root rot of small grains favored by reduced tillage in the Pacific Northwest. Plant Dis. 70:70-73.

39. Weller, D. M., Raaijmakers, J. M., McSpadden-Gardener, B. B., and Thomashow, L. S. 2002. Microbial populations responsible for specific soil suppressiveness to plant pathogens. Annu. Rev. Phytopathol.

40. 40:309-348. 40. Yibarek, S. M., Verma, P. R., and Morrall, R. A. A. 1987. Anastomosis groups, pathogenicity, and specificity of Rhizoctonia solani isolates from seedling and adult rapeseed/canola plants and soils in Saskatchewan. Can. J. Plant Pathol. 9:6-13.

41. Zaikin A. A., Young, D. L., and Schillinger, W. F. 2008. Economics of an irrigated no-till crop rotation with alternative stubble management practices versus continuous irrigated winter wheat with burning and plowing of stubble. Washington State Univ. Ext. Bull. EB2029E. Online: http://farm.mngt.wsu.edu/nonirr.htm.

15
CHAPTER

LONG-TERM CROP ROTATION, SOIL TILLAGE, AND FERTILITY ON THE SEVERITY OF RICE GRAIN SMUTS

Steven A. Brooks, Merle M. Anders, and Kathleen M. Yeater

False smut (Ustilaginoidea virens (Cooke) Takah.) and kernel smut (Neovossia horrida (Takah.) Padwick & A. Khan, syn. Tilletia barclayana (Bref.) Sacc. & P. Syd.) are important fungal diseases of rice (Oryza sativa L.) that reduce grain yield and quality (29). Little research has been conducted to date to understand the etiology, epidemiology, and host plant susceptibility to these diseases, and rice-breeding priorities have not included developing smut-resistant cultivars. This has led to the widespread cultivation of susceptible rice cultivars in the United States (27), and a concurrent lack of disease management in rice production, due to the sporadic occurrence of severe smut epidemics. However, recent epidemics have drawn attention to these diseases and the need for their control. Reports of false smut emerging as a disease of importance for the southern United States began in 1997 (8,23), although the disease was known to exist in the United States since 1906 (22). However, the historical distribution and full impact of the disease over the past century is unknown. In Arkansas, false smut was documented in all rice-producing regions in 1999 (7) and, since that time, multiple outbreaks have occurred, including significant epidemics in 2008 and 2009 (4,15). Kernel smut has been a chronic disease problem for U.S.-based rice production

Citation: Brooks, S. A., Anders, M. M., and Yeater, K. M. 2011. Long-Term Crop Rotation, Soil Tillage, and Fertility on the Severity of Rice Grain Smuts. doi:10.1094 / PDIS-09-10-0689

(6,9,10), and also has significant economic impact. For example, during the winter of 2008–09, the presence of kernel smut halted the export of infested U.S. rough rice in transit to rice mills in Mexico (17). Although the disease has the potential for significant yield losses (29), phytosanitary and export restrictions are also a concern for seedborne pathogens like N. horrida. False smut and kernel smut present unique challenges for disease management. Currently, all major inbred rice cultivars and hybrids grown in the southern United States are susceptible to false smut, and the inbred cultivars are also susceptible to kernel smut (16,27). Susceptible germplasm combined with a cultivation system (in rotation with soybean (Glycine max (L.) Merr.) on tilled soil, with high fertilizer input) that promotes high levels of smut severity (2) has created a situation where smut diseases are not being controlled. Fungicides can be used to control fungal diseases of rice (blast and sheath blight) in the United States. However, midseason fungicide applications to control smut diseases are a financial gamble for rice farmers for the following three reasons. (i) The visual symptoms of disease are not present until after grain fill; therefore, a timely fungicide application must be made without any knowledge of disease incidence in apparently healthy plants. (ii) There is the risk of significant yield loss if no action is taken in rice fields with a history of the diseases. (iii) Fungicides do not offer complete protection from smuts (18,26) and there has been no analysis of cost versus yield protection to show a financial gain from treatments. In order to control grain smuts on susceptible rice cultivars, fungicides are not the optimal method of disease control as they are for foliar diseases. Previously (2,3), we demonstrated that the severity of smut diseases could be reduced in susceptible rice cultivars without the use of fungicides. By changing crop-management practices to reducedinput systems, high levels of disease control were obtained without yield penalties. Herein, we have expanded upon this work to understand the influences from alternative crop management systems on smut disease severity. Alternatives to the common rice-soybean rotation are used in Arkansas, and are often driven by commodity prices where repeat rice crops or rotations to corn are used, because these options can be more profitable than soybean (28). For long-term sustainability of rice production, additional alternatives are needed to address soil conservation (11,13,19), limited irrigation water (14), greenhouse gas emissions (1,12), and waterway pollution from fertilizer runoff (5). Additional research is needed from an agronomic standpoint to address these issues of sustainable rice production. From a disease-management perspective, changes to rice production systems to address these pressing issues can also have positive or negative impacts on disease severity. Alternatives to standard rice cultivation practices can contribute to long-term sustainability of rice production in the United States but the effects on rice pathogen populations and disease severity are unknown, particularly in no-tillage where crop residue is not burned or incorporated into the soil. Crop rotation has a long history of use in disease management; however, its utility for control of rice smuts was unknown (2). The most common rotation for our region is rice-soybean and, less frequently, the same rotation with winter wheat grown after rice (rice-wheat-soybean). A corn crop can also be included in a rice rotation as a substitute for soybean (rice-corn) or integrated with rice and soybean as a three-phase rotation (rice-soybean-corn or rice-corn-soybean). The use of corn as a rotation crop is of particular interest because it is also a host of U. virens (30). Additional alternatives to the rice-soybean rotation are continuous rice and continuous rice with winter wheat (ricewheat). The primary objective of this work was to determine the effects of tillage management, nitrogen fertilizer rate, and crop rotation on the severities of false smut and kernel smut.

MATERIALS AND METHODS

Study location

A long-term rice cropping-systems study was established at the University of Arkansas Rice Research and Extension Center (Stuttgart; N 34°27′ latitude, W 91°24′ longitude) in 2000 (21). The location lies within the Grand Prairie region of eastern Arkansas, where rice is typically grown in rotation with soybean. False smut and kernel smut commonly occur on the experiment station and in production fields of the area (2–4). The experimental plots were known to be naturally infested with false smut and kernel smut prior to 2006, and no fungicides were applied since the inception of the study in 2000.

Crop rotation experiment

As part of the multi-year, multicropping-rotation study, the incidence of false smut and kernel smut were recorded each year for 4 years (2006 to 2009). The field experimental layout was a randomized split-strip plot design, where tillage type was the main plot, fertility level was the subplot, and rotation was randomly applied within each of four replicates at the beginning of the study (2,21). Tillage treatments were traditional tillage (TT) and no-tillage (NT). Fertility treatments for rice were "low" (L) (N at 112 kg ha^{-1} + P_2O_5 at 45 kg ha^{-1} + K_2O at 67 kg ha^{-1}) and "high" (H) (N at 168 kg ha^{-1} + P_2O_5 at 67 kg ha^{-1} + K_2O at 100 kg ha^{-1}) using Agrotain (Agrotain International, L.L.C.) coated urea pellets (46-0-0), triple super phosphate (0-46- 0), and muriate of potash (0-0-60). Phosphorus and potassium fertilizers were applied prior to planting and N was applied post emergence at the four- to five-leaf growth stage. The rotation treatments were continuous rice (RR), rice-soybean (RS), rice-corn (RC), rice-wheat (RW), rice-wheat-soybean-wheat (RWSW), ricesoybean-corn (RSC), and rice-corn-soybean (RCS; Table 1). Four rice cultivars that represent two maturity groups (standard and early) were grown each year. Standard maturity lines ('Wells', 'Cybonnet', and 'XL 723') were spring planted for full-season growth. Wells was grown in all years, whereas Cybonnet (2006) was replaced with hybrid 'XL 723' in 2007 through 2009.

Table 1. Summary of crop rotations and rice cultivars evaluated for false smut and kernel smut severity by year

	Rice cropping systems[x]							Rice cultivars			
Year	1	2/3[y]	4/5[y]	6	7/8[y]	9[z]	10[z]	V1	V2	V3	V4
2006	RR	RS	RC	...	...	...	...	Wells	Cybonnet	...	...
2007	RR	RS	RC	...	...	...	...	Wells	XL 723	...	...
2008	RR	RS	RC	RW	RWSW	RSC	RCS	Wells	XL 723	Spring	Spring
2009	RR	RS	RC	RW	RWSW	...	...	Wells	XL 723	Spring	RU0701124

[x] Crop rotation abbreviations indicate the rotation-years sampled for false smut and kernel smut. RR = continuous rice, RS = rice-soybean, RC = rice-corn, RSC = rice-soybean-corn, RCS = rice-corn-soybean, RW = rice-wheat, and RWSW = rice-wheat-soybean-wheat.

[y] Cropping systems 2/3, 4/5, and 7/8 were paired so that rice was grown every year for each rotation (i.e., when system 2 was planted in rice, system 3 was planted with soybean, but both are identical rice-soybean rotations).

[z] Cropping systems 9 and 10 were three-phase rotations where rice was grown once every 3 years. Rice occurred only once (2008) in these rotations during the 4 years of this study. These systems were not paired and differ because rice follows corn in system 9 but follows soybean in system 10.

Earlymaturity lines 'Spring' and 'RU0701124' were used in rotation with wheat, and were summer planted following the wheat harvest. Spring was grown in all years, and from 2006 through 2008 was entered twice in the experiment because no other early-maturing, locally adapted cultivar was available in sufficient quantity of seed. In 2009, experimental line RU0701124 was included with Spring in the wheat rotations. Seed of inbred cultivars Wells, Cybonnet, and Spring and experimental line RU0701124 were provided by the University of Arkansas. Seed of hybrid XL 723 was provided by RiceTec, Inc. (Alvin, TX). All entries were susceptible to both diseases, with the exception of hybrid XL723 that is moderately resistant to kernel smut (2; unpublished data). Rice was drill-seeded at the rates of 101 kg ha–1 for the inbred cultivars and 39 kg ha–1 for hybrid XL 723 using an Almaco grain drill (Almaco, Nevada, IA) and a 19-cm row spacing (6-by-19-m individual plot size). At the five-leaf growth stage, N fertilizer was applied and a permanent flood 10 cm deep was immediately established and maintained throughout the growing season. Mature rice was harvested with a John Deere 45 combine (Deere & Company, Moline, IL) retrofitted with a Carter weigh bucket (Carter Mfg. Co. Inc., Brookston, IN) at approximately 18% moisture for yield measurements. Grain samples for moisture determinations and disease evaluations were collected from the combine at harvest.

Sampling and disease rating

False smut and kernel smut disease ratings were performed according to Brooks et al. (2). Briefly, false smut sori were recovered by hand sieve from 1- to 1.5-kg grain plot-samples, and disease severity reported as sori number per kilogram of seed harvested. Subsequently, four 50-g subsamples of seed were used to rate kernel smut severity. Each subsample was soaked overnight in 0.27 M KOH to clear the hulls (palea and lemma; 20), and smutted kernels were identified visually over a light box. Kernel smut severity was reported as the number of symptomatic kernels per kilogram of seed harvested. Herein, disease severity is reported as least squares means (LSM) of logtransformed severity values.

Data analyses

Analyses of all dependent variables were performed using the GLIMMIX procedure in SAS (version 9.2). False smut and kernel smut were lognormally distributed. This probability distribution was specified in the model after the variables were coded by adding +1 to each value. Yield followed a Gaussian (normal) distribution. Tillage, fertility, rotation, and their interactions were considered fixed effects in the statistical models. The replicated block was random as well as the cultivar within the rotation. Cultivar was a random effect because we were inferring the range of all rice cultivars and we were not implicitly focused on the means results of the specific cultivars included in this study. The data were originally assessed on a year-to-year basis (not shown). The fixed effects showed similar trends in all years of the study. The full experiment over all years was analyzed by incorporating the random time effect (years) upon the repeated measurement of each experimental unit (cultivar plot within each rotation). A Toeplitz covariance structure was modeled for each cultivar within rotation across years. This approach allowed for more sampling and power to interpret the rotation and the variability of the random cultivar within rotation. The Kenward-Roger degrees of free dom method was used in all variable models. Best linear unbiased predictors and predicted standard errors (SEs) of cultivar were calculated using the SOLUTIONS option. Estimated means, SEs, and differences of means in the fixed effects were calculated using the LSMEANS option. Specific comparisons of fixed effects were calculated via ESTIMATE statements.

RESULTS

False smut and kernel smut disease severities were evaluated from 2006 through 2009 under a variety of crop-management systems. The statistical approach used detected high SEs which were attributed to the inherent variability of each disease, natural infestation of the research plots, environmental variability over the 4- year research period, and the frequency that individual treatment combinations were present over the course of the study. The latter was most important for the presentation of the results that follow, because the frequency in which treatment combinations were measured resulted in unbalanced treatment groups and unequal variances. Therefore, the three-phase rotations that appeared only once in the study had the highest SEs, followed by the wheat rotations (2 years), and the lowest SEs occurred in the rotations (RR, RS, and RC) that were present in all years (Tables 1 and 2). Because unequal variances were present, the letter groupings presented in the tables do not reflect all significant comparisons. To make the presentation of the data the least cumbersome, the additional significant comparisons (treatment combinations with relatively small SEs) are noted in the text and appropriate tables with P values. The statistical analysis of false smut results for the study shows that the main effects of tillage, fertility, and a rotation–tillage interaction significantly affected disease severity (Table 3). Although there were differences in disease severity between crop rotations (Table 2), these differences were not significant at $P < 0.05$ and were due to a high level of variance within each rotation (Table 2, SE).

Table 2. Rice false smut and kernel smut severity under different crop rotations[w]

Rotation[x]	LSM[y]	SE[y]	LSM grouping[z]
False smut			
RS	2.8104	0.6350	A
RC	2.1986	0.6349	A
RR	1.9929	0.6349	A
RW	1.8001	0.8029	A
RWSW	1.7830	0.7928	A
RSC	0.5823	0.8106	A
RCS	0.5522	0.8106	A
Kernel smut			
RWSW	4.6494	0.6581	A
RW	3.8843	0.6712	AB
RS	2.9669	0.5059	AB
RC	2.9070	0.5058	AB
RR	2.7165	0.5058	AB
RCS	1.0812	0.7027	B
RSC	0.8387	0.7027	B

[w] Disease severity = the number of infected kernels per kilogram of harvested seed. Disease severity values for each rotation are inclusive of both fertility and tillage treatments.

[x] RR = continuous rice, RS = rice-soybean, RC = rice-corn, RSC = rice-soybean-corn, RCS = rice-corn-soybean, RW = rice-wheat, and RWSW = rice-wheat-soybean-wheat.

[y] Least squares mean (LSM) of log-transformed disease severity values and standard error of the mean (SE).

[z] Conservative letter groupings of differences of LSMs were derived from log-transformed data for the multiple comparisons and use of the Tukey-Kramer *P* value adjustment. Within a treatment group, values followed by the same letter are not significantly different at $\alpha = 0.05$.

Table 3. Analysis of variance summary of P values for false smut and kernel smut for the fixed effects of rotation, tillage, fertility, and their interactions

	***P* values**	
Source of variation	**False smut**	**Kernel smut**
Rotation	0.3498	0.0180
Tillage	0.0002	0.2782
Fertility	<0.0001	0.0141
Tillage × fertility	0.5084	0.4066
Rotation × tillage	<0.0001	0.9888
Rotation × fertility	0.8189	0.9580
Rotation × tillage × fertility	0.3709	0.8782

The high SEs resulted from the inherent variability of a naturally infected field experiment conducted over 4 years (see covariance parameter estimates below). The only near-significant comparisons for false smut between rotations occurred when a corn crop was added to the RS rotation (RSC and RCS), which resulted in P values of 0.0550 and 0.0523, respectively (Table 4). The rotation–tillage interaction provided more informative comparisons between the paired effects of rotation and tillage (Table 5). The rotation by tillage interaction was the result of differing effects from tillage treatments on false smut severity among rotations (Table 5). For example, when data were pooled across fertility treatments, the effect of tillage on the two rotations containing a wheat phase (RW and RWSW) was opposite from those that did not contain wheat. RW/NT and RWSW/NT treatment combinations had higher levels of false smut than the tilled plots in the same rotations. This difference was 30% in the RW rotation (RW/NT > RW/TT; LSM diff. = 0.6242, SE = 0.3353, P = 0.0646), and 52% in the RWSW rotation (RWSW/NT > RWSW/TT; LSM diff. = 1.2571, SE = 0.2524, P < 0.0001). In the remaining two-phase rotation–tillage treatment combinations, NT resulted in lower levels of false smut than TT (RR/NT < RR/TT, P = 0.0181; RS/NT < RS/TT, P < 0.0001; RC/NT < RC/TT, P < 0.0001). The reduction in false smut severity from NT among these rotations was 21% in RR, 50% in RS, and 51% in RC. No significant differences were observed in the three-phase rotation by tillage treatment combinations (RSC/TT-RSC/NT and RCS/TTRCS/NT). The three-phase rotations had numerically less false smut in the NT treatment but this difference was not significant due to the comparatively high SEs relative to the low levels of disease (Table 5). Although the tillage effect in wheat rotations was opposite from the corresponding summer crop rotations, greater differences were observed in rotations containing a soybean phase (RS/TT > RS/NT = 50% and RWSW/NT > RWSW/TT = 52%) than when a winter wheat crop was added to an RR rotation (RR/TT > RR/NT = 21% and RW/NT > RW/TT = 30%).

Table 4. Comparisons of rice false smut and kernel smut severity between specific crop rotations[x]

Rotation comparison[y]	Difference[z]	SE[z]	*P* value
False smut			
RR, RR vs. RW	0.1928	1.0236	0.8544
RS, RS vs. RWSW	1.0274	1.0157	0.3363
RS vs. RC	0.6118	0.8980	0.5124
RC vs. RSC	1.6163	1.0297	0.1467
RC vs. RCS	1.6464	1.0297	0.1400
RS vs. RSC	2.2281	1.0297	0.0550
RS vs. RCS	2.2583	1.0297	0.0523
Kernel smut			
RR vs. RW	−1.1678	0.8305	0.1863
RS vs. RWSW	−1.6826	0.8199	0.0649
RS vs. RC	0.05986	0.7036	0.9340
RC vs. RSC	2.0684	0.8562	0.0307
RC vs. RCS	1.8259	0.8562	0.0521
RS vs. RSC	2.1282	0.8562	0.0269
RS vs. RCS	1.8857	0.8562	0.0458

[x] Disease severity = the number of infected kernels per kilogram of harvested seed. Disease severity values for each rotation are inclusive of both fertility and tillage treatments.

[y] RR = continuous rice, RS = rice-soybean, RC = rice-corn, RSC = rice-soybean-corn, RCS = rice-corn-soybean, RW = rice-wheat, and RWSW = rice-wheat-soybean-wheat.

[z] Difference = Difference in least squares means within individual comparisons as calculated via ESTIMATE statement. SE = standard error of the difference of means.

All comparisons of tillage within rotation for false smut, where additional observations were made, are listed in Table 5. Most notable are the RS/TT and RC/TT treatment combinations that had the highest levels of false smut. The lowest levels of false smut were observed in the three-phase rotations, where disease severities were reduced by 80% (RSC/TT) or more (in RCS/TT, RSC/NT, and RCS/NT treatment combinations) compared with RS/TT (RSC/TT < RS/TT; LSM diff. = 2.9866, SE = 1.0422, P = 0.0157). These data also show that incorporating wheat into an RS/TT treatment combination reduced the severity of false smut by 69% (RWSW/TT < RS/TT; LSM diff. = 2.5815, SE = 1.0273, P = 0.0304). However, this effect was lost with the NT treatment because false smut severities were similar in RWSW/NT and RS/NT. Finally, inclusion of winter wheat in continuous rice cropping had no effect on false smut severity regardless of rotation–tillage treatment comparisons (RW[TT or NT] versus RR[TT or NT]). The fertility treatment effect was consistent across all other treatments (no significant interactions, Table 3); therefore, a studywide value for the reduction in false smut severity attributable to reduced fertility was a relevant estimate. The low-fertility treatment reduced the severity of false smut by 20% compared with the high-fertility treatment across all rotation and tillage treatment combinations. Comparisons of fertility treatments within rotation are presented in Table 6. The fertility treatment was significant within all two-phase rotations. The reduction in false smut severity from the low fertility treatment was 27% in RW, 19% in RC, 18% in RS, and 15% in RR (P values = 0.0361, 0.0011, Covariance parameter estimates for false smut show that 30% of the total experimental variance was due to year, 44% due to cultivar within rotation, and 26% due to experimental error.

The annual variation in disease severity was a useful measure because all research plots were naturally infected with false smut. Plots of false smut severity values by year show consistent trends for rotations evaluated in all 4 years (RR, RS, and RC; Fig. 1A). Years 2006 and 2009 had the highest levels of false smut, whereas 2007 and 2008 were the lowest years tested. The low levels of false smut in 2008 were attributed to early planting and harvest, because severe outbreaks of false smut were recorded on late planted rice that year (data not shown). Kernel smut severity was also evaluated using the same study, time frame, and experimental treatments as above. The statistical analysis shows that the main effects of fertility and rotation were significant and there were no interactions (Table 3). Tillage was not detected as a significant main effect for kernel smut (Table 3); however; there was a consistent trend of lower disease severity in the NT treatment combinations when compared with the TT combinations (Table 5). Fertility effects for kernel smut were consistent across all rotation and tillage treatment combinations, and the study-wide value for the reduction in kernel smut due to low fertility was 16% (H > L; LSM diff. = 0.4831, SE = 0.1298, P = 0.0141). Comparisons of fertility treatments within rotation are presented in Table 6. Fertility comparisons were significant within RR (H/L = 19%; P = 0.0072), RS (H/L = 15%; P = 0.0165), and RC (H/L = 20%; P = 0.0029) rotations. The remaining rotations followed the same trend of reduced levels of kernel smut in the low-fertility treatment but these differences were not significant (Table 6). Therefore, variability in the magnitude of the fertility effect and corresponding SEs resulted in some significant comparisons, although the trend was consistent over the entire study. Overall comparisons of crop rotation effects on kernel smut severity, pooled over tillage and fertility treatments, were possible due to the consistent main effects and no interactions (Tables 2 and 4).

Table 5. Comparison of tillage by rotation treatment effects on disease severity within each crop rotation for rice false smut and kernel smut

Rotation[v]	Tillage[w]	LSM[x]	SE[x]	Group[y]
False smut[z]				
RS	TT	3.7360	0.6410	A
RC	TT	2.9516	0.6410	AB
RWSW	NT	2.4116	0.8028	ABC
RR	TT	2.2202	0.6410	ABC
RW	NT	2.1122	0.8034	ABC
RS	NT	1.8849	0.6411	ABC
RR	NT	1.7656	0.6410	ABC
RW	TT	1.4880	0.8367	ABC
RC	NT	1.4457	0.6410	BC
RWSW	TT	1.1545	0.8028	BC
RSC	TT	0.7494	0.8243	BC
RCS	TT	0.7145	0.8243	BC
RSC	NT	0.4152	0.8243	C
RCS	NT	0.3899	0.8243	C
Kernel smut				
RWSW	TT	4.8074	0.6740	A
RWSW	NT	4.4915	0.6740	AB
RW	TT	4.0345	0.7191	ABC
RW	NT	3.7341	0.6744	ABC
RS	TT	3.0785	0.5174	ABC
RC	TT	3.0331	0.5174	ABC
RS	NT	2.8553	0.5176	BCD
RR	TT	2.8109	0.5174	BCD
RC	NT	2.7809	0.5174	BCD
RR	NT	2.6221	0.5174	CD
RCS	NT	1.0854	0.7229	DE
RCS	TT	1.0769	0.7229	DE
RSC	TT	0.9751	0.7229	DE
RSC	NT	0.7023	0.7229	E

[u] Tillage by rotation values include both fertility treatments. Disease severity = the number of infected kernels per kilogram of harvested seed.

[v] RR = continuous rice, RS = rice-soybean, RC = rice-corn, RSC = rice-soybean-corn, RCS = rice-corn-soybean, RW = rice-wheat, and RWSW = rice-wheat-soybean-wheat.

[w] TT = traditional soil tillage and NT = no-tillage.

[x] Least squares mean (LSM) of log-transformed disease severity values and standard error of the mean (SE).

[y] LSM grouping. Conservative letter groupings of differences of LSM were derived from log-transformed data for the multiple comparisons and use of the Tukey-Kramer *P* value adjustment. Within a treatment group, values followed by the same letter are not significantly different at $\alpha = 0.05$.

[z] Additional significant comparisons for false smut: RS-TT, RS-NT: LSM diff. = 1.8512, SE = 0.1764, *P* value < 0.0001; RC-TT, RC-NT: LSM diff. = 1.5059, SE = 0.1758, *P* value < 0.0001; RWSW-NT, RWSW-TT: LSM diff. = 1.2571, SE = 0.2524, *P* value < 0.0001; and RR-TT, RR-NT: LSM diff. = 0.4546, SE = 0.1758, *P* value = 0.0181.

Table 6. Comparison of fertility by rotation treatment effects on disease severity within each crop rotation for rice false smut and kernel smut[t]

Rotation[u]	Fertility[v]	LSM[w]	SE[w]	Group[x]
False smut[y]				
RS	H	3.0809	0.6389	A
RS	L	2.5400	0.6389	AB
RC	H	2.4294	0.6388	AB
RR	H	2.1582	0.6388	AB
RW	H	2.0795	0.8138	AB
RC	L	1.9678	0.6388	AB
RWSW	H	1.9656	0.8011	AB
RR	L	1.8276	0.6388	AB
RWSW	L	1.6005	0.8011	AB
RW	L	1.5208	0.8138	AB
RCS	H	0.6891	0.8226	B
RSC	H	0.6458	0.8226	B
RSC	L	0.5189	0.8226	B
RCS	L	0.4152	0.8226	B
Kernel smut[z]				
RWSW	H	4.8758	0.6712	A
RWSW	L	4.4231	0.6712	AB
RW	H	4.1122	0.6879	ABC
RW	L	3.6564	0.6879	ABC
RC	H	3.2325	0.5137	ABC
RS	H	3.2130	0.5137	ABC
RR	H	3.0006	0.5137	BCD
RS	L	2.7208	0.5139	BCDE
RC	L	2.5815	0.5137	CDE
RR	L	2.4324	0.5137	CDEF
RCS	H	1.3020	0.7203	DEF
RSC	H	0.9988	0.7203	EF
RCS	L	0.8603	0.7203	EF
RSC	L	0.6785	0.7203	F

[t] Fertility by rotation values include both tillage treatments. Disease severity = the number of infected kernels per kilogram of harvested seed.

[u] RR = continuous rice, RS = rice-soybean, RC = rice-corn, RSC = rice-soybean-corn, RCS = rice-corn-soybean, RW = rice-wheat, and RWSW = rice-wheat-soybean-wheat.

[v] H = high fertility rate and L = low fertility rate.

[w] Least squares mean (LSM) of log-transformed disease severity values and standard error of the mean (SE).

[x] LSM grouping. Conservative letter groupings of differences of LSM were derived from log-transformed data for the multiple comparisons and use of the Tukey-Kramer *P* value adjustment. Within a treatment group, values followed by the same letter are not significantly different at $\alpha = 0.05$.

[y] Additional significant comparisons for false smut: RS-H, RS-L: LSM diff. = 0.5409, SE = 0.1411, *P* value < 0.0001; RC-H, RC-L: LSM diff. = 0.4616, SE = 0.1403, *P* value = 0.0011; RR-H, RR-L: LSM diff. = 0.3306, SE = 0.1403, *P* value = 0.0190; and RW-H, RW-L: LSM diff. = 0.5587, SE = 0.2656, *P* value = 0.0361.

[z] Additional significant comparisons for kernel smut: RC-H, RC-L: LSM diff. = 0.6510, SE = 0.1792, *P* value = 0.0029; RS-H, RS-L: LSM diff. = 0.4922, SE = 0.1799, *P* value = 0.0165; and RR-H, RR-L: LSM diff. = 0.5681, SE = 0.1792, *P* value = 0.0072.

Crop rotations exhibited distinct effects that were categorized by the crop components of the rotations. For example, both rotations containing a wheat phase (RW and RWSW) had the highest levels of kernel smut, and no other rotation–treatment combinations resulted in kernel smut severities that exceeded the wheat rotation values (Tables 5 and 6). Significant differences were observed between the RWSW and either of the three-phase rotations (RWSW-RCS; LSM diff. = 3.5683, SE = 0.9541, P = 0.0021 and RWSW-RSC; LSM diff. = 3.8108, SE = 0.9541, P = 0.0013), and the comparison of RWSW to the corresponding spring crop rotation RS was nearly significant (P = 0.0649; Table 4). In addition, no differences in kernel smut severities were observed when corn was substituted for soybean in two-phase rotations (RS versus RC), whereas the three-phase rotations significantly reduced kernel smut severity when compared with either of the corresponding twophase rotations (RSC and RCS versus RS or RC; Table 4). Covariance parameter estimates for kernel smut show that 55% of the total experimental variance was due to year, 18% due to cultivar within rotation, and 25% due to experimental error.

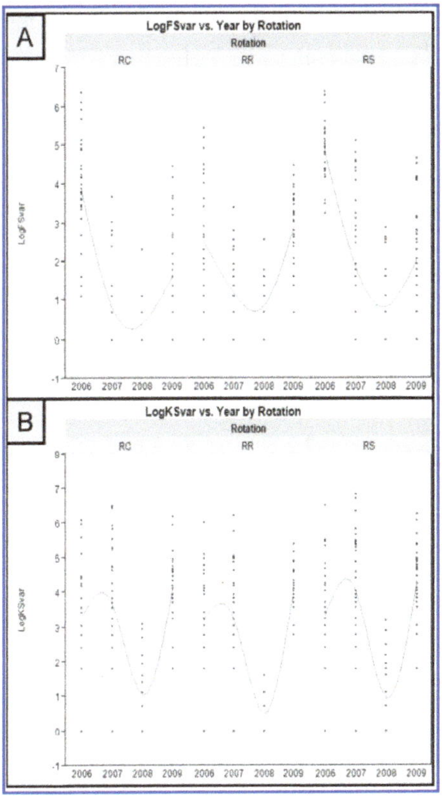

Figure 1. Log-transformed smut severity values (number of symptomatic kernels per kilogram of harvested seed; Y-axes) plotted by each year of the study (X-axes) for continuous rice (RR), rice-soybean (RS), and rice-corn (RC) rotations. A, False smut and B, kernel smut plots show the range of values obtained within rotation by year.

The annual variation in disease severity was an interesting measure because all research plots were naturally infected with kernel smut. Plots of kernel smut severity values by year show consistent trends for rotations evaluated in all 4 years (RR, RS, and RC; Fig. 1B). Years 2006, 2007, and 2009 were consistent and had the highest levels of kernel smut, whereas 2008 was the lowest year tested.

DISCUSSION

Previously (2), we reported an analysis of smut severity in RR and RS rotations. In that report, we focused our attention on continuous rice as an alternative to the standard RS rotation that is used on the majority of rice production acreage in the southern United States An analysis of the effect of continuous rice cropping on smut diseases was done, because there are circumstances where RS rotations are not used. For example, continuous rice is desirable when market prices make it attractive for farmers to repeat rice crops (28), or where poorly draining soil types on precision-leveled ground make soybean an unproductive rotation component for rice farmers. In these situations, understanding the effects of crop rotation on disease can predict future problems, such as expected increases or decreases in diseases due to alterations to rotation patterns. Or, if disease control is a primary concern, specific rotations can be chosen for field-level disease management for susceptible rice cultivars. The present study expands upon our previous research to examine additional crop rotation effects on smut severity. Corn is an example of an alternate crop that was not commonly grown in rotation with rice in the Mississippi Delta region of Arkansas but has gained acreage due to interest in biofuels and corresponding market prices in certain years. A high-value year for corn will not likely entice rice farmers to switch to an RC rotation but more likely include corn into an established RS rotation. In this case, a three-phase rotation is created but the crop immediately preceding rice cannot be predicted, and rice may follow soybean or corn (RCS or RSC) in rotation. Wheat is another crop that is not common in the rice-producing regions of Arkansas but is used as a winter crop particularly when dry fall weather permits ground preparation and planting and when wheat prices are acceptable. Winter wheat poses challenges for rice farmers due to the late harvest of the crop (usually in June), which is well after the normal rice-planting period for the region (usually late April to mid-May). This system (RW) requires short-duration cultivars due to the late planting, which changes both the germplasm and the environmental conditions that are relevant to disease severity. Due to limited availability of short-duration rice cultivars, soybean generally follows wheat in rotation (RWSW) due to a greater selection of cultivars and the ability to plant soybean onto the wheat beds without soil tillage. Previous reports (2,25,29) have shown that fertility is the most important agronomic factor influencing smut diseases, where high nitrogen fertilizer input positively affects smut severity. The present work is in agreement with these findings and, in our previous work (2), we suggested that the differences in disease severity from tillage and rotation treatments were caused by buffering N availability (NT and RR) or nitrogen enrichment of the soil (RS). Therefore, the effect from rotation was closely linked to the rotation's effect on N accumulation or availability. The same phenomenon holds true for the present work, where field history has a greater role in N accumulation and availability than fertilizer applied during the rice cultivation year. RS, RC, and RR ranked as the three highest rotations in terms of false smut severity. False smut severity was comparable in RS, RC, and RR within tillage groups and, in all three rotations, false smut was significantly higher with TT than NT. The high levels of false smut recorded in RS-TT may be attributed to nitrogen fixation by the soybean crop and in RC-TT by residual nitrogen in the soil from high N fertilizer levels used on corn. Soil organic matter accumulation buffers N availability to rice plants (2,21,24) and may be responsible for the reduction in false smut severity due to NT, but we did not collect soil samples and, thus, cannot confirm these hypotheses. In contrast, the three-phase rotations had the lowest smut severities regardless of fertility or tillage treatments. Although tillage and fertility within RSC and RCS followed the same trends as expected (H > L, TT > NT) for both diseases, these differences were not significant. This shows that fertility effects and N availability due to tillage were similar to those in RS and RC rotations. It also shows that high N fertility rates can be used in three-phase rice rotations without the risk of significant smut damage. For instance, on traditionally tilled soil, adding a year of corn into a soybean rotation will reduce both false smut and kernel smut (RS/TT versus RSC/TT or RCS/TT) even with high N input. Based upon this evidence, we hypothesize that rotating out of rice for 2 years may be reducing the amount of viable smut inoculum in the soil, because there is no reason to suspect limited N availability in the three-phase rotations. However, the RSC and RCS rotations occurred only once (2008) over the course of the 4-year study. False smut severity was least in 2008 but comparable with 2007 and 2009, and kernel smut severity in 2008 was noticeably lower than all other years tested (Fig. 1). Therefore, the apparent reduction of disease severity in the three-phase rotations may have been influenced by the overall lower values recorded in 2008. The effects from corn in rotation with rice are especially interesting for false smut because corn is an alternative host for U. virens. In rotations where corn was present, the alternative host did not appear to have any impact on disease severity in subsequent rice crops. Although the degree of susceptibility to U. virens in the corn hybrids used is unknown, very little false smut was observed in corn tassels over the course of the study (S. Brooks's field observations). U. virens is a minor disease of corn (30), and our data from rice plots immediately following corn in rotation (RC or RSC) show that an alternative host in rotation with rice had no detectable influence on false smut. Incorporating wheat into continuous rice (RW) or an RS rotation (RWSW) had interesting impacts on false smut severity. False smut severity was consistent in the NT system for RR, RS, and the corresponding wheat-containing rotations RW and RWSW, whereas false smut severity was not consistent among these rotations in TT; most notably, the 69% reduction of false smut in RWSW/TT compared with RS/TT. Moreover, TT reduced false smut in both rotations containing a wheat phase relative to NT (TT < NT; RW $P < 0.0646$ and RWSW $P < 0.0001$). At this time, we have no explanation for the interaction of wheat and tillage on false smut severity. Adding a wheat phase to rice rotations had a significant effect on kernel smut severity. In this case, the analysis was simpler because there were no significant interactions and disease severity was consistent across tillage treatments. It appears that the high kernel smut severities were the result of susceptibility in rice entry RU0701124 (on average, 238 smutted kernels per kilogram of seed). As stated above, rice grown in rotation with wheat requires shortduration rice cultivars that are grown later in the cropping season.

The combination of susceptibility (in the short-season Spring also) and, possibly, late-season conditions conducive to the disease produced high kernel smut values for the wheat rotations. Therefore, the high levels of kernel smut in wheat rotations may have been the result of germplasm selection and change in environment. There were several consistent trends within our results that are noteworthy. As mentioned above, the three-phase rotations present the best management strategy for both diseases, where the highest levels of disease control were achieved without confounding factors such as tillage or fertility. In addition, the three-phase rotations had remarkably consistent results for both diseases and both fertility and tillage treatments. Similarly, the RR system also produced consistent results throughout the study. Although disease severity was higher in RR than the three-phase rotations, severity was constant across other treatments. In general, the NT treatment also produced consistent results between rotation and fertility treatment combinations. Finally, the kernel smut disease severity was consistent, where tillage had little effect and fertility effects were similar across rotations. This contrasts with false smut, where disease severity was more erratic within and between rotation and tillage treatments. This was evident in the covariance parameter estimates, where variance due to cultivar was high for false smut (44%) and lower for kernel smut (18%). In contrast, there was greater variance due to time for kernel smut (55%) than for false smut (30%). This report concludes our research on smut disease control in susceptible rice cultivars using crop-management approaches. We have analyzed a number of agronomic factors that influence smut severity for three reasons: (i) to provide information on disease control for current economically important rice cultivars, because cultivars that occupy the majority of production acreage in the southern United States are all susceptible to grain smuts; (ii) to predict the positive or negative affects of alternate cropping systems on disease severity; and (iii) to use the most important factors for smut disease severity (rotation and fertility) to develop nurseries that maximize disease severity for resistance screening. The final reason is addressed in another report on smut resistance in rice. In this present work and our previous reports (2,3), we identified a number of conditions where smut diseases were controlled in rice cultivation without the use of fungicides or losses in yield. Incorporating NT will require a significant departure from standard practices in the southern United States. However, moderating the use of N fertilizer and three-phase rotations provide the highest levels of smut control and require little additional effort from rice producers. Finally, our results demonstrate the importance of field history influences on smut disease severity. Crop-management decisions are often based upon current-year conditions and, although current production practices are important, the previous rotation crops and tillage practices also have a significant influence on smut severity.

ACKNOWLEDGMENTS

We thank S. Agrama, J. Gibbons, A. Jackson, D. McCarty, A. McClung, T. Moss, B. Ottis, T. Sells, and K. Ueda for their contributions to this project; C. Isbell for providing the false smut spore cloud picture submitted for the journal cover; and RiceTec, Inc. for providing seed and financial support for this project.

REFERENCES

1. Bouwman, A. F. 1991. Agronomic aspects of wetland rice cultivation and associated methane emissions. Biogeochemistry 15:65-88.
2. Brooks, S. A., Anders, M. M., and Yeater, K. M. 2009. Effect of cultural management practices on the severity of false smut and kernel smut of rice. Plant Dis. 93:1202-1208.
3. Brooks, S. A., Anders, M. M., and Yeater, K. M. 2010. Effect of furrow irrigation on the severity of false smut in susceptible rice varieties. Plant Dis. 94:570-574.
4. Brooks, S. A., Anders, M. M., and Yeater, K. M. 2010. The Arkansas-2009 year in smut. (Abstr.) Proc. 33rd Annu. RTWG Meet. Biloxi, MS.
5. Carpenter, S. R., Caraco, N. F., Correll, D. L., Howarth, R. W., Sharpley, A. N., and Smith, V. H. 1998. Nonpoint pollution of surface waters with phosphorous and nitrogen. Ecol. Appl. 8:559-568.
6. Carris, L. M., Castlebury, L. A., and Goates, B. J. 2006. Nonsystemic bunt fungi—Tilletia indica and T. horrida: a review of history, systematics, and biology. Annu. Rev. Phytopathol. 44:113-133.
7. Cartwright, R. D., Lee, F. N., Beaty, T., Sutton, E. A., and Parsons, C. E. 1999. Reaction of rice cultivars/lines to false smut, stem rot, and black sheath rot disease. Pages 158-168 in: Arkansas Rice Research Studies 1999. B. R. Wells, ed. Res. Ser. 476. Ark. Agric. Exp. Stn., Fayetteville.
8. Cartwright, R. D., Lee, F. N., Chlapecka, R. M., and Vann, S. R. 1998. First report of false smut of rice in Arkansas. (Abstr.) Proc. 30th Annu. RTWG Meet. Reno, NV.
9. Cartwright, R. D., Lee, F. N., Crippen, D. L., and Templeton, G. E. 1993. Kernel smut research in Arkansas. Pages 71-78 in: Arkansas Rice Research Studies 1993. B. R. Wells, ed. Res. Ser. 439. Ark. Agric. Exp. Stn., Fayetteville.
10. Cartwright, R. D., Lee, F. N., Hunter, R. L., Crippen, D. L., and Templeton, G. E. 1994. Research in Arkansas on kernel smut of rice. Pages 115-123 in: Arkansas Rice Research Studies 1194. B. R. Wells, ed. Res. Ser. 446. Ark. Agric. Exp. Stn., Fayetteville.
11. Cassman, K. G., Dobermann, A., Sta Cruz, P. C., Gines, G. C., Samson, M. I., Descalsota, J. P., Alcantara, J. M., Dizon, M. A., and Olk, D. C. 1996. Soil organic matter and the indigenous nitrogen supply of intensive irrigated systems in the tropics. Plant Soil 182:267-278.
12. Cicerone, R. J., and Delwiche, C. C. 1993. Factors effecting methane production under rice. Global Biogeochem. Cycl. 7:143-155.
13. Dick, R. P. 1992. A review: long-term effects of agricultural systems on soil biochemical and microbial parameters. Agric. Ecosyst. Environ. 40:25-36.
14. Gillip, J. A., and Czarnecki, J. B. 2009. Validation of a ground-water flow model of the Mississippi river valley alluvial aquifer using water-level and water-use data for 1998–2005 and evaluation of water use scenarios. U.S. Geological Survey Scientific Investigations Report 2009-5040.

15. Ginn, W., and Cartwright, R. D. 2009. Arkansas rice hammered by false smut. Delta Farm Press. http://deltafarmpress.com/rice/Rice-disease-0325/

16. Gravois, K. A., and Bernhardt, J. L. 2000. Heritability and genotype x environment interactions for discolored rice kernels. Crop Sci. 40:314-318.

17. Guinn, J. 2008. U.S. paddy rice shipments halted at Mexican border. Page 1 in: USA Rice Federation Daily, December 22, 2008.

18. Hopkins, A., Bowman, R., Parsons, C. E., Clark, S. D., and Cartwright, R. D. 2002. Evaluation of foliar fungicides for control of false smut of rice, 2002. Fungic. Nematicide Tests 58:FC069.

19. Lal, R. 2004. Soil carbon sequestration impacts on global climate change and food security. Science 304:1623-1627.

20. Lee, F. N., Templeton, G. E., Cartwright, R. D., Long, D. H., and Hornbeck, J. M. 1992. Pathogenicity of Tilletia barclayana isolates. Pages 81-85 in: Arkansas Rice Research Studies 1992. B. R. Wells, ed. Res. Ser. 422. Ark. Agric. Exp. Stn., Fayetteville.

21. Olk, D. C., Anders, M. M., Filley, T. R., and Isbell, C. 2009. Crop nitrogen uptake and soil phenols accumulation under continuous rice cropping in Arkansas. Soil Sci. Soc. Am. J. 73:952-960.

22. Orton, W. A. 1907. Plant diseases of 1906. Pages 499-508 in: Yearbook U.S. Department of Agriculture. U.S. Government Printing Office, Washington, DC.

23. Rush, M. C., Shahjahan, A. K. M., Jones, J. P., and Groth, D. E. 2000. Outbreak of false smut of rice in Louisiana. Plant Dis. 84:100.

24. Schmid, B. T. 2008. Conservation tillage effects on soil physical and chemical properties in rice production in the Arkansas delta. M.S. thesis, University of Arkansas, Fayetteville.

25. Slaton, N. A., Gbur, E. E., Cartwright, R. D., DeLong, R. E., Norman, R. J., and Brye, K. R. 2004. Grain yield and kernel smut of rice as affected by preflood and midseason nitrogen fertilization in Arkansas. Agron. J. 96:91-99.

26. Tsuda, M., Sasahara, M., Ohara, T., and Kato, S. 2006. Optimal application timing of simeconazole granules for control of rice kernel smut and false smut. J. Gen. Plant. Pathol. 72:301-304.

27. University of Arkansas. 2007. Arkansas rice performance trials, 2005–2007. In: Rice Information, No. 166.

28. Watkins, K. B., Anders, M. M., and Windham, T. E. 2004. An economic comparison of alternative rice production systems in Arkansas. J. Sustain. Agric. 24:57-78.

29. Webster, R. K., and Gunnell, P. S. 1992. Compendium of Rice Diseases. American Phytopathological Society, St. Paul, MN.

30. White, D. G. 1999. Compendium of Corn Diseases, 3rd ed. American Phytopathological Society, St. Paul, MN.

INDEX

L

M

N

O

P

R

S

T

V

W

Z